Horizons in World Physics, Volume 259

Self Focusing of Relativistic Electron Bunches in Plasma

Horizons in World Physics Series

Volume 258 **New Topics in Theoretical Physics,** Edited by Henk F. Arnoldus and Thomas George

Volume 257 **Boson's Ferromagnetism and Crystal Growth Research,** Edited by Emerson D. Seifer

Volume 256 **Horizons in World Physics,** Edited by Victor H. Marselle

Volume 255 **Frontiers in General Relativity and Quantum Cosmology Research,** Edited by Victor H. Marselle

Volume 254 **Optics and Electro-Optics Research,** Edited by Albert V. Berzilla

Volume 253 **Trends in General Relativity and Quantum Cosmology,** Edited by Albert Reimer

Volume 252 **Chemical Physics Research Trends,** Edited by Benjamin V. Arnold

Volume 251 **Quantum Dots: Research Developments,** Edited by Peter A. Ling

Volume 250 **Quantum Gravity Research Trends,** Edited by Albert Reimer

Volume 249 **General Relativity Research Trends,** Edited by Albert Reimer

Volume 248 **Spacetime Physics Research Trends,** Edited by Albert Reimer

Volume 247 **New Developments in Quantum Cosmology Research,** Edited by Albert Reimer

Volume 246 **Quantum Cosmology Research Trends,** Edited by Albert Reimer

Volume 245 **Horizons in World Physics,** Edited by Tori V. Lynch

Volume 244 **Horizons in World Physics,** Edited by Albert Reimer

Volume 243 **Theoretical Physics 2002, Part 2,** Edited by Henk F. Arnoldus and Thomas F. George

Volume 242 **Models and Methods of High-T_c Superconductivity: Some Frontal Aspects, Volume 2,** Edited by J. K. Srivastava and S. M. Rao

Volume 241 **Models and Methods of High-T_c Superconductivity: Some Frontal Aspects, Volume 1,** Edited by J. K. Srivastava and S. M. Rao

Volume 240 **Horizons in World Physics,** Edited by Albert Reimer

Volume 239 **Theoretical Physics 2002, Part 1,** Edited by Henk F. Arnoldus and Thomas F. George

Volume 238 **Theoretical Physics 2001,** Edited by Henk F. Arnoldus and Thomas F. George

Volume 237 **Generalized functions in mathematical physics: Main ideas and concepts,** Edited by A.S. Demidov

Volume 236 **Edge Excitations of Low-Dimensional Charged Systems,** Edited by Oleg Kirichek

Volume 235 **On the Structure of the Physical Vacuum and a New Interaction in Nature (Theory, Experiment, Applications),** Edited by Yu. A. Baurov

Volume 234 **Electron Scattering on Complex Atoms (Ions),** Edited by V. Lengyel, O. Zatsarinny and E. Remeta

Volume 233 **An Introduction to Hot Laser Plasma Physics,** Edited by A.A. Andreev, A.A. Mak and N.A. Solovyev

Volume 232 **Electron-Atom Scattering: An Introduction,** Maurizio Dapor

Volume 231 **Theory of Structure Transformations in Non-equilibrium Condensed Matter**, Edited by Alexander L. Olemskoi

Volume 230 **Topics in Cosmic-Ray Astrophysics**, Edited by M. A. DuVernois

Volume 229 **Interactions of Ions in Condensed Matter**, Edited by A. Galdikas and L. Pranevièius

Volume 228 **Optical Vortices**, Edited by M. Vasnetsov and K. Staliunas

Volume 227 **Frontiers in Field Theory**, Edited by Quantum Gravity and Strings, R.K. Kaul, J. Maharana, S. Mukhi and S. K. Rama

Volume 226 **Quantum Groups, Noncommutative Geometry and Fundamental Physical Interactions**, Edited by D. Kastler, M. Russo and T. Schücker

Volume 225 **Aspects of Gravitational Interactions**, Edited by S.K. Srivastava and K.P. Sinha

Volume 224 **Dynamics of Transition Metals and Alloys**, Edited by S. Prakash

Volume 223 **Fields and Transients in Superhigh Pulse Current Devices** Edited by G.A. Shneerson

Volume 222 **Gaussian Beams and Optical Resonators,** Edited by A. N. Oraevsky

Volume 221 **Laser Cathode-Ray Tubes,** Edited by Yu. M. Popov

Volume 220 **Quantum Electrodynamics with Unstable Vacuum,** Edited by V.L. Ginzburg

Volume 219 **The Muon Method in Science,** Edited by V.P. Smilga and Yu.M. Belousov

Volume 218 **Radiative and Collisional Characteristics of Ions in Hot Plasmas,** Edited by I.I. Sobel'man

Volume 217 **Nonlinear and Quantum Optical Phenomena in Nonequilibrium Media** , Edited by V.A. Shcheglov

Volume 216 **Semiconductors and Insulators: Optical and Spectroscopic ...,** Edited by Yu.I. Koptev

Volume 215 **Relativistic Theory of Gravity,** Edited by A.A. Logunov

Volume 214 **Research in Quantum Field Theory**, Edited by V.I. Man'ko

Volume 213 **Nonlinear Theory of Strong Electromagnetic Wave-plasma...,** Edited by O.N. Krokhin

Volume 212 **Dynamics of Elementary Atomic-Molecular...**, Edited by O. S. Scheglov

Volume 211 **Optics and Lasers,** Edited by G.G. Petrash

Volume 210 **Theory and Technology of High Temperature Superconductors,** Edited by Yu.I. Koptev

Volume 209 **Theory of Interaction of Multilevel Systems...**, Edited by V.I. Man'ko

Volume 208 **Physical Properties of High Temperature Superconductors,** Edited by Yu.I. Koptev

Volume 207 **Generalized Mayer Series in Classical Statistical Mechanics**, Edited by Igor J. Ivanchik

Volume 206 **Mathematical Physics, Applied Mathematics and Informatics,** Edited by Yu.I. Koptev

Volume 205 **Squeezed and Correlated States of Quantum Systems**, Edited by M.A. Markov

Volume 204 **Sakharov Memorial Lectures in Physics**, Edited by L.V. Keldysh and V.Ya. Fainberg

Volume 203 **Fiber Optics: Research and Development,** Edited by Ye.M. Dianov

Volume 202 **Gas Dynamics**, Edited by Yu.I. Koptev

Volume 201 **Generation of Nonlinear Waves and Quasistationary ...**, Edited by L.M. Kovrizhnykh

Volume 200 **Laser Thermonuclear Targets and Superdurable Microballoons**, Edited by A.I. Isakov

Volume 199 **Spatiotemporal Characteristics of Laser Emission,** Edited by M.V. Pyatakhin and A.F. Suchkov

Volume 198 **Theory of Nonstationary Quantum Oscillators**, Edited by M.A. Markov

Volume 197 **Reliability Problems of Semiconductor Lasers**, P.G. Eliseev

Volume 196 **Pulsars,** Edited by A.D. Kuzmin

Volume 195 **Physical Characteristics and Critical Temperature of High Temperature Superconductors** , Edited by M.M. Sushchinskiy

Volume 194 **Biophysical Approach to Complex Biological Phenomena,** Edited by E. Volkov

Volume 193 **Research on Chemical Lasers** Edited by A.N. Orayevskiy

Volume 192	**Atomic and Ionic Spectra and Elementary Processes in Plasma** , Edited by I.I. Sobelman
Volume 191	**Polarization of the Vacuum and a Quantum Relativistic Gas ...**, A.Ye. Shabad
Volume 190 , (Supp. Vol.)	**Superconductors with A15 Lattice and Bridge Contacts Based ...**, Edited by M. M. Sushchinskiy
Volume 190	**Metal-Optics and Superconductivity**, Edited by A.I. Golovashkin
Volume 189	**Clusters of Galaxies and Extragalactic Radio Sources**, Edited by A.D. Kuz'min
Volume 188	**Integral Systems, Solid State Physics and Theory of Phase Transitions. Pt. 2. Symm. and Alg. Structures in Physics**, Edited by V.V. Dodonov and V.I. Man'ko
Volume 187	**Quantum Field Theory, Quantum Mechanics and Quantum Optics: Pt. 1. Symm. and Alg. Structures in Physics,** Edited by V.V. Dodonov and V.I. Man'ko
Volume 186	**Photoproduction of Pions on Nucleons and Nuclei**, Edited by A.A. Komar
Volume 185	**Injection Lasers in Optical Communication and Information Processing Systems**, Edited by Yu. M. Popov
Volume 184	**Electron Processes in MIS-Structure Memories**, Edited by A.F. Plotnikov
Volume 183	**Invariants and the Evolution of Nonstationary Quantum Systems**, Edited by M.A. Markov
Volume 182	**Luminescence of Wideband Semiconductors**, Edited by M.D. Galanin
Volume 181	**Pulsed Metal and Metal Halogenide Vapor Lasers**, Edited by G.G. Petrash
Volume 180	**Inelastic Light Scattering in Crystals**, Edited by M.M. Sushchinskiy
Volume 179 , (Supp. Vol. 2)	**Electron-Excited Molecules in Nonequilibrium Plasma**, Edited by N.N. Sobolev
Volume 179 , (Supp. Vol. 1)	**Interaction of Ultrashort Pulses with Matter** , Edited by M.D. Galanin
Volume 179	**X-Ray Plasma Spectroscopy and the Properties of Multiply-Charged Ions** , Edited by I.I. Sobel'man
Volume 178	**The Delfin Laser Thermonuclear Installation: Operational Complex ...**, Edited by G.V. Sklizkov
Volume 177	**Stoichiometry in Crystal Compounds and Its Influence on Their Physical Properties**, Edited by Yu. M. Popov
Volume 176 , (Supp. Vol)	**Quantization, Gravitation and Group Methods in Physics** Edited by A.A. Komar
Volume 176	**Classical and Quantum Effects in Electrodynamics** Edited by A.A. Komar
Volume 175 , (Supp. Vol.)	**Luminescence and Anisotropy of Zinc Sulfide Crystals** Edited by M.D. Galanin
Volume 175	**Luminescence Centers of Rare Earth Ions in Crystal Phosphors** , Edited by M.D. Galanin
Volume 174 , (Supp. Vol. 2)	**Electron Liquid Theory of Normal Metals** Edited by V.P. Silin
Volume 174 , (Supp. Vol. 1)	**Nonequilibrium Superconductivity** , Edited by V.L. Ginzburg
Volume 174	**Thermodynamics and Electrodynamics of Superconductors,** Edited by V.L. Ginzburg
Volume 173	**Quantum Mechanics and Statistical Methods** , Edited by M.M. Sushchinskiy
Volume 172	**Phase Conjugation of Laser Emission** , Edited by N.G. Basov
Volume 171	**Research on Laser Theory** , Edited by A.N. Orayevskiy
Volume 170	**The Theory of Target Compression by Longwave Laser Emission** , Edited by G.V. Sklizkov
Volume 169	**The Physical Effects in the Gravitational Field Black Holes**, Edited by M.A. Markov
Volume 168	**Issues in Intense-Field Quantum Electrodynamics**, Edited by V.L. Ginzburg
Volume 167	**Group Theory, Gravitation and Elementary Particle Physics**, Edited by A.A. Komar

Horizons in World Physics, Volume 259

SELF FOCUSING OF RELATIVISTIC ELECTRON BUNCHES IN PLASMA

VALERY B. KRASOVITSKIY

Nova Science Publishers, Inc.
New York

For permission to use material from this book please contact us:
Telephone 631-231-7269; Fax 631-231-8175
Web Site: http://www.novapublishers.com

LIBRARY OF CONGRESS CATALOGING-IN-PUBLICATION DATA
Available upon request

ISBN 13: 978-1-60021-529-2
ISBN 10: 1-60021-529-7

Published by Nova Science Publishers, Inc. ✦ New York

Contents

PREFACE

The main subject of this book is the nonlinear theory of the electrostatic focusing of an electron beam split into bunches under conditions when the plasma permittivity at the modulation frequency is negative and the effective Coulomb force acting on the electron bunches is reversed. Conditions for the spatial equilibrium between the bunch and plasma emission, as well as the dynamics of the formation of focused bunches, are confirmed by solving (both analytically and numerically) the self-consistent set of equations.

Results are also presented from investigations on nonlinear plasma theory, including the propagation of Langmuir and helicon solitons in a plasma with a charged particle beam under the conditions of the Cherenkov and cyclotron resonances, the collective acceleration of ions by a relativistic electron beam, and the propagation of Langmuir perturbations in a weekly ionized gas described by the Davydov kinetic equation.

The particle-field resonance interaction is frequently encountered in laboratory, geophysical and astrophysical plasma studies. Therefore, the book will be of interest to experts in plasma physics and relative sciences, as well as to all those who are interested in the problems of modern physics.

To solve the applied problems of plasma physics, the general methods of macroscopic electrodynamics are used. Most of the chapters are supplemented by appendices that contain detailed mathematical derivations and may be useful for lecturers and students.

INTRODUCTION

The idea of the possibility of using the strong electromagnetic self-fields of uncompensated beams and plasma for the effective acceleration of the charged particles was first put forward in [1-4]. However the practical realization of this idea depends on methods of self-focusing of the beam which makes possible to concentrate an electromagnetic energy into the region of particle acceleration. The well known wave guide system of such a kind is the equilibrium Bennet pinch [5] which arises, if the spatial charge of the relativistic electron beam is partially compensated by the ions and a focusing action of the static electromagnetic field is equalized by the gradient of kinetic pressure in the beam. A magnetostatic self-focusing of the relativistic electron beam occurs also in a dense collisional plasma, where plasma electrons fully neutralize the beam charge and only partially beam current (Ivanov and Rudakov [6]). More detailed bibliography on the problem of macroscopic equilibrium and stability of powerful relativistic beams can be found in Refs. [7-9].

As was shown by Budker [3], in the method of the self-stabilized beam discovered by him the radial pinching of a relativistic electron-ion stream occurs under the action of the radiative reaction force that acts on electrons oscillating in the radial potential well of an uncompensated ionic charge, while the energy of the transverse oscillations is continually replenished as a result of the electron-ion scattering processes which change the longitudinal motion of the electrons into transverse motion.

In order to reduce the beam-focusing time, which is determined by the reciprocal of the electron-ion binary collision rate and which out to be quite substantial, Fainberg [10] proposed the use of the coherent radiation of electrons for the radial pinching of beams.

The first computations carried out by Krasovitskii [11] in the linear and quasilinear approximation confirmed the principle possibility of radial beam self-focusing. It is shown that this possibility exists when a bounded beam interacts with the plasma under conditions of the two-stream instability. The radial beam compression occurs under the effect of high-frecuency pressure forces due to the surface plasma waves generated by the beam. The focusing time is of the order of the reciprocal growth rate for the two-stream instability. This effect was predicted theoretically in [11] and experimentally realized in [12].

The possibility of beam self-focusing under the action of electrostatic oscillations, excited by the beam in the unbounded thermonuclear plasma, was proposed by Winterberg [13]. This effect arises as the beam reaction on the field of slow wave which is exited by the beam in a plasma and then transforms into electromagnetic wave and leaves the beam through its boundary [14].

The nonlinear theory of self-compression of a relativistic electron beam exposed to coherent electromagnetic radiation (fast electromagnetic wave) is developed by Bliokh et al.

in papers [15].

It was shown by Krasovitskii in [16] that the compression of a bunch by the field of the plasma wave exited by the beam is balanced by the gradient of the kinetic pressure in the bunch. Self-focusing occurs if the beam modulation frequency is lower than the plasma frequency and plasma permittivity is negative. Under the plasma resonance condition the focusing polarization field greatly exceeds the Coulomb repulsion field for the electrons in the bunches which allows the gradient of the kinetic pressure in the beam to be balanced.

The self-focusing of negative ionic and electron bunches in a plasma has observed in the experiments [17] and [18], respectively.

In Chapters 1-3 the spatial equilibrium of a strongly modulated electron beam (or of a train of electron bunches) in a dense plasma is analyzed theoretically in the self-focusing case. But the mathematical difficulties arising in the search for analytical models allow only the conditions under which beam equilibrium can occur in the plasma and the shape of the electron bunches to be determined. That is why, the process by which bunches form and decay, including the stage in which potential wells develop, the compression of relativistic bunches to high density in the wave field as it reaches large amplitude, and the approach to the steady state of beam transport in the plasma is analyzed numerically (Chap. 3, Ref. [13]).

The second part of the monograph studies the nonlinear steady-state Langmuir nad electromagnetic waves (solitons) in a plasma with a charged particle beam.

The conditions of the Langmuir and electromagnetic envelope soliton propagation along a monoenergetic electron beam in a plasma is discussed in Chaps. 4 and 5. The carrier wave number and the carrier frequencies of these waves satisfy the conditions of the Cherenkov and anomalous or normal Doppler resonances. These two-resonance solutions in the criss-crossing of beam and plasma branches of oscillations that are of interest for transporting electromagnetic energy along the beam in the plasma.

Ion acceleration by relativistic electron beam is discussed in Chap. 6. Studied is the dynamics of ion bunches in the wave field of the charge density, excited in the electron beam under the conditions of the electron-ion instability and an electron bunch acceleration mechanism is put forward by a wave with modulated phase. The nonlinear theory presented here yields the characteristic parameters of the process, viz., the energy acquired by the ions, the acceleration time, and the maximum density of the accelerated ion beam.

The possibility of using a tubular relativistic electron beam, whose charge is partially neutralized by the charge of the ions captured by the electrostatic field of the beam to form a background plasma, for the slow waveguide with conducting walls is discussed in the case of the beam current closed to the limit current.

The attenuation of the electrostatic disturbances in a weakly ionized plasma is analyzed in final Chap. 7, as a result of elastic collisions of electrons and neutrals. It is shown that the frequency and the imaginary increment of small Langmuir oscillations are determined by the ratio of the elastic collision frequency to the plasma frequency. The interaction of a plasma with a short electric pulse field includes the stages of heating and cooling-off of electrons. The computations are made on the base of the Davydov kinetic equations in the approximation of applied field and in the self-consistent approximation.

This monograph presents results of theoretical research, conducted with the author's participation at Rostov State University and Keldysh Institute of Applied Mathematics (KIAM) during 1973-2000.

A great contribution to elaboration of theory and numeric methods was made by researchers and post-graduate students of the Chair of Theoretical and Nuclear Physics of Rostov State University: I.V. Bachin, V.G. Dorofeenko, G.V. Fomin, D.V. Krasovitskii and V.V. Prudskikh. The numeric simulations of plasmas and the beam, using the particle method, were made by O.V. Batischev and Yu.S Sigov at the KIAM.

CHAPTER 1

ELECTROSTATIC EQUILIBRIUM OF MODULATED ELECTRON BEAM IN A PLASMA

The foundation of electrostatic self-focusing of previously broken up into distinct bunches electron beam in a dense plasma is the inversion of the Coulomb force sign under the condition that the beam modulation frequency is lower than plasma Langmuir frequency $\omega_m < \omega_p$ [1]. In this case the plasma permittivity $\epsilon = 1 - \omega_p^2/\omega_m^2$ is negative and a spatial self-focusing of the electron bunches occurs. The electromagnetic radiation emitted by the beam in a plasma is a superposition of the vacuum Coulomb field of moving bunches and of the polarization wave exited by this field in a plasma. Under the resonance condition with $|\epsilon| \ll 1$, the polarization field greatly exceeds the Coulomb repulsion field for the electrons in the bunches which allows the gradient of the kinetic pressure in the beam to be balanced. In other words, an electron beam previously broken up into clumps repels plasma electrons from the interior of the clumps and the uncompensated ion background creates moving potential wells for the beam electrons maintaining the wave-beam equilibrium.

§1.1. Self-focusing of moving electron disks in a plasma.

Following the work [1], let us analyze the process of laminated electron beam self-focusing in a dense plasma. The beam is supposed to be preliminarily divided into a number of bunches. In accordance with the charged disk model, let us consider the electron disks of surface density $\sigma_e(r)$ moving with the velocity $\mathbf{v}$. We suppose the distance between the disks equals l. The current density, produced in the plasma by the system of these disks, can be written as

$$\mathbf{j} = (0, 0, j_z), \quad j_z = ev \sum_{s=1}^{\infty} \sigma_{es}(r) \delta(z - z_{es} - vt - sl), \tag{1.1.1}$$

where z_{es} are the coordinates of the electron disks in the beam frame of reference.

The self-consistent electric field $\mathbf{E}$ in the plasma at the presence of the beam should satisfy the system of Maxwell equations [2]:

$$\text{grad}\,\text{div}\mathbf{E} - \Delta\mathbf{E} + \frac{1}{c^2}\frac{\partial^2}{\partial t^2}\hat{\epsilon}\mathbf{E} = -\frac{4\pi}{c^2}\frac{\partial \mathbf{j}}{\partial t}, \tag{1.1.2}$$

where $\hat{\epsilon}$ is a plasma dielectric function matrix and $\mathbf{j} = (0, 0, j_z)$. We direct z axis along the beam velocity and consider the azimuthally symmetrical solution $\mathbf{E} = (E_r, 0, E_z)$ which satisfy the following equations:

$$\frac{\partial^2 E_z}{\partial r \partial z} - \frac{\partial E_r^2}{\partial z^2} = -\frac{1}{c^2}\frac{\partial^2}{\partial t^2}\hat{\epsilon} E_r, \tag{1.1.3}$$

$$\hat{\epsilon}\left[\frac{1}{r}\frac{\partial}{\partial r}(rE_r) + \frac{\partial E_z}{\partial z}\right] = \frac{4\pi j_z}{v}. \tag{1.1.4}$$

Taking into account that the beam field is a periodic function of the coordinate, we represent the solution of Eqs. (1.1.3) and (1.1.4) as the Fourier series in this variable

$$\mathbf{E}(t, \mathbf{r}) = \sum_{n=-\infty}^{\infty} \mathbf{E}^n(t, r)\exp(inkz),$$

where $k = 2\pi/l$, and ℓ is the spatial period of the beam. As the result, we get the system of equations for the components of the electric field

$$ink\frac{\partial E_z^n}{\partial r} + (nk)^2 E_r^n = -\frac{1}{c^2}\frac{\partial^2}{\partial t^2}\epsilon E_r^n, \tag{1.1.5}$$

$$\hat{\epsilon}\left[\frac{1}{r}\frac{\partial}{\partial r}(rE_r^n) + inkE_z^n\right] = 2ke\sigma_e(r)\exp[-in(\omega_m t + kz_e], \tag{1.1.6}$$

where $\omega_m = kv$ is the beam modulation frequency.

As the time dependence is defined by $e^{-in\omega_m t}$ one can use the relation

$$\hat{\epsilon}e^{-in\omega_m t} = \epsilon(n\omega_m)e^{-in\omega_m t},$$

(derived in [2]) to represent Eq. (1.1.5) in the form

$$E_r^n(r) = \frac{i}{nk}\left[\frac{v^2}{c^2}\epsilon(n\omega_m) - 1\right]^{-1}\frac{dE_z^n}{dr}. \tag{1.1.7}$$

For the case of thin disks, i.e., when the beam radius r_e is small compared to the interdisk spacing, $r_e \ll l$, it follows from Eq. (1.1.7), that $E_z^n \simeq nkr_e E_r^n$. Consequently, for the components of moderate wave number values, $n \ll l/r_e$, the contribution of which in the Fourier series expansion is fairly small, it is possible to neglect the longitudinal component of the field in Eq. (1.1.6). After multiplying of both equation sides by the $\exp(iknz)$ of and taking the summation with respect to all harmonics, one can obtain the equation for the radial component of field

$$\frac{1}{r}\frac{d}{dr}(rE_r) = 2ke\sigma_e(r)\sum_{n=-\infty}^{\infty}\frac{1}{\epsilon(n\omega_m)}\exp[in(kz - kz_e - \omega_m t)],$$

where

$$\epsilon(n\omega_m) = 1 - \frac{\omega_p^2}{n^2\omega_m^2}$$

is the plasma permittivity with respect to the frequency $n\omega_m$.

The electric fields that act upon the electron bunches at the points $z = sl + z_e + vt$ satisfy the following equation:

$$\frac{1}{r}\frac{d}{dr}(rE_r) = 2ke\sigma_e(r)\sum_{n=-\infty}^{\infty}\frac{1}{\epsilon(n\omega_m)}, \tag{1.1.8}$$

which contains the divergent series expansion because of the longitudinal bunch dimension of a bunch is equal zero. According to Ref. [2], this problem can be corrected using the substitution $1/\epsilon \to 1/\epsilon - 1$. Under the resonance conditions, $|\omega_m - \omega_p| \ll \omega_p$, the harmonic with the frequency ω_m dominates over the whole spectrum. Thus, only the term $|n| = 1$ should be kept in the right-hand side of Eq. (1.1.8)

$$\sum_{n=-\infty}^{\infty} \left[\frac{1}{\epsilon(n\omega_m)} - 1\right] = -\frac{\alpha\pi\cos(\alpha\pi)}{\sin[(1-\alpha)\tau]} \quad \to \quad \frac{2}{\epsilon(\omega_m)},$$

$$\alpha \to 1$$

where $\alpha = \omega_p/\omega_m$.

Making use of the relation $E_r = -d\varphi/dr$ in the above equation, let us consider that the focusing electric force $e\sigma E_r$ and the gradient of the kinetic pressure $-T\, d\sigma/dr$ are balanced. On this assumption, we find the surface density σ as a function of the potential

$$\sigma(r) = \sigma(0)\exp\left(-\frac{2e\varphi}{mv_T^2}\right), \tag{1.1.9}$$

where $v_T = \sqrt{2T/m}$ is the thermal velocity of beam electrons.

As a result, we get an ordinary differential equation:

$$\frac{1}{r}\frac{d}{dr}\left(r\frac{d\phi}{dr}\right) = \frac{8}{r_e^2}\exp(-\phi),$$
$$\phi = \frac{2e\varphi}{mv_T^2}, \quad r_e^2 = \frac{2v_T^2}{\omega_b^2}|\epsilon|, \quad \omega_b^2 = \frac{4\pi e^2\sigma(0)}{ml}, \tag{1.1.10}$$

where $\epsilon(\omega_m) = -|\epsilon|$.

Integrating Eq. (1.1.10), we find the distribution of potential and electron density over the cross section of the beam

$$\phi = 2\ln\left(1 + \frac{r^2}{r_e^2}\right), \quad \sigma(r) = \sigma(0)\left(1 + \frac{r^2}{r_e^2}\right)^{-2}. \tag{1.1.11}$$

Note that in comparison with the magnetostatic self-focusing [3], the effective beam radius r_e is proportional to a small parameter $\sqrt{|\epsilon|}$ so that the efficiency of this method increases near a plasma resonance.

The above condition $kr_e \ll 1$ allowing us to neglect a longitudinal component of the electric field can be shown up as

$$\left(\frac{v_T}{v}\right)^2 |\epsilon| \ll \frac{\sigma(0)}{n_p l}. \tag{1.1.12}$$

Correspondingly, the beam current is small in comparison with its maximum value

$$I_b = \frac{2\pi e v}{l}\int_0^\infty \sigma(r) r\, dr = \frac{mv_T^2 v|\epsilon|}{e} \ll \frac{\sigma(0)mv^3}{eln_p}. \tag{1.1.13}$$

The two-dimensional solution of the problem of the electrostatic equilibrium of a modulated electron beam in a plasma, taking account of capture of electrons by the field of the wave in both the longitudinal and transverse directions when condition (1.1.12) violates, have obtained in the succeeding sections.

§1.2. Longitudinal equilibrium of moving monoenergetic bunches.

It is shown above that the transport of small radius disks in a plasma, $r_e \ll l$, ($l \simeq \lambda$, $\lambda = 2\pi v/\omega_p$), is accompanied by the radial compression of the beam. When the one-dimensional modulated electron beam, $r_e \gg l$, propagates in a plasma under the condition $\epsilon < 0$ a periodic potential well occurs, capturing the electron bunches, and the longitudinal electrostatic equilibrium is established in the beam coordinate system [3].

The emission by an electron beam with the density ρ and the velocity $\mathbf{v}$ in a dense (linear) plasma is described by the set of inhomogenious hydrodynamic equations

$$\frac{\partial \mathbf{v}_p}{\partial t} = -\frac{e}{m}\frac{\partial \varphi}{\partial \mathbf{r}}, \quad \frac{\partial \tilde{n}_p}{\partial t} + \operatorname{div}(n_p \mathbf{v}_p) = 0,$$
$$\Delta\varphi = -4\pi e(\tilde{n}_p + \rho),$$

where n_p is the background plasma density and Δ is the Laplace operator. Eliminating the quantities $\mathbf{v}_p$ and $\tilde{n}_p$ from the above equations, we find

$$\left(\frac{\partial^2}{\partial t^2} + \omega_p^2\right)\Delta\varphi = -4\pi e\frac{\partial^2 \rho}{\partial t^2}. \tag{1.2.1}$$

Assuming that z axis of a cylindrical system of coordinates is directed along the beam axis, we consider that a steady-state axisymmetrical solution of Eq. (1.2.1) can be written as follows

$$\varphi(\mathbf{r}) = \sum_{n=-\infty}^{\infty} \varphi^n(r)\exp(inkz'),$$

where $z' = z - vt$, $k = 2\pi/l$, and l is the distance between bunches. The usual Fourier procedure may be followed to show that

$$\Delta\varphi = -4\pi e\sum_{n=-\infty}^{\infty}\frac{n^2}{n^2-\alpha^2}\rho^n(r)\exp(inkz'),$$
$$\rho^n(r) = \frac{1}{l}\int_{-l/2}^{l/2}\rho(r,z')\exp(-inkz')\,dz', \tag{1.2.2}$$

where $\alpha = \omega_p/\omega_m$, ω_p, and ω_m are the Langmuir plasma frequency and the beam modulation frequency, respectively.

Summing the series of the right-hand side of Eq. (1.2.2), we find the following integro-differential equation [4]:

$$\frac{1}{r}\frac{d}{dr}\left(r\frac{d\varphi}{dr}\right) + \frac{\partial^2\varphi}{\partial z'^2} = -4\pi e\tilde{\rho}(r,z'),$$
$$\tilde{\rho}(r,z') = \rho(r,z') - \frac{\alpha\pi}{\sin\alpha\pi}\frac{1}{l}\int_{-l/2}^{l/2}\rho(r,x)\cos\left[2\pi\alpha\left(\frac{|z'-x|}{l} - \frac{1}{2}\right)\right]dx. \tag{1.2.3}$$

It is essential that the self-consisting potential is a periodic function of z'. Therefore the effective electron density $\tilde{\rho}(r,z')$ which is a superposition of a beam density $\rho(r,z')$ and an induced plasma electron density satisfies the charge neutrality condition

$$\int_{-l/2}^{l/2}\tilde{\rho}(r,z')\,dz' = 0.$$

The steady-state solution exists, if the resultant force which acts on the beam electrons equals zero. Under this assumption, the condition of transverse and longitudinal equilibrium has the form:

$$e\frac{\partial\varphi}{\partial r}+\frac{T_{\perp}}{\rho}\frac{\partial\rho}{\partial r}=0,\quad e\frac{\partial\varphi}{\partial z'}+\frac{T_{||}}{\rho}\frac{\partial\rho}{\partial z'}=0, \tag{1.2.4}$$

where $T_{||}$ and $T_{\perp}$ are the components of electron temperature.

The set of equations (1.2.3) and (1.2.4) defines the self-consistent potential in a plasma. In the case of monoenergetic beam $T_{||}=T_{\perp}=0$, the equalities (1.2.4) realize into the bunch volume $|z|<a/2$ in the only case when the internal electrostatic field is absent. Proceeding on the assumption that $\varphi_i=0$, we obtain from Eq. (1.2.3) that $\tilde{\rho}=0$.

To find a self-consistent longitudinal dimension of the bunch, we assume that the charge density of the beam depends on the coordinates as:

$$\rho(r,z)=\begin{cases}\rho(r) & |z'|\leq a/2,\\ 0 & a/2\leq|z'|\leq l/2.\end{cases} \tag{1.2.5}$$

Substituting Eq. (1.2.5) into of Eq. (1.2.3) and evaluating the integral, we find

$$\tilde{\rho}=\frac{\sin[\alpha\pi(1-a/l)]}{\sin\alpha\pi}\cos(\alpha kz')\rho(r)=0. \tag{1.2.6}$$

The solution of the equation (1.2.6) is $\alpha(1-a/l)=n$, where n is a whole number. Thus, the result of Ref. [3] realizes in the special case of $n=1$,

$$a_c=(1-\omega_m/\omega_p)l\equiv\delta l. \tag{1.2.7}$$

According to Eqs. (1.2.3) and (1.2.5), the potential in a plasma $a/2<|z|\leq l/2$ satisfies the equation

$$\Delta\varphi_e=-4\pi e\rho(r)\cos\left[2\pi\alpha\left(\frac{|z'|}{l}-\frac{1}{2}\right)\right]. \tag{1.2.8}$$

The solution of the above equation has the form

$$\begin{aligned}\varphi_e(r,z')=4\pi e\Big[K_0(k_{\perp}r)\int_0^r I_0(k_{\perp}r')\rho(r')r'\,dr'\\ +I_0(k_{\perp}r)\int_r^{\infty}K_0(k_{\perp}r')\rho(r')r'\,dr'\Big]\cos\left(k|z'|-\alpha\pi\right),\end{aligned} \tag{1.2.9}$$

where I_0 and K_0 are modified Bessel functions, and $k_{\perp}=\alpha k$.

The radial component of the electric field equals zero at the boundaries of bunches, $\partial\varphi_e(\mp a/2)/\partial r=0$. Substituting (1.2.8) into this boundary condition, we find the following integral equation:

$$K_1(k_{\perp}r)\int_0^r I_0(k_{\perp}r')\rho(r')\,r'dr'=I_1(k_{\perp}r)\int_r^{\infty}K_0(k_{\perp}r')\rho(r')\,r'dr',$$

To proceed, let us differentiate this equality with respect to r and use the recurrence relations (well known from the theory of Bessel functions [5]) to transform it to the form

$$rK_1(k_{\perp}r)\frac{d\rho}{dr}=0.$$

It is easy to see that the above condition can be fulfilled only for $\rho = \text{const}$ (over a range $r > 0$). In the actual case $\rho \neq \text{const}$ the radial component of the electric field has a discontinuity at the points $z' = \pm a/2$, and the two-dimensional analog of the solutions of Ref. [3] does not exist.

Let us now consider the physical mechanism causing the effect of electrostatic self-focusing in the simple case of the one-dimensional solution (1.2.7). We assume that $\delta =\ll 1$ and the collective interaction with the plasma occurs basically through the field of this mode $E(t, z)$. We find from the Eqs. (1.2.2) and (1.2.5) the field of the first harmonic

$$E(t,z) = E_0 \sin(kz - \omega_m t), \quad E_0 = \frac{8e\rho_0}{k} \sin\frac{ka}{2},$$

where $k = 2\pi/l$ and $\omega_m = kv$.

Due to the field in the plasma the polarization wave E_p arises

$$\frac{\partial \tilde{v}_p}{\partial t} = \frac{e}{m}(E + E_p), \quad \frac{\partial E_p}{\partial t} = -4\pi e n_p \tilde{v}_p, \tag{1.2.10}$$

whose field E_p exceeds the vacuum field of the bunch, $E(t, z)$, by the factor $\delta^{-1} \gg 1$. It follows from Eq. (1.2.10) that

$$\begin{aligned} \tilde{v}_p(t,z) &= \frac{eE_0}{m\omega_m \epsilon} \cos(kz - \omega_m t), \\ E + E_p &= \frac{E}{\epsilon}, \quad \epsilon = 1 - \frac{\omega_p^2}{\omega_m^2}. \end{aligned} \tag{1.2.11}$$

Consequently, the phase velocity of speed $\tilde{v}_p$ and the density $\tilde{n}_p = n_p \tilde{v}_p / v$ of plasma electrons equals to the beam velocity and the electric field of perturbation is static in the frame of reference of the beam.

The number of plasma electrons displaced from the volume of bunches by the polarization field E_p is

$$N_p = -\int_{-a/2}^{a/2} n_p \, dz' = \frac{2\rho_0}{\pi \delta k} \sin^2 \frac{ka}{2}. \tag{1.2.12}$$

As a result the beam electrons find themselves in ion potential wells moving with the velocity v, and the Coulomb repulsion in the bunches is compensated by the charge of plasma ions. A complete charge neutralization of the bunches occurs if the number of electrons $N_b = \rho_0 a_c$ equals the number of free plasma ions N_p. Assuming that the bunches are sufficiently thin, $a_c \ll l$, we find using Eq. (1.2.12) the equilibrium longitudinal size $a_c = \delta l$, which coincides with Eq. (1.2.7).

We note that the effect of electrostatic self-focusing occurs in the case of plasma with the negative permittivity, $\epsilon < 0$, because of the effective electric field $E_\Sigma = E + E_p$ acting on the beam electrons changes sign.

According to Eq. (1.2.9), the radial component of the electric field equals

$$E_r(r) = 4\pi e \left[K_1(k_\perp r) \int_0^r I_1(k_\perp r') \frac{d\rho}{dr'} r' dr' + I_1(k_\perp r) \int_r^\infty K_1(k_\perp r') \frac{d\rho}{dr'} r' \, dr' \right]. \tag{1.2.13}$$

Since the electron density in the beam reaches a maximum value on the axis and drops off towards the periphery, it follows that $d\rho/dr < 0$ and the field exerts a focusing action on the beam. Therefore, the transverse equilibrium exists only under the condition that the focusing electric force eE_r is balanced by the gradient of kinetic pressure.

§1.3. Resonant anisotropic bunches.

Under the resonant condition $\sin(\alpha\pi) \to 0$, the polarization plasma field greatly exceeds the Coulomb field of bunches so that the first term of the right-hand side of Eq. (1.2.3) can be discarded. Accordingly, the second term is simplified to the form:

$$\frac{1}{r}\frac{d}{dr}\left(r\frac{d\varphi}{dr}\right) + \frac{\partial^2\varphi}{\partial z'^2} = \frac{4ek}{\delta}\cos kz' \int_0^{l/2} \rho(r,x)\cos kx\,dx, \tag{1.3.1}$$

where

$$\delta = 1 - \omega_m/\omega_p, \quad \rho(r,z') = \rho(r,-z').$$

Let us represent Eqs. (1.2.4) as the following equation

$$-\frac{d\rho}{\rho} = \frac{e}{T_\perp}\frac{\partial\varphi}{\partial r}dr + \frac{e}{T_{||}}\frac{\partial\varphi}{\partial z}dz', \tag{1.3.2}$$

which determines the beam density as a function of potential. For the non-isotropic temperature, $T_\perp \neq T_{||}$, the right-hand side of Eq. (1.3.2) can be represented as a total differential only in the special case $\varphi(r,z) = \varphi_\perp(r) + \varphi_{||}(z)$. Choosing the potential in this form and integrating Eq. (1.3.2), we find

$$\rho(r,z') = \rho_0 \exp\left(-\frac{e\varphi_\perp}{T_\perp} - \frac{e\varphi_{||}}{T_{||}}\right). \tag{1.3.3}$$

It is easy to see that density distribution (1.3.3) is compatible with Eq. (1.3.1), when the inequality $k|z'| \ll 1$ holds and the right-hand side of the latter equation does not depend on z'. We take into account that $\varphi \sim z'^2$ and put

$$\varphi_{||}(z') = \frac{T_{||}}{e}\frac{4z'^2}{a^2}, \tag{1.3.4}$$

where a is the longitudinal dimension of a bunch, $a \ll l$. Substituting (1.3.3) and (1.3.4) into (1.3.1), we find

$$\frac{1}{r}\frac{d}{dr}\left(r\frac{d\varphi_\perp}{dr}\right) + \frac{8T_{||}}{ea^2} = \frac{4\pi e}{\delta}\langle\rho\rangle_0 \exp\left(-\frac{e\varphi_\perp}{T_\perp}\right), \tag{1.3.5}$$

where

$$\langle\rho\rangle_0 = \frac{2\rho_0}{l}\int_0^{l/2} \exp\left(-\frac{4z'^2}{a^2}\right)dz' \simeq \frac{\sqrt{\pi}}{2}\frac{\rho_0 a}{l}.$$

Note that the first term omitted from the right-hand side of Eq. (1.2.3) is small if the width of the bunch a greatly exceeds the equilibrium longitudinal dimension a_c [see Eq. (1.2.7)]. Under this condition, $a \gg a_c$, the compression of bunches exceeds the Coulomb's repulsion and balances the gradient of kinetic pressure. The simple one-dimensional solution of Eq. (1.3.5), $\varphi_\perp = 0$, corresponds to the infinite layer

$$a_T = \frac{1}{\sqrt{\pi}}\left(\frac{4T_{||}\delta l}{\rho_0 e^2}\right)^{1/2} \tag{1.3.6}$$

and realizes if the longitudinal temperature of the bunch, $T_{||}$, equals the depth of potential well,

$$W_{||} = e\Delta\varphi = \pi e^2 a^2\langle\rho\rangle_0/2\delta,$$

initiated by the collective field in a plasma.

The appearance of $W_{||}$ has a simple physical explanation. According to Eq. (1.2.12), the density of excessive plasma ions which acts as a focusing lens on cold beam electrons equals $\Delta n_p = \rho_0 a/\delta l$. Therefore, the voltage drops across the points $z_0 = 0$ and $z_{\mp} = \mp a/2$ for the Gaussian distribution of beam electrons $\langle \Delta n_p \rangle = \langle \rho \rangle_0 / \delta$ can be found from Poisson's equation: $\Delta\varphi = \pi e a^2 \langle \rho \rangle_0 / 2\delta$.

To proceed on condition that $a \neq a_T$, we introduce the following dimensionless variables

$$\phi = \frac{e\varphi_\perp}{T_\perp}, \quad \xi = \frac{r}{r_e}, \quad \mu = \frac{T_{||}}{W_{||}}, \quad r_e^2 = \frac{T_\perp \delta}{4\pi \langle \rho \rangle_0 e^2},$$

and rewrite Eq. (1.3.5) in the form:

$$\frac{1}{\xi}\frac{d}{d\xi}\left(\xi \frac{d\phi}{d\xi}\right) + \mu = \exp(-\phi). \tag{1.3.7}$$

Let us analyze the solution of Eq. (1.3.7) in the limit case $|1-\mu| \ll 1$, when the inequality $|\phi| \ll 1$ holds, and this equation is simplified to the Bessel equation [5]. Integrating (1.3.7), we obtain the solution which satisfies $\phi(0) = \phi'(0) = 0$ at the beam axis,

$$\begin{aligned} \phi &= (1-\mu)\left[1 - J_0(\xi)\right], \\ \langle \rho \rangle &= \langle \rho \rangle_0 \left[\mu + (1-\mu) J_0(\xi)\right], \end{aligned} \tag{1.3.8}$$

where $J_0(\xi)$ is the Bessel function [5]. For $\mu < 1$ the potential is positive and the density of electrons reaches its maximum on the beam axis. In the opposite case $\mu > 1$, the self-consistent electric field defocuses beam electrons.

As it is shown above, the self-focusing solution exists when the longitudinal temperature of electrons is less than the depth of the potential well on the beam axis: $T_{||} < W_{||}$. Since the electron density of the beam $\langle \rho(r) \rangle$ drops off towards the periphery, it follows that the potential energy

$$W_{||}(\xi) = \frac{\langle \rho \rangle}{\langle \rho \rangle_0} W_{||}$$

decreases and the capture condition for fast beam electrons is violated.

The numerical solution of Eq. (1.3.7) in the case of arbitrary μ is shown in Fig. 1.3.1. The shape of curves is qualitatively invariable because of the potential and beam density asymptotics equal $\phi_\infty = -\ln\mu$ and $\langle \rho \rangle_\infty = \mu \langle \rho \rangle_0$, respectively.

The integration of the beam density over the cross section in the case of $\mu > 0$,

$$N = 2\pi \int_0^\infty \rho(r)\, r dr,$$

leads to the diverging integral. Therefore, the solution (1.3.8) shows that the self-focusing occurs at a short distance near the beam axis, but it is not correct at the periphery of the bunch.

In the limit $\mu \to 0$, the oscillations of the beam density and the potential disappear because of the radial well infinitely increases (see Fig. 1.3.1). For $\mu = 0$ we have the solution

$$\phi = 2\ln(1 + \xi^2/8), \quad \langle \rho \rangle = \frac{\langle \rho \rangle_0}{(1 + \xi^2/8)^2}, \tag{1.3.9}$$

which coincides with (1.1.11) that found from the asymptotic model of think disks.

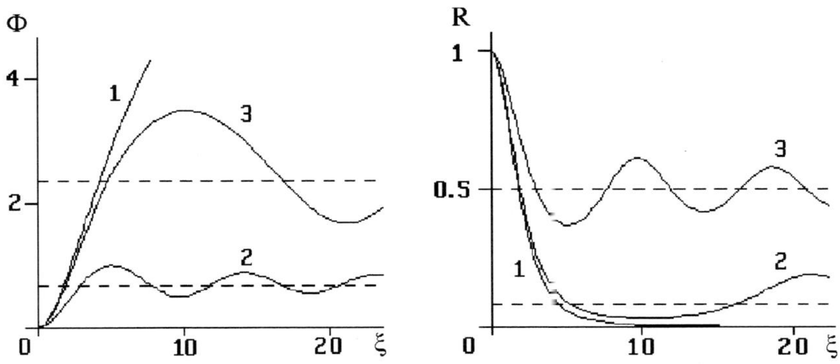

Fig. 1.3.1. the potential ϕ and the average bunch density R= $\langle\rho\rangle/\langle\rho\rangle_0$ as a function of the radial coordinate $\xi = 2\pi r/l$ for different μ: (1) 0; (2) 0.1; (3) 0.5.

Thus, the formulas (1.3.6) and (1.3.9) represent the solutions of Eq. (1.3.5) in two limiting cases of the longitudinal ($\varphi_\perp$=0) and transverse ($\varphi_\| = 0$) potential wells, respectively.

One can use an external potential wave to ensure the additional longitudinal focusing and remove radial oscillations of a beam density. Substituting

$$\varphi_{||}(z') \to \tilde{\varphi}_{||}(z') = \varphi_{||}(z') - \varphi_0 \cos(k_0 z')$$

into Eqs. (1.3.3) and (1.3.4), and using the inequality $k_0 z \ll 1$, we obtain the additional term $-k_0^2 \varphi_0$, which equalizes the gradient of kinetic pressure

$$\varphi_0 = \frac{8T_{||}}{ek_0^2 a^2}.$$

As a result the solution of Eq. (1.3.5) is the same as Eq. (1.3.9) which found above for cold thin disks.

§1.4. Self-focusing of the Boltzmann bunches.

For the isotropic heated bunches $T_\perp = T_{||} = T$, the right-hand side of Eq. (1.3.2) becomes a total differential. Therefore after the integration, we obtain the Boltzmann distribution function

$$\rho(r, z') = \tilde{\rho}_0 \exp\left[-\frac{e\varphi(r, z')}{T}\right] \tag{1.4.1}$$

which together with Eq. (1.3.1) determines a self-consistent potential in a plasma with a beam and the shape of equilibrium bunches.

We seek a driven solution of Eq. (1.3.1) in the form

$$\begin{aligned} \varphi(r, z') &= \frac{T}{e}\phi(r) \cos kz', \\ \rho(r, z') &= \rho_0 \exp\left[-\phi_0 - \phi\,(r) \cos kz'\right], \end{aligned} \tag{1.4.2}$$

where $z' = z - vt$, $\delta = 1 - \omega_m/\omega_p \ll 1$, and $k = 2\pi/l$ and $\omega_m = kv$ are the wave number and the modulation frequency; $\rho(0, 0) = \rho_0$ is the density at the center of bunch, and $\phi(0, 0) = \phi_0$.

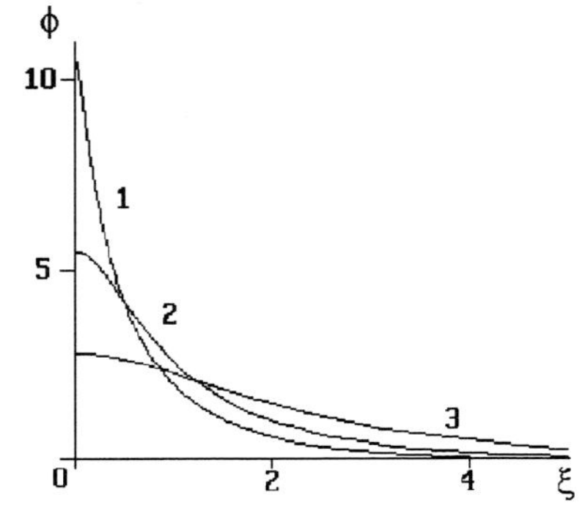

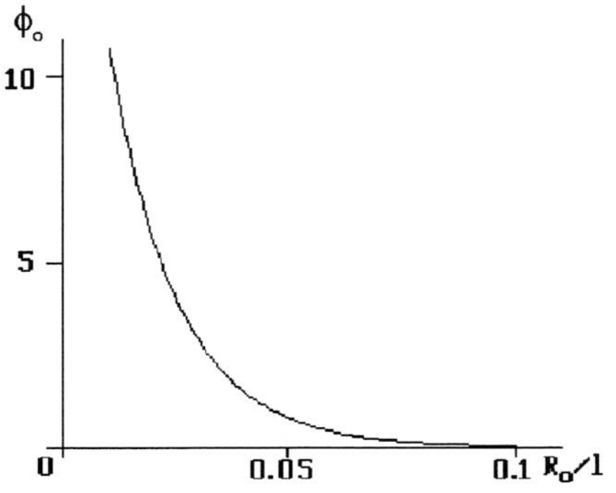

Fig. 1.4.1. Potential ϕ as a function of radial variable $\xi = 2\pi r/l$ for different ϕ_0: (1) 10.5; (2) 5.5; (3) 2.8.

Fig. 1.4.2. Depth of the potential well ϕ_0 as a function of R_0/l for soliton solutions of Eq. (1.4.3).

Substituting Eq. (1.4.2) in Eq. (1.3.1), averaging over the spatial period of the beam l, and using the relations

$$I_1(x) = \pi^{-1} \int_0^{\pi} \cos t \exp(x \cos t)\, dt, \quad I_1(-x) = -I_1(x),$$

we find the ordinary differential equation

$$\frac{1}{r}\frac{d}{dr}\left(r\frac{d\phi}{dr}\right) - k^2\phi = -\frac{1}{R_0^2}\exp(-\phi_0)I_1(\phi), \tag{1.4.3}$$

where $R_0^2 = T\delta/4\pi e^2\rho_0$, and $I_1(\phi_1)$ is a modified Bessel function [5].

A limit passage to Eq. (1.3.5) realizes in the case of a deep potential well

$$\phi(r) = \phi_0 - \tilde{\phi}(r), \quad \phi_0 = 2(l/\pi a)^2 \gg \tilde{\phi}(r)$$

when the longitudinal size a of bunch is small in comparison with the modulation period l of the beam. Replacing the Bessel function in Eq. (1.4.3) by its asymptotic form

$$I_1(\phi) \simeq (2\pi\phi_0)^{-1/2}\exp\left(\phi_0 - \tilde{\phi}\right), \tag{1.4.4}$$

we get Eq. (1.3.5). Note that in comparison with Eq. (1.3.5), the density distribution (1.4.2) is a Gaussian function only near the beam axis, $r = 0$, and for $k|z'| \ll 1$.

The numerical calculation shows that the equation (1.4.3) has a soliton-like solution satisfying the conditions, $\phi(0) = \phi_0$, and $\phi'(0) = 0$, and monotonically decreasing in the region $\phi \ll 1$,

$$\phi(r) \sim K_0(\sigma r), \quad r \to \infty, \tag{1.4.5}$$

where $K_0(\sigma r)$ is a Mcdonald function (a modified Bessel function of the second kind) and

$$\sigma^2 = k^2 - \frac{1}{2R_0^2}, \exp(-\phi_0) > 0.$$

Examples of these solitons for different ϕ_0 are presented in Fig. 1.4.1.

The depth of potential well ϕ_0 as a function of R_0/l is given in Fig. 1.4.2. It follows from this figure that in the limit $R_0/l \to 0$, the quantity ϕ_0 becomes infinite. However, in the latter case the linearity condition of the equation of motion for plasma electrons,

$$\phi_0 \ll (kR_0)^2 \frac{mv^2}{T} \sqrt{\delta}. \tag{1.4.6}$$

may be violated. By analogy with Eq. (1.3.7), the radial oscillations are observed, if the function $\phi_0(R_0/l)$ deviates from its optimal value represented by the curve in Fig. 1.4.2. In actuality, the condition for the capture of the electrons in the potential well holds only for $\delta > 0$.

The results of numerical integration of Eqs. (1.4.2) and (1.4.3) are illustrated in Fig. 1.4.3 and Fig. 1.4.4, respectively.

Note that the Maxwell-Boltzmann distribution function used above causes the beam density to have a tail of fast electrons. Because of this, a part of particles are not captured by the wave and the non-zero asymptotic, $\rho_\infty = \rho_0 e^{-\phi_c}$, occurs as $r \to \infty$. However this value becomes small exponentially if the potential at the center of bunch is large enough, $\phi_0 \gg 1$.

§1.5. Captured particles.

It is well known that the distribution function of the beam satisfy the kinetic equation

$$\frac{\partial f}{\partial t} + \mathbf{v}\frac{\partial f}{\partial \mathbf{r}} - \frac{e}{m}\frac{\partial \varphi}{\partial \mathbf{r}}\frac{\partial f}{\partial \mathbf{v}} = 0 \tag{1.5.1}$$

which solution in the most general form is an arbitrary function of the electron energy

$$f(\mathbf{v}, \varphi) = f_0(m\mathbf{v}^2/2 + e\varphi)$$

and depends on the initial distribution function, $f_0(v)$.

It is shown in §1.4 that Maxwell-Boltzmann distribution function

$$f_M(\mathbf{v}, \varphi) = \rho_0 \left(\frac{m}{2\pi T}\right)^{3/2} \exp\left(-\frac{mv^2/2 + e\varphi}{T}\right)$$

does not take to account the capture of the fast beam electrons by the wave. Therefore, following [7], we shall assume

$$f(\mathbf{v}, \varphi) = \rho_0 \left(\frac{m}{\pi e \varphi_m}\right)^2 \sqrt{V^2(\varphi) - v^2}\, \theta(\varphi), \tag{1.5.2}$$

where ρ_0 is the electron density at the bottom of the potential well, φ_m, $V^2(\varphi) = -2e\varphi/m$ $(e < 0)$. The discontinuous function on the right-hand side of Eq. (1.5.2),

$$\theta(\varphi) = \begin{cases} 1 & \varphi \geq 0, \\ 0 & \varphi < 0. \end{cases}$$

reflects the condition for capture of the electrons by the wave.

Integrating (1.5.2) with respect to velocities with weight $4\pi v^2$ in the range from zero to V, we find the density of the captured particles as a function of potential

$$\rho(\varphi) = \rho_0 \left(\frac{\varphi}{\varphi_m}\right)^2 \theta(\varphi). \tag{1.5.3}$$

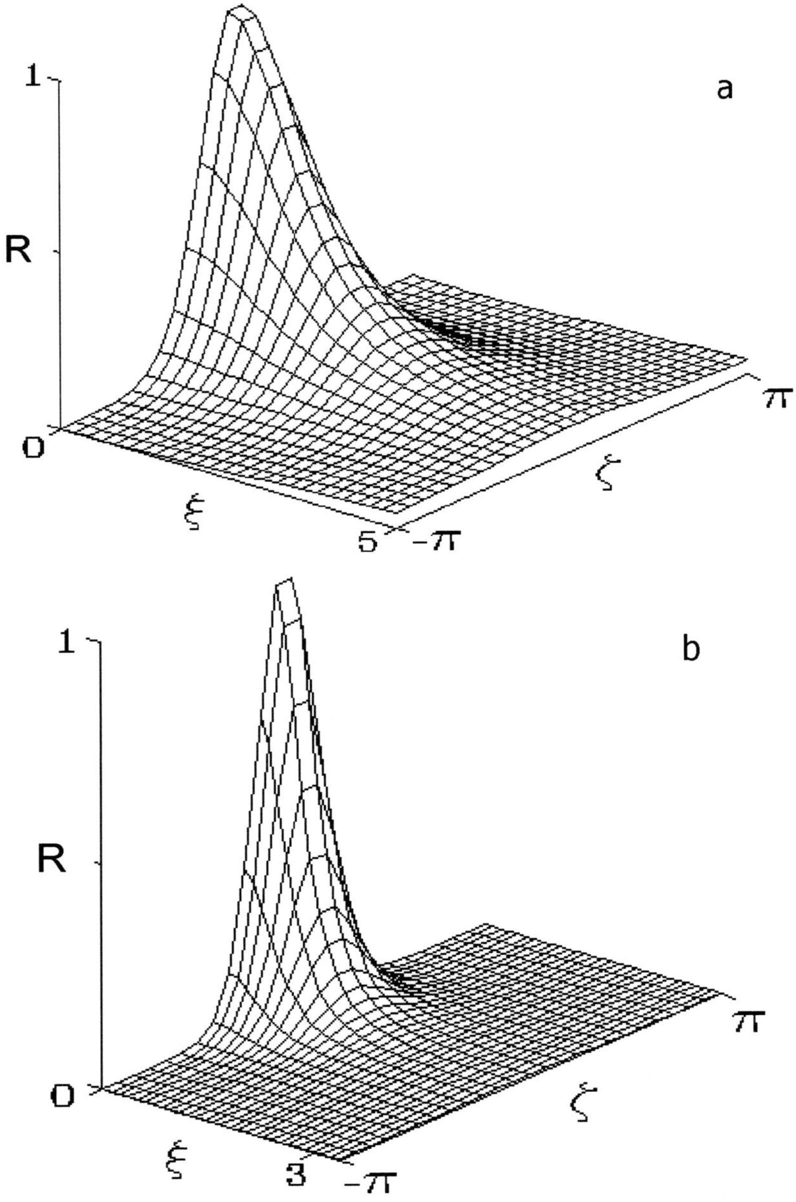

Fig. 1.4.3. Density of Boltzmann's bunches R=ρ/ρ_0 for (a) ϕ_0=2.8 and (b) ϕ_0=5.5; $\xi=kx$, and $\zeta=kz'$.

Substituting (1.5.3) into (1.3.1), and taking into account the form of the right side of the equation, we look for a solution in the form

$$\varphi(r, z') = \varphi_m \Phi(r) \cos kz', \quad k = 2\pi/l. \tag{1.5.4}$$

As a result we obtain a nonlinear equation for the radial part of the potential:

$$\xi^{-1}(\xi\Phi')' - \Phi = -q\Phi^2, \tag{1.5.5}$$

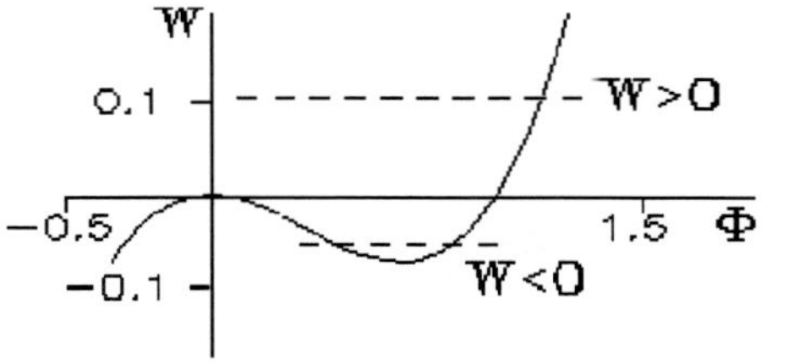

Fig. 1.5.1. Radial potential well for q=1.5.

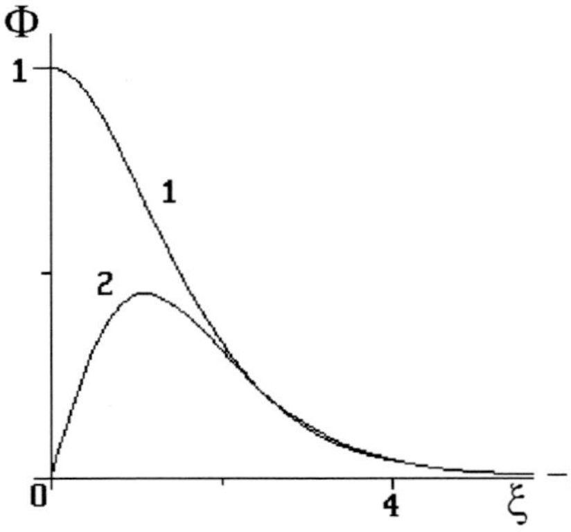

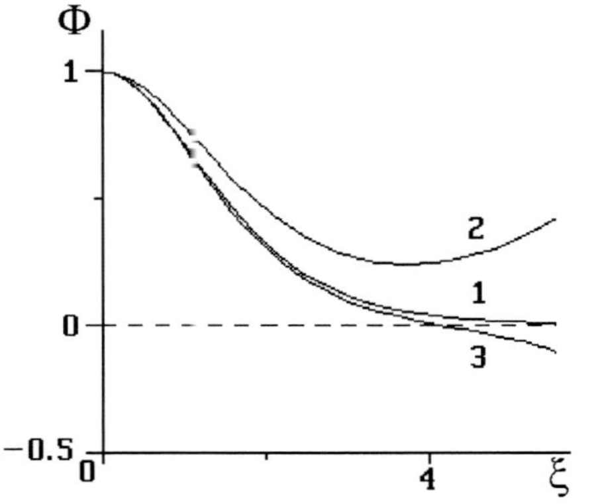

Fig. 1.5.2. Soliton solution q=2.39.: (1) shape of the potential and (2) radial electric field as a function of ξ,

Fig. 1.5.3. Potential as a function of coordinate for various values of the parameter q: (1) 2.39; (2) 2; (3) 2.45.

where $q = 2|e|\rho_0 l^2/3\pi^2\delta\varphi_m$, $\delta = \omega_m/\omega_p > 0$ and the prime denotes differentiation with respect to the variable $\xi = 2\pi r/l$. In calculating the integral in Eq. (1.5.5) we take account of the condition for capture of electrons by the wave, which for $\Phi > 0$ corresponds to the region $|z| \leq l/4$, and

$$\int_0^{\pi/2} \cos^3 t\,\theta(\cos t)\;dt = \frac{2}{3}.$$

For the problem considered the positive solution of Eq. (1.5.5) is physically meaningful, and reaches a maximum $\Phi(0) = 1$ and $\Phi'(0) = 0$ on the beam axis, and falls monotonically with increase of ξ in proportion to the MacDonald function $K_0(\xi)$.

Since we cannot find the integral of Eq. (1.5.5), we shall investigate this equation qualitatively, using the analogy with the problem of motion of a mechanical particle in a potential field [8], and omitting the "friction force" $\xi^{-1}\Phi'$. We represent the right side of Eq. (1.5.5) in the form $dW/d\Phi$. Carrying out the integration, we find the energy conservation law

$$\frac{\Phi'^2}{2} + W(\Phi) = E, \tag{1.5.6}$$

where the constant E corresponds to the total energy of the particle in the potential well $W(\Phi) = q\Phi^3/3 - \Phi^2/2$. The graph of this function whose zero points are $\Phi_1 = 0$ and $\Phi_2 = 3/2$ is shown in Fig. 1.5.1. For $E > 0$, the motion is infinite, and for $E < 0$ it is periodic. In the case $E = 0$ there is a soliton solution with the properties mentioned above, and the point Φ_2 corresponds to the beam axis, where by definition $\Phi_1 = 1$, and therefore, $q = 3/2$.

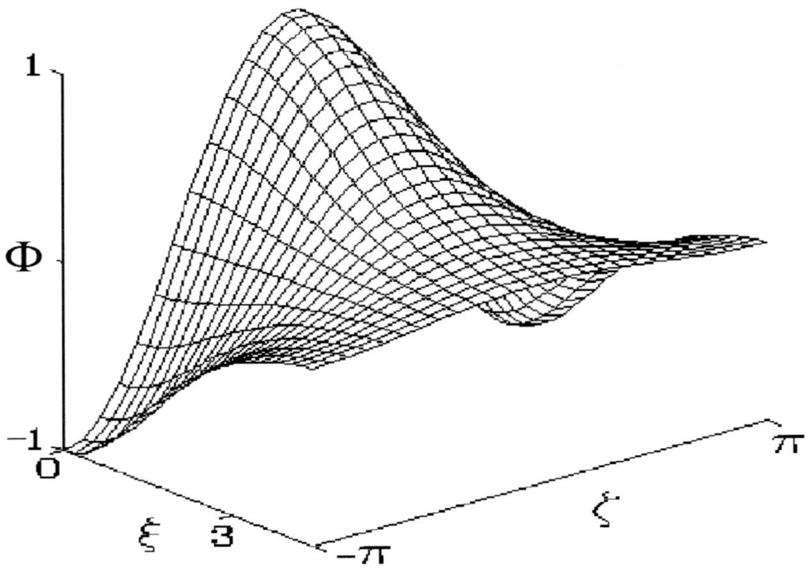

Fig. 1.5.4. Spatial shape of the wave potential.

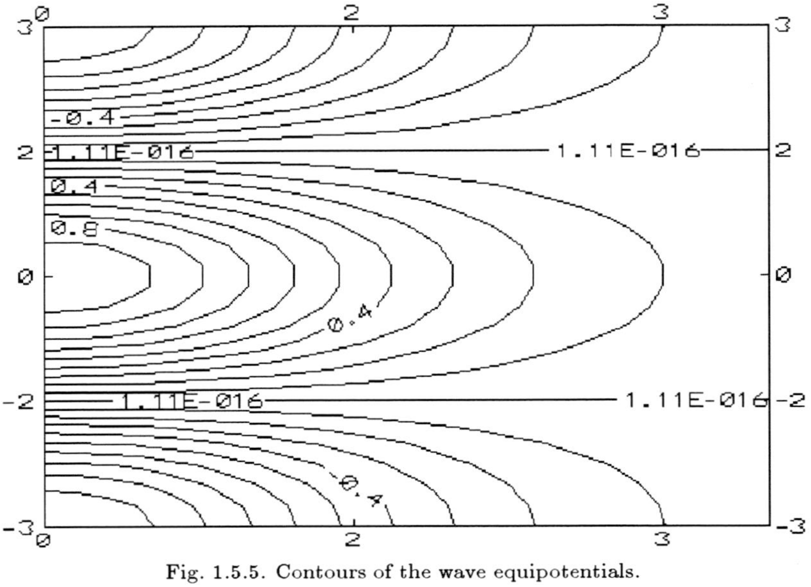

Fig. 1.5.5. Contours of the wave equipotentials.

Numerical integration of Eq. (1.5.5) (allowing for $\xi^{-1}\Phi'$) shows that the solitonic solution corresponds to $q_{\rm sol} = 2.392$, so that the depth of the potential well for each bunch is

$$\varphi_m \simeq 0.028\frac{|e|\rho_0 l^2}{\delta} \simeq 0.1\frac{\rho_0}{\rho_p\delta}\frac{mv^2}{|e|} \tag{1.5.7}$$

(see Fig. 1.5.2). When the parameter q deviates from the optimal value the nature of the solutions of changes, and the curves do not have the required shape (see Fig. 1.5.3). In actuality, the condition for capture of the electrons in the potential well, Eq. (1.5.2), holds only for $\delta > 0$.

Thus, an electron beam modulated with a frequency ω_m close to Langmuir plasma

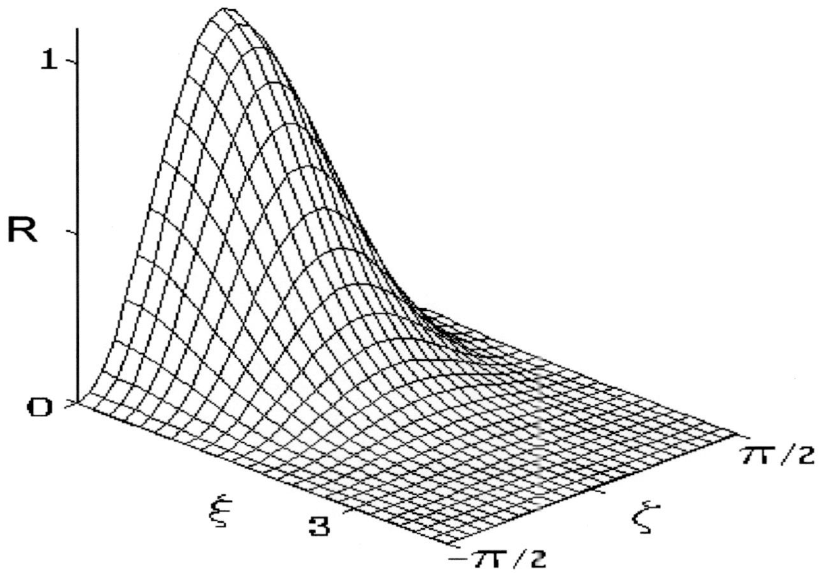

Fig .1.5.6. Density of captured electrons R(ξ,ζ)=ρ/ρ_0 (ρ_0 is the density at the center of bunch, $\xi=kx$ and $\zeta=kz'$).

frequency, with $\delta = 1 - \omega_m/\omega_p \ll 1$, creates a spatial potential wells whose characteristic dimensions in the longitudinal and transverse directions are $l/2$ and l/π, respectively. Figures 1.5.4 - 1.5.6 illustrates the shape of the equipotential surfaces.

It should be noted that in contrast to the Maxwell function when the temperature of beam electrons is an arbitrary value, the kinetic energy of captured electrons is determined by the depth of the potential well, $T \simeq |e|\varphi_m$.

§1.6. Collisional dissipation in the plasma.

The presence of collisional dissipation in the plasma is accompanied by the appearance of phase detuning of the beam with the wave and collapse of the self-focusing regime when the modulating frequency coincides with the Langmuir plasma frequency. Therefore, the frequency detuning is limited to low values of the inequality $\delta \gg \nu_c/\omega_p$, where ν is the frequency of collisions of plasma electrons with ions, taking into account energy dissipation in the plasma.

The singularity which appears in right-hand side of Eq. (1.5.5) in the limit $\delta \to 0$ can be removed if the friction force $-m\nu_c\mathbf{v_p}$ is taken into account in the equation of motion for plasma electrons. Accordingly, the additional term appears in the wave equation

$$\left(\frac{\partial^2}{\partial t^2} + \nu_c\frac{\partial}{\partial t} + \omega_p^2\right)\Delta\varphi = -4\pi e\,\frac{\partial^2\rho}{\partial t^2}. \qquad (1.6.1)$$

As was shown above, under resonance conditions in the plasma-modulated beam system one field harmonic dominates. At the same time for $\delta \to 0$ one must calculate the wave phase $\vartheta(r)$, as well as the amplitude $\Phi(r)$:

$$\varphi(t,r) = \varphi_m\Phi(r)\cos\left[kz' + \vartheta(r)\right],$$

$$z' = z - vt, \quad k = 2\pi/l, \quad \delta = 1 - \omega_m/\omega_p \ll 1,$$

and the order of the system of equations is increased in comparison with Eq. (1.6.1) [7]:

$$\begin{aligned} \xi^{-1}(\xi\Phi')' - (1+\vartheta'^2)\Phi &= -q\Phi^2, \\ \xi^{-1}(\xi\Phi^2\vartheta')' &= -q\Gamma\Phi^3, \end{aligned} \tag{1.6.2}$$

where

$$q = \frac{2|e|\rho_0 l^2}{3\pi^2\varphi_m\delta(1+\Gamma^2)}, \qquad \Gamma = \frac{\nu_c}{2\omega_m\delta}.$$

The prime denotes differentiation with respect to variable $\xi = 2\pi r/l$, and the following equality

$$\int_\pi^\pi \rho(\cos t)\exp(-it)\,dt = \int_\pi^\pi \rho(\cos t)\cos t\,dt$$

is used in deriving Eq. (1.6.2) so that the right-hand side of the first Eq. (1.6.2) coincides with the Eq. (1.5.5).

The presence of energy dissipation in the plasma in principle alters the form of the solution for the potential, since for $\Gamma > 0$ system (1.6.2) has no solitonic solution. This can be verified by a simple qualitative analysis, using the relation

$$\vartheta' = -\frac{q\Gamma}{\xi\Phi^2}\int_0^\xi \Phi^3\xi d\xi \tag{1.6.3}$$

which follows from the second part of Eq. (1.6.2).

Clearly, for $\Gamma \ll 1$ and $\Phi(0) = 1$, $\Phi'(0) = \vartheta'(0) = 0$ the solution of Eq. (1.6.2) is asymptotically close to the solitonic solution of Eq. (1.5.5) as long as $\vartheta'^2 \ll 1$ and the corresponding term in the second equation can be neglect. However, far enough from the beam axis we have $\xi \to \infty$ where $\Phi \to 0$, and the integral in (1.6.3) remains finite, this condition breaks down for $\xi \simeq \Gamma\Phi^2$, and the loss of phase of the wave appears as changes of its amplitude. Correspondingly, in the first equation of (1.6.2) there is a "reversal" point where the potential reaches a minimum (see Fig. 1.6.1), and then increases with increase of the coordinate. Since the asymptotic form $\xi_{\min}$ is valid near

$$\Phi \simeq K_0(\xi) \simeq \xi^{-1/2}\exp(-\xi),$$

it follows from the condition $|\vartheta'| \ll 1$ that

$$\xi_{\min} \simeq \frac{1}{2}\ln\Gamma. \tag{1.6.4}$$

Thus, in a plasma with energy dissipation $\Gamma > 0$ there is no solitonic solution corresponding to the spatial potential well. However, far enough from the resonance

$$\Gamma = \frac{\nu_c}{2\omega_m\delta} \ll 1, \tag{1.6.5}$$

there is an asymptotical solution close to solitonic. This inequality limits the depth of the potential well of Eq. (1.5.7) on the low value side of the parameter δ.

§1.7. Equilibrium of laminated electron-ion beam in a plasma.

A bunched electron beam creates a potential distribution, which presents alternating peaks and valleys. The bunches are trapped in the valleys, while the excess of electrons is carried within the peaks. In turn, the peaks represent potential wells for the positron

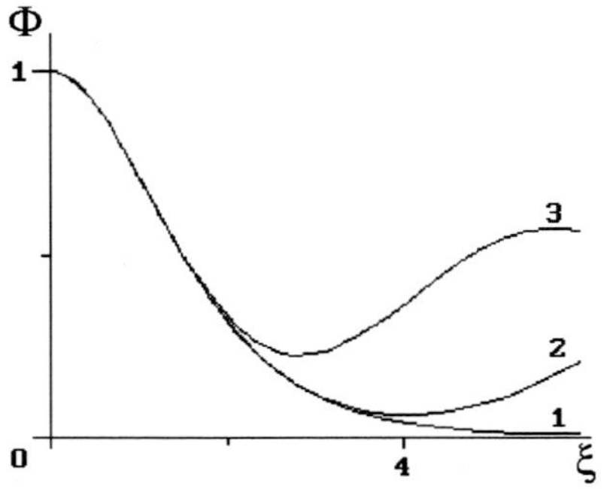

Fig. 1.6.1. Curves illustrating deviation of the solution from solitonic for different values of the parameter Γ: (1) 0.001; (2) 0.01; (3) 0.1.

bunches that move synchronously with electrons. Principally, this circumstance allows the realization of a laminated electron-ion beam in a dense plasma [8].

Following the §1.1, let us analyze the process of laminated electron-ion beam self-focusing in a dense plasma. The beam is supposed to be preliminarily divided into a number of bunches. In accordance with the charged disk model, let us consider the electron and ion disks of surface charge density $\sigma_e(r)$ and $\sigma_i(r)$, respectively, moving with velocity $\mathbf{v}$. We suppose the spacing between the disks is equal l. The current density, produced in the plasma by the system of these disks, can be written as

$$j_z = v \sum_{s=-\infty}^{\infty} \big[-|e|\sigma_e s(r)\delta(z - z_{es} - vt - sl) + q_i\sigma_i s(r)\delta(z - z_{is} - vt - sl)\big], \tag{7.1.1}$$

where z_{es} and z_{is} are the coordinates of the electron and ion disks in a frame of reference of the beam, respectively. Under the conditions, $|\omega_m - \omega_p| \ll \omega_p$ and $r_{ei} \ll l$, we find the equation for the radial component of the field (by analogy with Eq. (1.1.6)):

$$\frac{1}{r}\frac{d}{dr}(rE_r) = \frac{4k}{\epsilon(\omega_m)}\big[-|e|\sigma_e(r)\cos k(z' - z_e) + q\sigma_i(r)\cos k(z' - z_i)\big], \tag{1.7.2}$$

where $z' = z - vt$.

Making use of the relation $E_r = -d\varphi/dr$ in the above equation, one can express the surface densities σ_e and σ_i as a function of potentials $\varphi_e \equiv \varphi(r, z_e)$ and $\varphi_p \equiv \varphi(r, z_i)$, the corresponding potentials being applied to the electron and ion disks:

$$\sigma_e(r) = \sigma_e(0)\exp\left(\frac{|e|\varphi_e}{T_e}\right), \quad \sigma_i(r) = \sigma_i(0)\exp\left(-\frac{q\varphi_i}{T_i}\right). \tag{1.7.3}$$

Taking account of Eq. (1.7.2) and Eq. (1.7.3) one readily obtains the following nonlinear equations for the potentials:

$$\Delta_r\varphi_e = \frac{4k}{|\epsilon|}\left[-|e|\sigma_{0e}\exp\left(\frac{e\varphi_e}{T_e}\right) + q\sigma_{0i}\exp\left(-\frac{q\varphi_i}{T_i}\right)\cos\eta\right],$$
$$\Delta_r\varphi_i = \frac{4k}{|\epsilon|}\left[-|e|\sigma_{0e}\exp\left(\frac{e\varphi_e}{T_e}\right)\cos\eta + q\sigma_{0i}\exp\left(-\frac{q\varphi_i}{T_i}\right)\right], \tag{1.7.4}$$

where the following notations used:

$$\Delta_r = r^{-1}d/dr\,(rd/dr)\,, \quad \eta = k(z_e - z_i),$$
$$\epsilon(\omega_m) \equiv -|\epsilon|.$$

To proceed, let us first consider the case of $\cos\eta = 0$. This condition implies, that the electron disk displacement with respect to ion disk is equal to a quarter of wavelenth. As it can be readily seen, the system Eqs. (1.7.4) is equivalent to the system of two similar equations, any of which describes the radial equilibrium in the field of both electron and ion beams. As was shown in §1.1, the solutions of these equations can be written as Eq. (7.1.1).

To find the solution for the general case $\cos\eta \neq 0$ it is convenient to represent the system Eqs. (1.7.4) in the following manner:

$$(\xi\Phi_e')' = 8\xi[-\exp(\Phi_e) + Q\cos\eta\exp(-\Phi_i)],$$
$$(\xi\Phi_i')' = 8R\xi[-\cos\eta\exp(\Phi_e) + Q\exp(-\Phi_i)], \tag{1.7.5}$$

here we use dimensionless variables

$$\Phi_e = \frac{|e|\varphi_e}{T_e}, \quad \Phi_i = \frac{q\varphi_i}{T_i}, \quad \xi = \frac{r}{r_{0e}},$$
$$Q = \frac{q}{|e|}\frac{\sigma_{0i}}{\sigma_{0e}}, \qquad R = \frac{q}{|e|}\frac{T_e}{T_i}.$$

If the radii of the electron and ion disks are equal, the following relations should be fulfilled:

$$\Phi_i = -\Phi_e = 2\ln(1 + \xi^2/\xi_0^2),$$
$$\xi_0^{-2} = 1 - Q\cos\eta = R(Q - \cos\eta). \tag{1.7.6}$$

Under these condition the distributions of beam particle density can be determined with the help of Eq. (1.1.11), where one should replace

$$r_{e,i}^2 \to r_{ei}^2 = \frac{r_e^2}{1 - Q\cos\eta}. \tag{1.7.7}$$

(This solution does exists provided the inequality $\xi_0^2 > 0$).

In the phase domain $\eta < \pi/2$ ($\cos\eta > 0$), when the charges of the electron and ion disks completely or partially neutralize each other, the radius r_{ei} of the laminated beam exceeds the radius r_e of the electron beam: $r_{ei} > r_e$. In the phase domain $\eta > \pi/2$ ($\cos\eta < 0$) the fields of the electrons and protons enhance each other, resulting in the decrease of the laminated beam radius as compare to the electron beam radius: $r_{ei} < r_e$.

For the case of the charge neutralized beam with $Q = R = 1$ the minimum value of radius is equal to

$$r_{ei} = r_e/\sqrt{2}, \quad \cos\eta = -1. \tag{1.7.8}$$

Let us suppose that both electron and ion are in the Boltzmann equilibrium with the wave of potential. Using this condition one can express the beam charge density as

$$\rho(t,\mathbf{r}) = q_e n_e(r,z') + q_i n_i(r,z'),$$
$$n_{e,i}(r,z') = n_{0e,i}\exp\left\{\frac{q_{e,i}}{T_{e,i}}[\varphi(0,z_{e,i}) - \varphi(r,z')]\right\}, \tag{1.7.9}$$

where z_e and $z_i = z_e + \pi/k$ are points at the centers of electron and ion bunches, respectively.

Under the resonance condition, when $\alpha \to 1$, the wave of potential tends asymptotically to the harmonic wave and, similar to the beam of the electron bunches §1.4, the solution can be written as follows:

$$\begin{aligned} \varphi &= (T_e/e)\phi(r) \cos kz' \\ n_e(r, z') &= n_{0e} \exp\left[-\phi_0 + \phi(r) \cos kz'\right], \\ n_p(r, z') &= n_{0i} \exp\left[-R\phi_0 - R\phi(r) \cos kz'\right], \end{aligned} \tag{7.1.10}$$

where $q_e = -|e|$, $q_i = q$, and $\phi_0 = \varphi(0, z_e) = -\varphi(0, z_i)$.

The substitution of equation Eq. (7.1.10) into Eq. (1.3.1) and averaging over the spatial period of the beam allows us to obtain the following equation:

$$\begin{aligned} &\frac{1}{r}\frac{d}{dr}\left(r\frac{d\phi}{dr}\right) - k^2\phi \\ &= -\frac{1}{R_0^2}\left[\exp(-\phi_0)I_1(\phi) + Q\exp(-R\phi_0)I_1(R\phi)\right], \end{aligned} \tag{7.1.11}$$

where $I_1(x)$ is a modified Bessel function and $R_0^2 = T_e\delta/4\pi n_{0e}e^2$. In the limiting case $Q \to 0$, this equation describes the equilibrium of the electron bunches in a plasma (see Eq. (1.4.2)).

Compared to the similar system Eq. (1.7.5), which is dependent upon the relative electron and ion disk coordinate (the parameter $\cos\eta$) as well as upon the dimensionless parameters Q, R and R_0, the above Eq. (7.1.11) determines the equilibrium of the electron and ion bunches with the wave potential, the longitudinal dimensions of the bunches being finite and conjugated with the value of the field. Note, that in this case the distance between the maximums of the electron and ion densities Eq. (1.7.9) is equal to half the wavelength. This value corresponds to the minimum beam radius under the approximation of electron and ion disks.

§1.8. Instability of modulated electron-ion beam in a plasma.

The transport of electron-ion beam in a dense plasma accompanies by an excitation of the Langmuir oscillations. To analyze a plasma-beam instability, we assume that beam radius (1.7.8) greatly exceeds a distance between bunches l, $r_{e,i} \gg l$, and we consider that a set of one-dimensional equations has the form

$$\begin{aligned} &\frac{\partial^2 E}{\partial t^2} + \omega_p^2 E = -4\pi\frac{\partial j}{\partial t}, \\ &j = |e|\sigma \sum_{s=-\infty}^{\infty} \left\{\dot{z}_+^{(s)}\delta\left[z - z_+^{(s)}(t)\right] - \dot{z}_-^{(s)}\delta\left[z - z_-^{(s)}(t)\right]\right\}, \\ &\ddot{z}_\pm^{(s)} = \pm\frac{|e|}{m_\pm}E\left[t, z_\pm^{(s)}(t)\right]. \end{aligned} \tag{1.8.1}$$

Here E is the self-consistent electric field acting upon the electron $(-)$ and ion $(+)$ bunches at the points $z_\mp^{(s)}(t)$ [9].

We seek a solution of Eqs.(1.8.1) in the form

$$E(t, z) = \mathrm{Re}\left[E(t)\exp\left(i\omega t - ikz\right)\right], \tag{1.8.2}$$

where $\omega = kv_0 = 2\pi v_0/l$ is the modulation frequency of the beam closed to the plasma frequency ω_p, $|\omega - \omega_p| \ll \omega_p$, and E is a slowly varying function, $\dot{E} \ll E$.

Substituting (1.8.2) into (1.8.1) and averaging over the longitudinal period of the beam l, we obtain the following set of ordinary differential equations

$$\begin{aligned} \dot{E} + i(\omega - \omega_p)E &= -\frac{4\pi|e|\sigma v_0}{l}\left(e^{ik\delta z_+} - e^{ik\delta z_-}\right), \\ \delta\ddot{z}_\pm &= \pm\frac{|e|}{m_\pm}\mathrm{Re}\left\{E(t)\exp\left[-ik\delta z_\pm(t)\right]\right\}, \end{aligned} \tag{1.8.3}$$

where the dot means the derivative with respect to time. We take into account a spatial periodicity of the beam and put $z_\pm^{(s)}(t) = sl + v_0 t + \delta z_\pm(t)$ so that the displacement of the electron or ion layer $\delta z_\pm(t)$ does not depend on index s.

Energy conservation law which follows from the Eq. (1.8.3) has the form

$$2\frac{\sigma}{l}v_0\left(m_+\delta\dot{z}_+ + m_-\delta\dot{z}_-\right) + \frac{|E|^2}{4\pi} = \text{const.} \tag{1.8.4}$$

The first term in the left-hand side of (1.8.4) is the density of beam energy and second one is the density of Langmuir oscillations in a plasma. The velocity variation is small $|\delta\dot{z}_\pm| \ll v_0$, if the inequality $\sigma/l \ll n_p$ holds. Therefore the terms of order $\delta\dot{z}_\pm^2$ are omitted in Eq. (1.8.4).

Since an ion mass m_+ greatly exceeds an electron one m_-, the contribution of beam ions is essential only in the energy integral. Putting $\delta z_+ = 0$ in Eq. (1.8.3), we rewrite the equations as

$$\begin{aligned} &\left[\xi^{(4)} + \frac{q^2}{2}\dot{\xi}\sin\xi\right]\left(\Delta + \dot{\xi}\right) - \ddot{\xi}\left[\xi^{(3)} + \frac{q^2}{2}(1-\cos\xi)\right] \\ &+ \left(\Delta + \dot{\xi}\right)^3\ddot{\xi} + \frac{q^2}{2}\left(\Delta + \dot{\xi}\right)^2\sin\xi = 0, \end{aligned} \tag{1.8.5}$$

where the dimensionless variables are

$$\tau = \omega t, \quad \xi = k\delta z_-, \quad \Delta = 1 - \frac{\omega_p}{\omega},$$

$$q^2 = \frac{8\pi e^2\sigma}{m\omega^2 l} \ll 1.$$

1. Linear theory.

Stationary points of differential equation (1.8.5) are defined by the solutions of the algebraic equation $\sin\xi_0 = 0$. Using the linear approximation $\xi = \xi_0 + \xi_1$, $|\xi_1| \ll 1$, we get the equation

$$\xi_1^{(4)} + \left(\Delta^2 - q^2\sin^2\frac{\xi_0}{2}\right)\ddot{\xi}_1 + \frac{\Delta q^2}{2}\cos\xi_0\,\xi_1 = 0. \tag{1.8.6}$$

To proceed, let us first consider the case of $\xi_0 = 0$. This condition implies that the initial electron plane displacement from the ion one is equal zero. Substituting $\xi_1 = \xi_1(0)\exp(\lambda\tau)$ into Eq. (1.8.6), we get the dispersion relation

$$\lambda^2 = -\frac{\Delta^2}{2} \pm \sqrt{\frac{\Delta^4}{4} - \frac{\Delta q^2}{2}}. \tag{1.8.7}$$

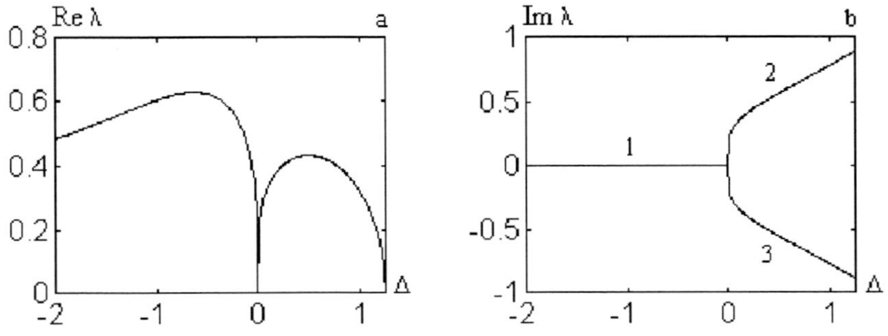

Fig. 1.8.1. (a) Growth rate and (b) detuning of the phase velocity as a function of deviation (in units of $q^{2/3}$). Curve 1 - 3 correspond to Cherenkov, anomalous, and normal Doppler effect, respectively.

We are interested in the roots of Eq. (1.8.7) which have a positive real part, $\mathrm{Re}\,\lambda > 0$, that corresponds to the increase in the energy of the plasma oscillations. These roots occur for both signs of the frequency detuning. However, for $\Delta < 0$ the increase in the energy of the Langmuir oscillations occurs irrespective of the relation between the parameters Δ and q, whereas for $\Delta > 0$ there is a threshold value $q_c^2 = \Delta^3/2$ and the plasma-beam system is unstable only for $q^2 > q_c^2$. The curves $\mathrm{Re}\,\lambda$ and $\mathrm{Im}\,\lambda$ are shown in Fig. 1.8.1.

A dependence of the growth rate $\mathrm{Re}\,\lambda > 0$ of the sign of the detuning Δ has the following physical explanation. The spatial displacement of electrons relatively ions leads to the appearance of electrostatic field in a frame of reference of the beam. This field acts upon the electrons depending on the sign of the plasma permittivity,

$$\epsilon = 1 - \frac{\omega_p^2}{\omega^2} \simeq 2\Delta.$$

The collective electric field attracts an electron layer to ion one for $\omega > \omega_p$ when the plasma permittivity is positive. To proceed, we use first equation (1.8.3) in asymptotic case when the condition $q^2 \gg \Delta^3$ is fulfilled,

$$\mathcal{E} = -\frac{q^2}{\epsilon}\,\xi, \quad \mathcal{E} = \frac{eEl}{2\pi m_- v_0^2}.$$

Thus, the electron layer becomes an oscillator which frequency $\Omega_b = q/\sqrt{\epsilon}$ is the same as the frequency of the longitudinal beam oscillations in a plasma at $\epsilon > 0$. It results from the inequality $q^2 \ll \Delta^3$ that the frequency Ω_b is small in comparison with the plasma frequency Ω_p in a frame of reference of the beam.

Langmuir oscillations become unstable when the inequality $\Omega_b \ll \Omega_p = \Delta$ is violated and the resonance plasma-beam interaction occurs. For $q > q_c$, Eq. (1.8.7) has two roots with a positive real part,

$$\lambda = \left(\frac{q^2\Delta}{8}\right)^{1/4}\left(\sqrt{1-\sqrt{\frac{\Delta^3}{2q^2}}} \pm i\sqrt{1+\sqrt{\frac{\Delta^3}{2q^2}}}\right). \tag{1.8.8}$$

Carrying out the phase velocities of this waves, $v_{\mathrm{ph}}^{\pm} = k^{-1}\left(\omega \pm \mathrm{Im}\,\lambda\right)$, one can find that signs "+" and "-" correspond to the anomalous and normal Doppler effects, respectively. The maximum growth rate of instability follows from (1.8.8),

$$\delta_{\mathrm{m}+} = \frac{\sqrt{3}}{4}\,q^{2/3}\,\omega, \quad \Delta_{\mathrm{m}+} = \frac{1}{2}\,q^{2/3}\,\omega. \tag{1.8.9}$$

When $\Delta \to 0$, the asymptotic solution of algebraic equation (1.8.8) has the form $\lambda \sim (\Delta q^2)^{1/4}$.

In the range $\omega < \omega_p$ ($\Delta = -|\Delta|$), the plasma permittivity is negative and the collective electric field repulses beam electrons from ions and the amplitude of the small perturbations increases exponentially with respect to time. Since the phase velocity of the wave is synchronized with the velocity of the beam, this mechanism of the instability is associated with the Cherenkov resonance. By analogy with Eq. (1.8.9) we get

$$\delta_{m-} = \left(\frac{q}{2}\right)^{2/3}\omega, \quad \Delta_{m-} = -\left(\frac{q}{2}\right)^{2/3}\omega. \tag{1.8.10}$$

The dispersion equation for $\xi_0 \neq 0$ can be found from Eq. (1.8.9) by means of the transformation

$$\Delta \to \Delta\cos\xi_0, \quad \Delta^2 \to \Delta^2 - q^2\sin^2\frac{\xi_0}{2}.$$

It is easy to see that in the range $|\Delta| \simeq q^{2/3}$, which corresponds to maximal growth rate of instability the additional term is small.

2. Nonlinear phase oscillations.

The goal of the present study is to obtain analytical solutions of nonlinear equation (1.8.5) in two asymptotic cases when the growth rate of instability is small in comparison with its maximum values (1.8.9) and (1.8.10).

We assume that in the range $\omega < \omega_p$, the inequality $q^2 \ll |\Delta|^3$ holds. Under this condition, the frequency detuning $|\Delta|$ greatly exceeds the growth rate $\delta_- = q/\sqrt{2|\Delta|}$ and Eq. (1.8.5) coincides with a known equation

$$\ddot{\xi} - \delta_-^2 \sin\xi = 0. \tag{1.8.11}$$

which determines the oscillations of a nonlinear pendulum with the equilibrium position at the topmost point. This equation appears arises from a coordinate dependence of the electric field in Eq. (1.8.3), $\mathcal{E} = \delta_-^2 \sin\xi$.

The first integral of Eq. (1.8.11)

$$\dot{\xi}^2 = \dot{\xi}_0^2 + 2\delta_-^2\left(\cos\xi_0 - \cos\xi\right) \tag{1.8.12}$$

allows us to determine a displacement of the electron layer relatively ion one. In the general case the solution of Eq. (1.8.12) can be represented in terms of elliptic functions. However for $\dot{\xi}_0^2 \ll 2\delta_-^2$ and $\xi_0 \ll 1$, the following simple asymptotic solution exists

$$\tan\frac{\xi}{4} = \exp\psi, \quad \mathcal{E} = 2\delta_-^2\frac{\sinh\psi}{\cosh^2\psi}, \quad \psi = \delta_-\tau + \ln\frac{\xi_0}{4}. \tag{1.8.13}$$

In the range of frequencies greater than the resonance frequency, $\Delta > 0$, an analytic solution of Eq. (1.8.5) can be obtained near the instability threshold, $q - q_c \ll q_c$, where the growth rate

$$\delta_+ = \frac{\Delta}{2^{3/2}}\sqrt{q^2/q_c^2 - 1}, \quad q - q_c \ll q_c = \sqrt{\Delta^3/2}$$

remains small compared with Δ. In the limit $|\xi| \ll 1$, expanding the left-hand side of Eq. (1.8.5) in a series and omitting the terms of order $\sim \xi^4$, we obtain the following asymptotic equation

$$\begin{aligned} &\xi^{(4)} + \Delta^2\ddot{\xi} + \frac{1}{2}\Delta q^2\xi + \frac{3}{4}\Delta^3\xi\dot{\xi} + 3\Delta\dot{\xi}\ddot{\xi} \\ &+ \frac{1}{\Delta}\left[\dot{\xi}\xi^{(4)} - \ddot{\xi}\xi^{(3)}\right] + \frac{1}{2}\Delta^2\xi\dot{\xi}^2 - \frac{1}{8}\Delta^2\xi^2\ddot{\xi} + 3\ddot{\xi}\dot{\xi}^2 - \frac{\Delta^4}{24}\xi^3 = 0. \end{aligned} \tag{1.8.14}$$

The solution of above Eq. (1.8.14) can be sought in the form

$$\xi = x(\tau)\cos\left[\frac{\Delta\tau}{\sqrt{2}} + \vartheta(\tau)\right] + y(\tau)\sin\left[2\left(\frac{\Delta\tau}{\sqrt{2}} + \vartheta(\tau)\right)\right],$$
$$\dot{x} \ll \Delta x, \quad \dot{y} \ll \Delta y, \quad \dot{\vartheta} \ll \Delta\vartheta. \tag{1.8.15}$$

Substituting (1.8.15) into (1.8.14), averaging the resulting expressions over the period $2^{3/2}\pi/\Delta$, and discarding terms of order $\delta_+/\Delta \ll 1$, we find the set of slow equations

$$\ddot{x} - \delta_+^2 x + 2\left(\frac{3\Delta}{16}\right)^2 x^3 + \frac{\Delta^3}{2^{5/2}}\dot{\vartheta} = 0, \quad \ddot{\vartheta} = \frac{5\dot{x}}{2x}\dot{\vartheta}, \quad y = \frac{x^2}{6\sqrt{2}}. \tag{1.8.16}$$

Expressing $\dot{\vartheta}$ in terms of x from second Eq. (1.8.16), we can rewrite the first equation in the form

$$\ddot{x} - \delta_+^2 x + 2\left(\frac{3\Delta}{16}\right)^2 x^3 + \Delta^3\dot{\vartheta}_0\left(\frac{x}{2x_0}\right)^{5/2} = 0. \tag{1.8.17}$$

Integrating Eq. (1.8.16), we find an energy integral

$$\dot{x}^2 - \delta_+^2 x^2 + \left(\frac{3\Delta}{16}\right)^2 x^4 + \frac{\Delta^3}{7\sqrt{2}}\dot{\vartheta}_0 x_0\left(\frac{x}{x_0}\right)^{7/2} = \mathrm{C}. \tag{1.8.18}$$

The solution of Eq. (1.8.12) can be represented in terms of elliptic functions. In the limit $\dot{\vartheta}_0 \to 0$, the asymptotic soliton-like solution exists

$$x = x_m \cosh^{-1}\left(\operatorname{arccosh}\frac{x_m}{x_0} - \delta_+\tau\right),$$
$$x_0 \ll x_m, \quad \dot{x}_0 \ll \delta_+ x_m, \quad x_m = 16\delta_+/3\Delta. \tag{1.8.19}$$

According to Eq. (1.8.3), the electric field satisfies the equation

$$\dot{\mathcal{E}} + i\Delta\mathcal{E} - iq^2\xi/2 = 0 \tag{1.8.20}$$

when the inequality $\xi \ll 1$ is fulfilled. Substituting (1.8.15) into (1.8.20) and carrying out the integration, we find

$$\mathcal{E} = \sqrt{\frac{3}{2}\frac{q^2}{\Delta}}\, x(\tau)\cos\left[\frac{\Delta\tau}{\sqrt{2}} + \vartheta(\tau) + \arctan\frac{1}{\sqrt{2}}\right], \tag{1.8.21}$$

where $x(\tau)$ and $\vartheta(\tau)$ are slow functions, $\Delta \simeq (2q^2)^{1/3}$. Taking into account (1.8.19) and (1.8.21), one can find the following estimation: $\mathcal{E}_{\mathrm{m}} \simeq q^{2/3}\delta_+$.

3. Numerical integration.

Numerical integration of Eq. (1.8.3) described below enable us to get the solution for arbitrary parameters q and δ. To proceed, we introduce a slow dimensionless variables

$$Y = \mathcal{E}/q^{4/3}, \quad D = \Delta/q^{2/3}, \quad T = q^{2/3}\tau,$$
$$\xi_e = k\delta z_-, \quad \xi_i = k\delta z_+, \quad \mu = m/M.$$

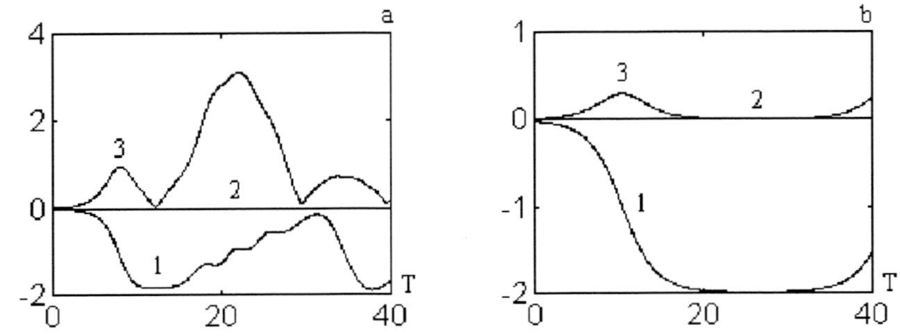

Fig. 1.8.2. Coordinates of (1) electron, (2) ion bunches, and (3) field amplitude versus time for deviation (a) $D=-2^{-1/3}$ and (b) $D=-3$ in case of Cherenkov resonance.

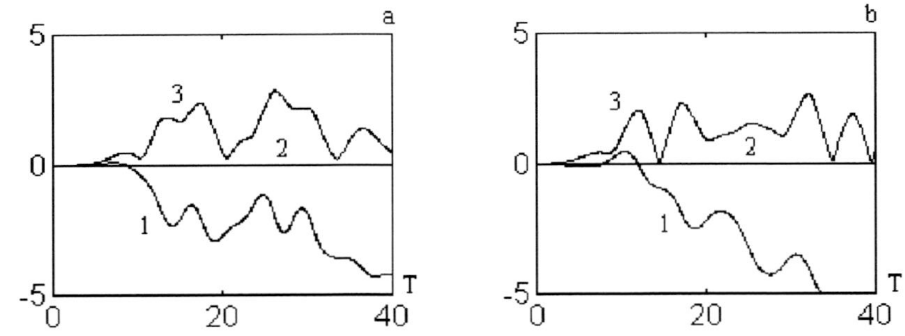

Fig. 1.8.3. Coordinates of (1) electron, (2) ion bunches, and (3) field amplitude versus time for deviation D=0.5 under the conditions of (a) normal and (b) anomalous Doppler effects.

As a result, the set of equations takes the form:

$$Y' + iDY = \frac{q^2}{2}\left[\exp\left(i\xi_e\right) - \exp\left(i\xi_i\right)\right],$$
$$\xi_e'' = -\mathrm{Re}\left[Y\exp\left(-ik\xi_e\right)\right], \qquad \xi_i'' = \mu\,\mathrm{Re}\left[Y\exp\left(-ik\xi_i\right)\right], \tag{1.8.22}$$

where the prime denotes differentiation with respect to variable T.

Numerical integration shows that in the case of Cherenkov resonance, $D < 0$, the repulsion of the electrons from the ions (moving with practically constant velocity, $\mu = 10^{-3}$) leads to an excitation of the nonlinear beam oscillations represented in Fig. 1.8.2a. Curves in Fig. 1.8.2b correspond to the limiting case of Eq. (1.8.11).

For $D > 0$, when an ionic layer attracts electrons, the instability in the presence of the anomalous or normal Doppler effect occurs (see Fig. 1.8.3).

According to energy conservation law (1.8.4), the excitation of plasma-beam oscillations for both signs of detuning D accompanies with the slowing of the laminated electron-ion beam.

CHAPTER 2

ELECTRON BUNCHES IN A HEATED NONLINEAR MAGNETIZED PLASMA

A practical implementation of the electrostatic self-focusing of modulated electron beam in a plasma essentially depends on the conditions of plasma creating and confinement. The mathematical difficulties arising in the search for analytical solutions of nonlinear partial differential equations allow only the conditions under which beam equilibrium can occur in the plasma and the shape of the electron clumps to be determined. More realistic parameters of the plasma. i.e., a finite temperature of plasma electrons, a dependence of plasma permittivity on the field amplitude, and an external magnetic field are taken into consideration in this chapter.

§2.1. Unbounded plasma modes in a finite magnetic field.

In order to account for the thermal motion of plasma electrons and the external magnetic field, we include the gradient of the kinetic pressure and the Lorentz force in the equation of motion:

$$\frac{\partial \mathbf{v}_p}{\partial t} - \omega_B[\mathbf{v}_p \mathbf{n}] = \mathbf{E} - \frac{T_p}{m n_p} \nabla \tilde{n}_p, \tag{2.1.1}$$

where n_p and T_p are the background plasma density and temperature, $\mathbf{v}_p$ and $\tilde{n}_p$ are the variable components of electron velocity and density with $\tilde{n}_p \ll n_p$, $\omega_B = eB/mc$ is the gyrofrequency, and $\mathbf{n} = \mathbf{B}/B$, $\mathbf{B}$ is the external magnetic field [1,2].

Using the continuity equation and the Poisson equation

$$\frac{\partial \tilde{n}_p}{\partial t} = -\mathrm{div}(n_p \mathbf{v}), \qquad \mathrm{div}\mathbf{E} = 4\pi e(\tilde{n}_p + \rho), \tag{2.1.2}$$

and eliminating the quantities $\mathbf{v}_p$ and $\tilde{n}_p$ from (2.1.1) and (2.1.2), we represent an equation for the electric field in the form

$$\begin{aligned}
&\left[\frac{\partial^2}{\partial t^2}\left(\frac{\partial^2}{\partial t^2} - V_T^2\Delta + \omega_p^2 + \omega_B^2\right) - V_T^2\omega_B^2(\mathbf{n}\nabla)^2\right]\mathrm{div}\mathbf{E} \\
&\quad + \omega_p^2\omega_B\frac{\partial}{\partial t}\mathbf{n}\,\mathrm{curl}\,\mathbf{E} + \omega_p^2\omega_B^2(\mathbf{n}\nabla)(\mathbf{n}\mathbf{E}) \\
&= 4\pi e\left[\frac{\partial^2}{\partial t^2}\left(\frac{\partial^2}{\partial t^2} + \omega_B^2 - V_T^2\Delta\right) - V_T^2\omega_B^2(\mathbf{n}\nabla)^2\right]\rho,
\end{aligned} \tag{2.1.3}$$

where $V_T = \sqrt{T_p/m}$ is the thermal velocity of plasma electrons, ρ is the beam density, and Δ is the Laplacian.

To simplify Eq. (2.1.3), we assume that the inequalities

$$V_T^2\Delta \simeq V_T^2/l^2 \ll \partial^2/\partial t^2 \simeq \omega_p^2,\ \omega_B^2 \ll \omega_p^2,$$

are fulfilled. In the above approximation, the terms of order V_T^2 and ω_B^2 are essential only close to the plasma resonance. Incorporating these terms in the operator on the left-hand side of Eq. (2.1.3), we find the equation for the self-consistent potential $\mathbf{E}$=-$\nabla\varphi$:

$$\frac{\partial^2}{\partial t^2}\left(\frac{\partial^2}{\partial t^2} - V_T^2\Delta + \omega_p^2 + \omega_B^2\right)\Delta\varphi + \omega_p^2\omega_B^2(\mathbf{n}\nabla)^2\varphi = -4\pi e\,\frac{\partial^4\rho}{\partial t^4}. \tag{2.1.4}$$

In the special case of $\omega_B = V_T = 0$, equation (2.1.4) is the same as analogous Eq. (1.2.1).

We direct the z axis of the cylindrical coordinate system along the beam velocity, and we seek a resonance solution of Eq. (2.1.4) in the form

$$\varphi(t,\mathbf{r}) = \mathrm{Re}\,\left[\varphi(x,y)\exp(ikz')\right], \tag{2.1.5}$$

where $z' = z - vt$, v is the beam velocity, and l is the spatial period of beam modulation.

Substituting (2.1.5) in (2.1.4) and averaging over the space, we find an equation for the wave amplitude

$$\begin{aligned}&\left\{-V_T^2\Delta_\perp^2 + \left(\omega_L^2 - \omega_m^2 + \omega_B^2 + k^2V_T^2\right)\frac{\partial^2}{\partial x^2}\right.\\ &+\left[\omega_L^2 - \omega_m^2 + \omega_B^2\left(1 - \frac{B_y^2}{B^2}\right) + k^2V_T^2\right]\frac{\partial^2}{\partial y^2}\\ &\left.-k^2\left[\omega_L^2 - \omega_m^2 + \omega_B^2\left(1 - \frac{B_z^2}{B^2}\right)\right] - \frac{4ikB_yB_z}{B^2}\frac{\partial}{\partial y}\right\}\varphi(x,y)\\ &= 4ke\omega_m^2\int_{l/2}^{l/2}\rho(r,z')\exp(-ikz')\,dz',\end{aligned} \tag{2.1.6}$$

where $\omega_L^2 = \omega_p^2 + k^2V_T^2$, $\Delta_\perp = \partial^2/\partial x^2 + \partial^2/\partial y^2$, $\omega_m = kv$ is the beam modulation frequency, and $\mathbf{B} = (0, B_y, B_z)$.

§2.2. Ribbon bunches in a heated plasma.

In the present section the results obtained in §1.5 for a cold plasma with $T_p = 0$ are generalized to the case of finite electron temperature $T_p > 0$. To proceed, we consider Eq. (2.1.6) in the limiting case of a weakly magnetic field,

$$\omega_B^2 \ll k^2V_T^2, \tag{2.2.1}$$

at which terms $\sim \omega_B^2$ can be discarded.

We assume that the beam electron density as a function of the potential is determined by formula (2.1.6). In the case of a ribbon beam $\partial/\partial y = 0$, we find a nonlinear ordinary differential equation for the amplitude of the wave potential:

$$\mu\Phi^{(4)} - (1+\mu)\Phi'' + \Phi - q\Phi^2 = 0, \tag{2.2.2}$$

where the dimensionless variables are

$$\Phi = \frac{\varphi}{\varphi_m}, \quad q = \frac{2|e|\rho_0 l^2}{3\pi^2 \varphi_m \delta}, \quad \mu = \frac{V_T^2}{2v^2\delta}, \quad \delta = 1 - \frac{\omega_m}{\omega_L} > 0,$$

and the prime means the derivative with respect to $\xi = kx$.

A first integral of Eq. (2.2.2) is

$$\mu(2\Phi'\Phi''' - \Phi''^2 - \Phi'^2) = \Phi'^2 + (2q/3)\Phi^3 - \Phi^2. \tag{2.2.3}$$

In a cold plasma $\mu = 0$ the necessary condition for the existence of self-focusing is that the solution of Eq. (2.2.3) satisfying the conditions $\Phi(0) = 1$ and $\Phi'(0) = 0$ on the axis has the form of soliton with $q = 3/2$:

$$\Phi = \cosh^{-2}(\xi/2). \tag{2.2.4}$$

In a warm plasma $\mu > 0$ the order of the equation increases and it is necessary to find $\Phi''(0)$. From Eq. (2.2.3) it follows that this quantity is related to the parameter q by

$$\Phi''(0) = -\sqrt{\mu^{-1}(1 - 2q/3)} \tag{2.2.5}$$

which must be taken into account in resorting to a numerical solution.

In the limiting case of a weakly heated plasma $\mu \ll 1$, regarding the left-hand side of (2.2.3) as small perturbation, we find

$$\Phi'^2[1 - (1 - 4q\Phi)] = \Phi^2[1 - (2q/3)\Phi - \mu(q\Phi - 1)^2], \tag{2.2.6}$$

The solution (2.2.6) has a monotonically decreasing solution which is asymptotically close to a soliton for $q = (3/2)(1 - \mu/4)$.

Equation (2.2.3) is invariant under the transformation:

$$\xi_1 = \frac{\xi}{\sqrt{\mu}}, \quad \mu_1 = \frac{1}{\mu}.$$

Hence the asymptotic solution for $\mu \gg 1$ follows from Eq. (2.2.4):

$$\Phi = \cosh^{-2}(\xi/2\sqrt{\mu}), \quad q = (3/2)(1 - 1/4\mu). \tag{2.2.7}$$

The results of integrating Eq. (2.2.2) numerically are illustrated in Fig. 2.2.1. As the parameter μ increases the solution which is soliton for a cold plasma $\mu = 0$ spreads out in space, retaining the shape of a solitary pulse in accordance with the asymptotic form (2.2.7).

The evolution of the depth of the radial potential well

$$\varphi_m(\mu) = \frac{2|e|\rho_0 l^2}{3\pi^2\delta}\frac{1}{q(\mu)} \tag{2.2.8}$$

as a thermal velocity of the plasma electrons increases is shown in Fig. 2.2.2, in which the function $q(\mu)$ is plotted. The minimum corresponds to $q_{min} \simeq 1.453$, and the asymptotes at $\mu \to 0$ and $\mu \to \infty$ coincide: $q(0) = q(\infty) = 3/2$.

Thus, the study described above shows that the self-focusing effect of a deeply modulated electron beam persists in a plasma with finite electron temperature. The cold-plasma approximation $\mu \ll 1$ holds if the Debye length λ_d is small in comparison with $l\delta^{1/2}$. As λ_d increases, the focusing effect of the plasma on the beam decreases and the electron clumps

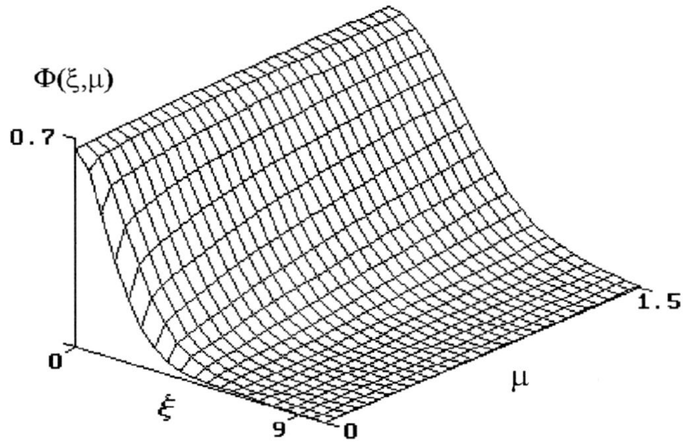

Fig. 2.2.1. Shape of the radial potential well $\Phi(\xi,\mu)$ (in units of $2|e|\rho_0 l^2/3\pi^2\delta$) as a function of the plasma temperature.

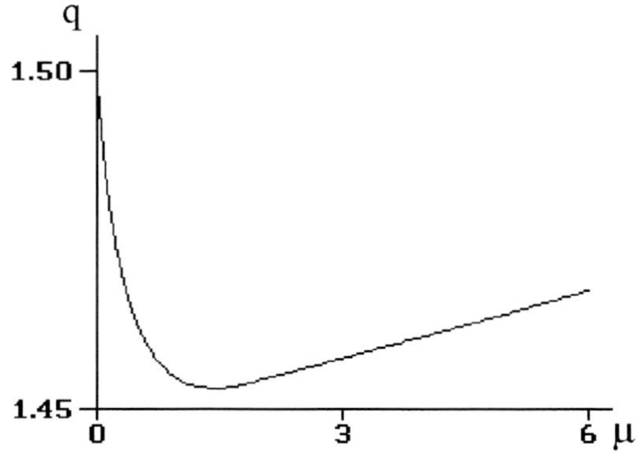

Рис.2.2.2. Depth of the potential well versus plasma temperature.

spread out, while the beam width increases in proportion to $l\sqrt{\mu}$ for $\mu \gg 1$, in accordance with the asymptotic formula (2.2.7).

The depth of the radial potential well for the beam electrons in a heated plasma $\mu > 0$ increases, reaching a maximum value in the region $\mu \simeq 1$. However numerical integration reveals that the relative increase in the parameter $\varphi(\mu)$ is slight (see Fig. 2.2.2).

§2.3. Ribbon bunches in a nonlinear plasma.

The propagation of electromagnetic wave in a plasma is accompanied by the appearance of a ponderomotive force [5]

$$\mathbf{f} = -\nabla\phi, \quad \phi = \frac{e^2|\mathbf{E}|^2}{4m\omega^2}, \tag{2.3.1}$$

where $|\mathbf{E}|$ and ω are the electric field amplitude and the frequency of the wave, respectively.

The density of plasma electrons as a function of a field amplitude satisfies the equilibrium condition

$$(\mathbf{f} - e\nabla\varphi_s)\, n_e - T_p \nabla n_e = 0, \quad n_e = n_i,$$

where $n_i = n_p \exp(e\varphi_s/T_p)$ is the density of plasma ions, φ_s is the electrostatic potential, and n_p and T_p are the equilibrium plasma density and the temperature. Carrying out the integration, we find [6]:

$$n(|\mathbf{E}|) = n_p \exp\left(-\frac{|\mathbf{E}|^2}{E_T^2}\right), \quad E_T^2 = \frac{8m\omega^2 T_p}{e^2}. \tag{2.3.2}$$

This formula permits to generalize Eq. (2.1.2) obtained in a linear approximation to the case of nonlinear plasma:

$$\operatorname{div}\left[\left(\frac{\partial^2}{\partial t^2} - V_T^2\Delta + \omega_p^2(|\mathbf{E}|)\right)\mathbf{E}\right] = 4\pi e\left(\frac{\partial^2}{\partial t^2} - V_T^2\Delta\right)\rho, \tag{2.3.3}$$

where $V_T^2 = T_p/m$ is the thermal velocity of plasma electrons and $\omega_p^2(|\mathbf{E}|) = 4\pi e^2 n(|\mathbf{E}|)/m$.

Switching to the potential $\mathbf{E} = -\nabla\varphi$, we consider the resonance solution of Eq. (2.3.3),

$$\varphi = \varphi(x)\cos[k(z - vt)], \quad k = 2\pi/l, \quad \omega_p - kv \ll \omega_p,$$

and we write down it for the case of small amplitude expanding exponential function (2.3.2) in the power series.

Averaging (2.3.3) over the spatial period l, we find a nonlinear differential equation for the wave amplitude:

$$\frac{d}{dx}\left[\left(|\epsilon(0)| - \frac{|\mathbf{E}|^2}{E_T^2} - \frac{V_T^2}{\omega^2}\frac{d^2}{dx^2}\right)\frac{d\varphi}{dx}\right]$$
$$-k^2\left(|\epsilon(0)| - \frac{|\mathbf{E}|^2}{E_T^2} - \frac{V_T^2}{\omega^2}\frac{d^2}{dx^2}\right)\varphi = 4ke\int_{-l/2}^{l/2}\rho(x,z')\cos kz'\,dz', \tag{2.3.4}$$
$$\epsilon(0) = 1 - \frac{\omega_L^2}{\omega_m^2}, \quad \omega_L^2 = \omega_p^2 + (kV_T)^2, \quad |\mathbf{E}|^2 = \varphi'^2 + (k\varphi)^2.$$

Carrying out the integration in the right-hand side, we get the equation

$$\mu\Phi^{(4)} - [(1 + \mu - \chi\Phi^2 - \chi\Phi'^2)\Phi']' + (1 - \chi\Phi^2 - \chi\Phi'^2)\Phi - q\Phi^2 = 0, \tag{2.3.5}$$

at which $\chi = k^2\varphi_m^2/2\delta E_T^2 > 0$. In the limit $\chi \to 0$, Eq. (2.3.5) does not differ from analogous Eq. (2.2.2).

Multiplying equation (2.3.5) by $6\chi\Phi'$, using the expression

$$(1 + \mu - \chi\Phi^2 - 3\chi\Phi'^2)' = -2\chi\Phi\,(\Phi + 3\Phi''),$$

and integrating, we represent the first integral in the form:

$$\begin{gathered} 6\mu\chi(2\Phi'\Phi''' - \Phi''^2) + (1 + \mu - \chi\Phi^2 - 3\chi\Phi'^2)^2 \\ + 8\chi[(1 + \mu/4)\Phi^2 - \chi\Phi^4/2] - 4q\chi\Phi^3 = (1 + \mu)^2, \end{gathered} \tag{2.3.6}$$

which satisfies the soliton condition

$$\Phi^{(n)}(\xi \to \infty) = 0, \quad n = 0, 1, 2, 3,$$

and the boundary condition at the beam axis,

$$\begin{gathered} \Phi(0) = 1, \quad \Phi'(0) = 0 \\ \Phi''(0) = -\sqrt{\mu^{-1}(1 - \chi/2 - 2q/3)}. \end{gathered} \tag{2.3.7}$$

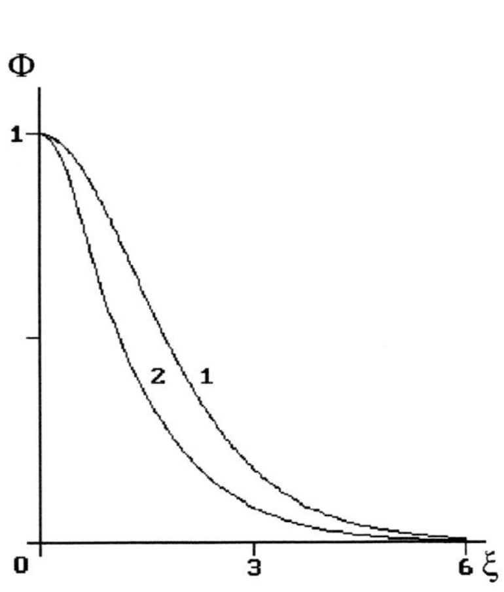

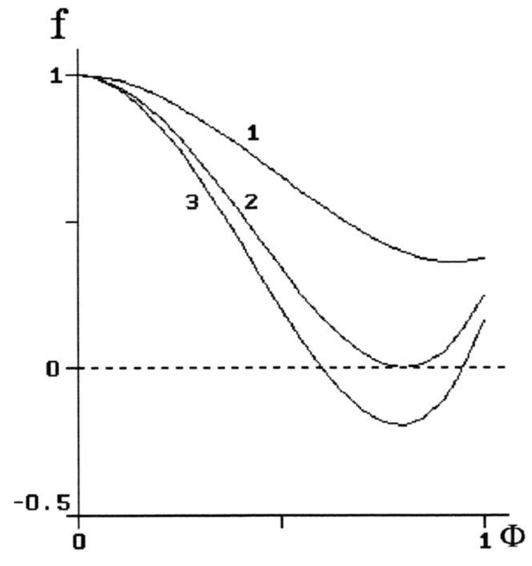

Fig. 2.3.1. Solution of Eq. (2.3.8) for $\chi = 0$ (1) and $\chi = 0.5$ (2).

Fig. 2.3.2. Displacement of $f(\Phi, \chi) < 0$ in the range of $f(\Phi_{min}, \chi) < 0$ for various values of $\chi = 0.25$ (1); 0.5 (2) и 0.6 (3).

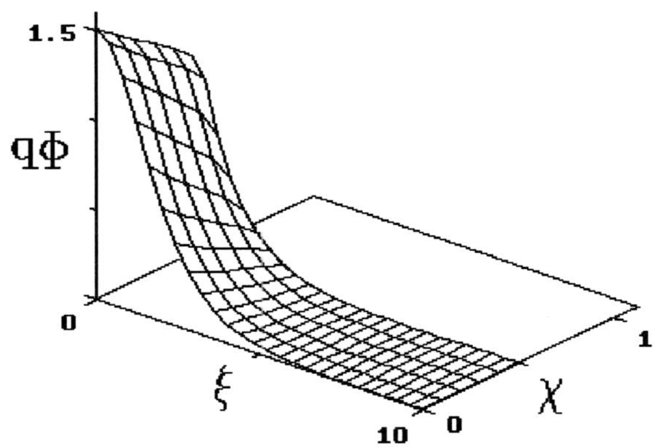

Fig. 2.3.3. Profiles of potential in cold plasma $\mu = 0$.

It is easy to see that equation (2.3.6) is substantially simplified in the limiting case of a cold nonlinear plasma, $\mu \ll \chi$, because of the higher derivatives be can discarded,

$$
\begin{aligned}
3\chi\Phi'^2 &= 1 - \chi\Phi^2 - \sqrt{f(\Phi, \chi)}, \\
f(\Phi, \chi) &= 1 - 8\chi\Phi^2(1 - \chi\Phi^2/2) + 4q\chi\Phi^3,
\end{aligned}
\tag{2.3.8}
$$

where $q = (3/4)(2 - \chi)$.

A solution of Eq. (2.3.8) in the form of soliton exists for $\chi = 0$. A numerical study shows that the soliton-like solution represented in Fig. 2.3.1 also exists in the case of nonlinear plasma, $\chi > 0$, if the inequality $f(\Phi, \chi) > 0$ is fulfilled. But this function

$$
\begin{aligned}
f(\Phi_{\min}, \chi) &= (\sqrt{7}/2 - 1 - \chi\Phi_{\min}^2)(\sqrt{7}/2 + 1 + \chi\Phi_{\min}^2), \\
\chi\Phi_{\min}^2 &= \sqrt{(9/16)^2(1 - \chi/2)^2 + \chi} - (9/16)(1 - \chi/2),
\end{aligned}
\tag{2.3.9}
$$

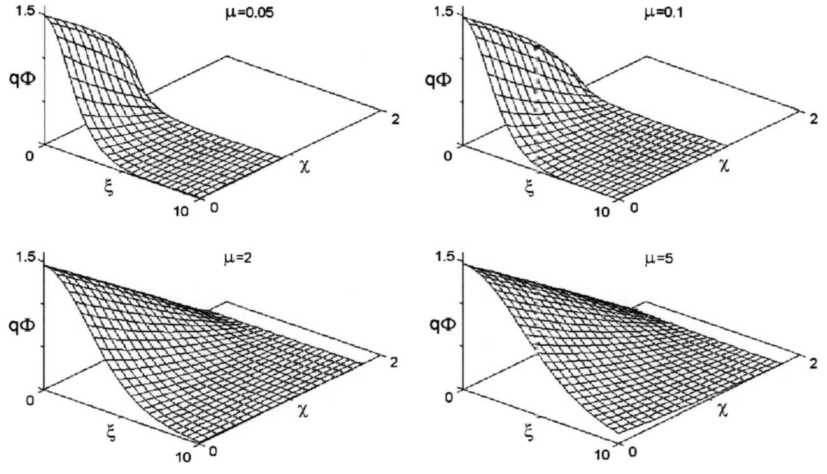

Fig. 2.3.4. Profiles of potential in a heated plasma $\mu>0$.

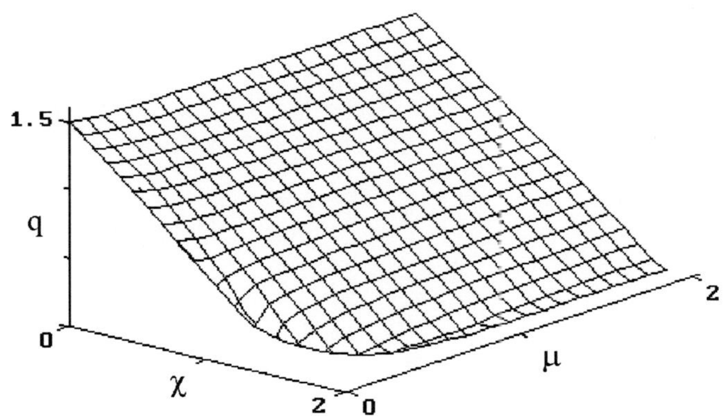

Fig. 2.3.5. Maximal depth of potential well $q(\chi,\mu)$.

becomes negative, $f(\Phi_{\min},\chi) < 0$ on the condition that $\chi\Phi_{\min}^2 > \sqrt{7}/2 - 1$ (Fig. 2.3.2) so that the inequality

$$\frac{\chi}{(2-\chi)^2} < \frac{81}{128}\frac{\sqrt{7}-2}{(4-\sqrt{7})^2} \tag{2.3.10}$$

restricts the region of definition of the soliton-like solution by the inequality $\chi \leq \chi_{\max} \simeq 0.501$. The distribution of potential over the beam cross section found by means of numerical integration of Eq. (2.3.8) is shown in Fig. 2.3.3.

A region of parameter χ corresponding to soliton-like solution extends if the thermal motion of plasma electrons is taking into account, $\mu > 0$. The results of numerical integration of Eq. (2.3.5) is shown in Fig. 2.3.4. The function $q(\chi\ \mu)$ which determines the depth of potential well,

$$\varphi_m(\chi,\mu) = \frac{2}{3\pi^2}\frac{|e|\rho_0 l^2}{\delta}\frac{1}{q(\chi,\mu)}, \tag{2.3.11}$$

is represented in Fig. 2.3.5.

According to Figs. 2.3.5, the threshold $\chi \leq \chi_{\max}$ also exists in a heated nonlinear

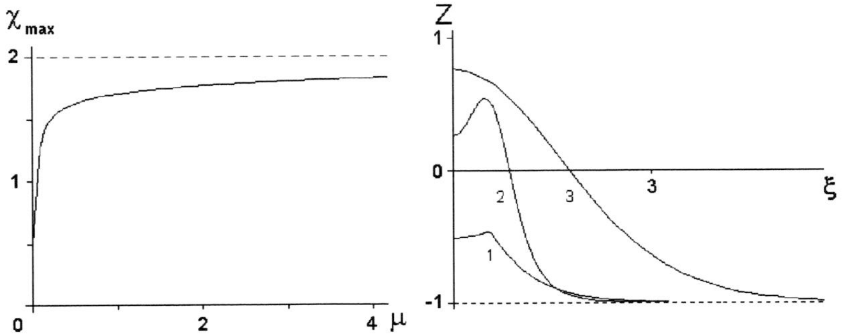

Fig. 2.3.6. Threshold value $\chi_{\max}$ versus parameter μ.
Fig. 2.3.7. Nonlinear plasma permittivity $Z{=}\epsilon_{NL}(|\mathbf{E}|)/|\epsilon(0)|$ along the radial variable ξ for various values of $\mu = 0$: $\chi_{\max} = 0.501$ (1); $\mu = 0.1$, $\chi_{\max} = 1.264$ (2); $\mu = 2$, $\chi_{\max} = 1.769$ (3).

plasma $\mu > 0$. But in distinction to the case of a cold plasma $\mu = 0$, this restriction arises from Eq. (2.3.7). The boundary function $\chi_{\max}(\mu)$ is shown in Fig. 2.3.6.

Let us now consider the physical mechanism causing the effect of electrostatic self-focusing in more general case of Eq. (2.3.5) taking into account a plasma nonlinearity. Due to the field in the linear plasma, a polarization wave displaces the part of plasma electrons from the volume of bunches [see Eq. (1.2.12)], and the Coulomb repulsion in the bunches is compensated by the charge of plasma ions in moving with beam velocity ionic potential wells.

The nonlinear ponderomotive force (2.3.1) repels the plasma ions together with the electrons from the interior of bunches, and a background plasma density decreases. As a result, the nonlinear permittivity of plasma,

$$\epsilon_{NL}(|\mathbf{E}|) = \epsilon(0) + \frac{|\mathbf{E}|^2}{E_T^2} = |\epsilon(0)|(-1 + \chi\Phi^2 + \chi\Phi'^2) \tag{2.3.12}$$

increases at the centers of bunches.

In the case of a cold plasma $\mu = 0$, the plasma permittivity

$$\epsilon_{NL}(|\mathbf{E}|) = -\frac{|\epsilon(0)|}{3}\left[\, 2(1 - \chi\Phi^2) + \sqrt{\mathrm{f}}\,\right],$$

is negative over the whole range $\chi \le \chi_{\max}$ where a self-focusing occurs (Fig. 2.3.7, curve 1).

A part of plasma particles turn back at the beam axis, if the thermal motion, $\mu > 0$, compensates partial losses of plasma electrons in this region. Therefore under an action of strong electric field exited by a powerful electron beam the plasma permittivity becomes positive near the beam axis, $\epsilon_{NL}(|\mathbf{E}|) > 0$, if the condition $\chi\Phi > 1$ is fulfilled (Fig. 2.3.7, curve 2 and 3).

§2.4. Longitudinal magnetic field.

In this section we consider the effect of an external magnetic field on the conditions for transport of a modulated beam in a cold plasma $V_T = 0$. In the special case of a longitudinal magnetic field $B_y = 0$, $B_z = B$, the equation (2.1.6) takes the form:

$$\begin{aligned}&\left[(\omega_p^2 - \omega_m^2 + \omega_B^2)\Delta_\perp - k^2(\omega_p^2 - \omega_m^2)\right]\varphi(x,y)\\ &= 4ke\omega_p^2 \int_{-l/2}^{l/2} \rho(r,z')\exp(-kz')\,dz',\end{aligned} \tag{2.4.1}$$

where $\Delta_\perp = \partial^2/\partial x^2 + \partial^2/\partial y^2$, ω_p, ω_m, and $\omega_B = eB/mc$ are the Langmuir frequency, the modulation frequency, and the gyrofrequency, respectively.

Since a bunch distribution function depends on a beam energy integral, the density of captured particles as a function of potential does not depend on a magnetic field. Therefore we transform Eq. (2.4.1) into the analogous equation obtained in the absence of the magnetic field in §1.3 by a change of the transverse scale

$$x_1 = \frac{x}{\sqrt{1+\eta}}, \quad y_1 = \frac{y}{\sqrt{1+\eta}}, \quad \eta = \frac{{\omega_B}^2}{\omega_p^2 - \omega_m^2}.$$

It follows from the latter relations that the longitudinal magnetic field $\eta > 0$ deforms the spatial potential well, changing the beam radius, $r_B = (l/2\pi)\sqrt{1+\eta}$, but the nature of the solution remains unchanged [7].

A defocusing action of a sufficiently weak longitudinal magnetic field $\omega_B \ll \omega_p$ on the beam arises because of an increasing of transverse plasma frequency $\Omega_p = \sqrt{\omega_p^2 + \omega_B^2}$. As a result, the transverse component of plasma permittivity,

$$\epsilon_\perp = 1 - \frac{\omega_p^2}{\omega_m^2 - {\omega_B}^2} = -(1+\eta)\epsilon_{||}, \quad \epsilon_{||} = 1 - \frac{\omega_p^2}{\omega_m^2},$$

exceeds the longitudinal one, $(|\epsilon_\perp > \epsilon_{||}|)$, and the focusing transverse field decreases [8].

To explain the physical mechanism of this effect, we assume that $\delta = 1 - \omega_m/\omega_p \ll 1$ and that the collective interaction with the plasma occurs basically via the field of this resonance mode

$$\varphi_b = \varphi_0(r)\exp(i\omega_m t - ikz),$$

where $k = \omega_p/v$. The amplitudes of radial R and longitudinal Z oscillations of plasma electrons are determined by the following equations:

$$\begin{aligned}(\omega_B^2 - \omega_m^2)R &= -\frac{e}{m}\frac{\partial}{\partial r}(\varphi_0 - \varphi_p), \\ \omega_m^2 Z &= -\frac{e}{m} ik(\varphi_0 + \varphi_p),\end{aligned} \tag{2.4.2}$$

where φ_p is the amplitude of the wave which satisfies the equation

$$\frac{1}{r}\frac{\partial}{\partial r}\left[r\left(\frac{\partial\varphi}{\partial r} - 4\pi e\rho_p R\right)\right] + ik\left(ik\varphi_p + 4\pi e\rho_p Z\right) = 0. \tag{2.4.3}$$

Using (2.4.2) and (2.4.3), we can express φ_p in terms of φ_o:

$$\begin{aligned}\varphi_p = \left(\epsilon_\perp^{-1} - 1\right)\varphi_0 + \left(\epsilon_\perp - \epsilon_{||}\right)\epsilon_\perp^{-2}k^2 \Big[& K_0(k_\perp r)\textstyle\int_0^r I_0(k_\perp r')\varphi_0(r')\, r'dr' \\ & + I_0(k_\perp r)\textstyle\int_r^\infty K_0(k_\perp r')\varphi_0(r')\, r'dr'\Big], \qquad k_\perp = (\epsilon_{||}/2\epsilon_\perp)k.\end{aligned} \tag{2.4.4}$$

It follows from (2.4.4) that

$$E_{rp}/E_{zp} \simeq R/Z \simeq (\epsilon_{||}/\epsilon_\perp)^{1/2} \simeq (1+\eta)^{-1/2},$$

and the ratio of the focusing radial field to the longitudinal one decreases as the magnetic field (parameter η) increases. Thus the longitudinal magnetic field suppresses transverse motion of the plasma electrons and the potential distribution in space becomes anisotropic.

For $\eta \gg 1$, when the transverse motion of the plasma electrons is "frozen", the potential well is close to one-dimensional longitudinal form, since the magnetic field does not influence the process of drawing out plasma electrons by the collective field along the beam.

§2.5. Transport of electron bunches across magnetic field.

We investigate now the effect of an arbitrary oriented magnetic field on the conditions for transport of a bunched beam in a plasma. The magnetic field manifests itself in changes in the wave guide properties of the plasma and the action of the Lorentz force $\mathbf{F} = e[\mathbf{vB}]/c$ on the electrons of the bunches. In a longitudinal field, $\mathbf{F} = 0$, the plasma anisotropy produces an increase in the bunch radius without changing the longitudinal dimensions. The appearance of a transverse magnetic field component $\mathbf{B}_\perp$ is accompanied by a displacement of the beam electrons due to the force $\mathbf{F}_\perp$, i.e., an analog of the Hall effect in the conductor. But since the boundaries of the conductor at which charge accumulates are absent here and the role of the walls is played by the smoothly decaying potential of the bunch at its periphery, the result of a constant force acting on the current is to empty out part of the electrons from the potential wells [7].

Numerical investigation shows that in a finite transverse magnetic field there are no physically well-posed steady solutions for which the density of the bunches attain a minimum on the beam axis and decays monotonically toward the periphery. However it is clear that in the limit $\mathbf{B}_\perp \to 0$ there exists solutions which are asymptotically close to those found in §1.4 in the absence of a magnetic field. The following analysis of these solutions enables us to find a bound on the value of the transverse component of the field below which this self-focusing effect of a modulated electron beam in a plasma can be observed.

The density ρ of the beam as a function of the potential is determined from the condition for the bunches to be in equilibrium,

$$e\rho\left(-\nabla\varphi + [\mathbf{vB}]/c\right) - T\nabla\rho = 0,$$

where T is the beam temperature and $\mathbf{n} = \mathbf{B}/B$. Integrating this equation we obtain

$$\rho(\mathbf{r}) = \tilde{\rho}_0 \exp\left[-\frac{e\varphi(\mathbf{r})}{T} + \frac{\omega_B}{T}\mathbf{nM}\right], \tag{2.5.1}$$

where $\tilde{\rho}_0$ is a constant of integration, $\mathbf{M} = m[\mathbf{rv}]$, and $\mathbf{r}$ is the radius vector.

We choose the z axis parallel to the beam velocity, and following §1.4 and §2.4 we consider the resonanse solution in a weak magnetic field:

$$\varphi(t, \mathbf{r}) = (T/e)\phi(x, y)\cos kz', \quad z' = z - vt, \quad k = 2\pi/l, \quad \omega_B \ll \omega_p. \tag{2.5.2}$$

In a longitudinal magnetic field we have $\mathbf{n}\nabla = \partial/\partial z$ and $\mathbf{M} = 0$, and substituting (2.5.1) and (2.5.2) in Eq. (2.1.6) we find an equation for the wave amplitude:

$$\left[(1+\eta)\left(\frac{\partial^2}{\partial x^2} + \frac{\partial^2}{\partial y^2}\right) - k^2\right]\phi = -\frac{1}{R_0^2}\exp(-\phi_0)I_1(\phi), \tag{2.5.3}$$

where $I_1(\phi)$ is modified Bessel function, $\eta = {\omega_B}^2/2\omega_m^2\delta$, $R_0^2 = T\delta/4\pi e^2\rho_0$, and $\delta = 1 - \omega_m/\omega_p \ll 1$. In passing from (2.5.1) to (2.5.3) we have set $\tilde{\rho}_0 = \rho_0\exp(-\phi_0)$, where $\phi_0 = \phi(0, 0)$ is the potential at the center $x = y = 0$ of a bunch.

According to §2.4, by a change of the transverse scale

$$x_1 = x\,(1+\eta)^{-1/2}, \quad y_1 = y\,(1+\eta)^{-1/2},$$

we can convert Eq. (2.5.3) into the analogous equation of §1.4, obtained in the absence of a magnetic field and having a solution in the form of a solitary pulse. It follows that the longitudinal magnetic field increases the beam radius by a factor of $(1+\eta)^{1/2}$, but does not change the physics of the self-focusing effect.

This circumstance enables us to break up the problem for an arbitrary oriented magnetic field into "longitudinal" and "transverse" parts and, without loss of generality, to set **vB** =0. If we choose the y axis in the direction of **B**, from Eqs. (2.5.1) and (2.5.2) we find

$$\rho(\mathbf{r}) = \rho_0 \exp\left[-\phi_0 + \phi(x,y)\cos kz' + \frac{mv|\omega_B|x}{T}\right]. \tag{2.5.4}$$

Substituting (2.5.4) in right-hand side of Eq. (2.1.6), we find an equation for the wave amplitude:

$$\left[(1+\eta)\frac{\partial^2}{\partial x^2} - k^2\right]\phi + \frac{\partial^2\phi}{\partial y^2} = -\frac{1}{R_0^2} I_1(\phi)\exp\left(-\phi_0 + \frac{mv|\omega_B|x}{T}\right), \tag{2.5.5}$$

analogous to (2.5.3). The exponential on the right-hand side of (2.5.5) results from the projection of the electron momentum $M_y = -mvx$ in the direction of magnetic field.

Following §1.3 and §1.4, we consider a deep potential well

$$\phi_0 = \frac{8}{(ka)^2} \gg 1, \rho(0,0,z') = \rho_0 \exp\left[-\left(\frac{2z'}{a}\right)^2\right],$$

where the longitudinal dimension a of the bunch is small in comparison with the modulation period l of the beam. Replacing the Bessel function in (2.5.5) by its asymptotic form

$$I_1(\phi) \simeq (2\pi\phi_0)^{-1/2}\exp(\phi_0 - \phi_1). \quad \phi_0 \gg \phi_1$$

and going over to the new variable

$$\tilde{\phi} = \phi_1 - \frac{mv|\omega_B|x}{T},$$

we find the equation

$$(1+\eta)\frac{\partial^2\phi}{\partial x^2} + \frac{\partial^2\phi}{\partial y^2} + \frac{\mu_B}{r_0^2} = \exp(-\phi), \tag{2.5.6}$$

$$r_0^2 = R_0^2\frac{\rho_0}{\langle\rho\rangle_0}, \quad \langle\rho\rangle_0 = \sqrt{\pi}\rho_0\frac{a}{2l}, \quad \mu_E = \frac{8r_0^2}{a^2}(1+\eta),$$

which differs from analogous Eq. (1.3.7) in the presence of the parameter $1+\eta$, which can be removed by a change of scale $x_1 = x/\sqrt{1+\eta}$. Thus a sufficiently weak transverse magnetic field $\eta \simeq 1$ does not interfere with self-focusing. At the same time the cross section of the bunch is transformed into an ellipse whose semiahes are in the ratio $\sqrt{1+\eta}$, with the longer being in the x direction (the direction in which the Lorentz force acts).

Equation (2.5.6) has less complete information than the original equation (2.5.5), which we consider for the case of a planar beam, $\partial/\partial y = 0$:

$$\frac{d^2\phi}{d\xi^2} - \phi = -\frac{1}{A} I_1(\phi)\exp(-\phi_0 + h\xi), \tag{2.5.7}$$

where

$$\xi = kx, \quad A = (kR_0)^2(1+\eta), \quad h = mv|\omega_B|/kT.$$

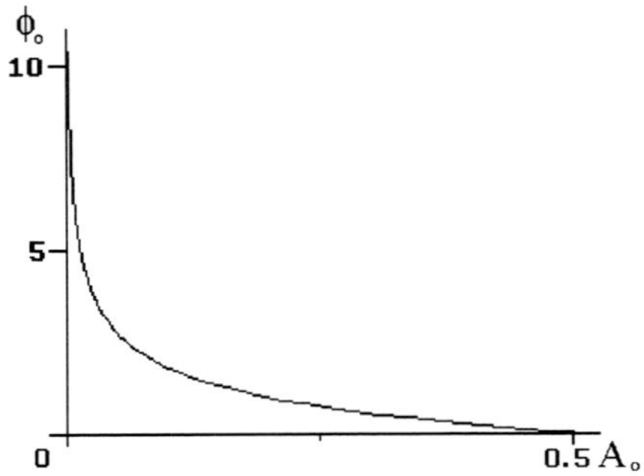

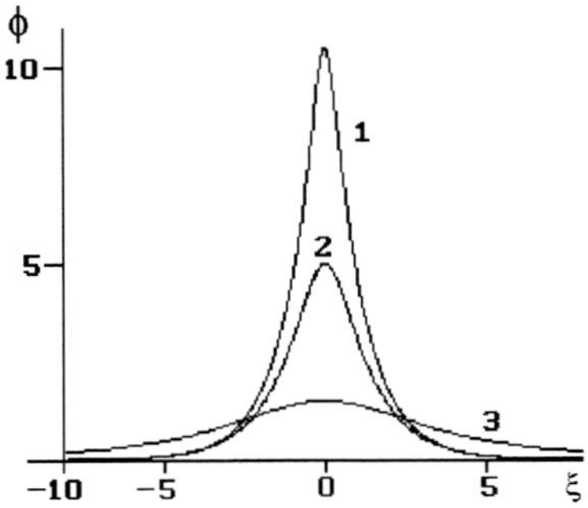

Fig. 2.5.1. Maximum soliton amplitude ϕ_0 as a function of A_0.

Fig. 2.5.2. Profiles of the potential for various values of ϕ_0 = 10.5 (1), 5 (2) и 1.5 (3).

If the magnetic field vanishes $h = \eta = 0$, Eq. (2.5.7) has a solution in the form of a solitary pulse, which vanishes exponentially for $|\xi| \to \infty$:

$$\phi_\infty \simeq \exp\left[-|\xi|\left(1 - \frac{1}{2A_0}\exp(-\phi_0)\right)^{1/2}\right], \tag{2.5.8}$$

$A_0 = (kR_0)^2$. The condition for it to exist is the relation

$$A_0\phi_0^2 = 2\exp(-\phi_0)[I_0(\phi_0) - 1], \tag{2.5.9}$$

which determines the maximum value of the potential $\phi_0(A_0)$ (see Fig. 2.5.1). The asymptotic form of this function is

$$\phi_0(A_0) = \begin{cases} (2/\pi A_0^2)^{1/5}, & A_0 \ll 1, \\ 1 - 2A_0, & 1 - 2A_0 \ll 1. \end{cases} \tag{2.5.10}$$

For small values $\phi_o \ll 1$ the analytical solution of (2.5.7) takes the form of a soliton

$$\phi = (1 - 2A_0)\cosh^{-1}[(1 - 2A_0)\xi/4].$$

Plots of $\phi(\xi)$ and $\rho(\xi)$ for different values of

$$A_0 = (kR_0)^2 \simeq \frac{T}{mv^2}\,\frac{n_p\delta}{\rho_0}$$

are shown by curves 1 in Fig. 2.5.3. A strong compression of the bunches takes place for $A_0 \ll 1$, which corresponds to a cold beam with a high energy density. Equilibrium (self-focusing) of a bunch occurs for $A_0 < 1$.

The deformation of the bunches under the action of the transverse magnetic field is illustrated by the numerical solution of Eq. (2.5.7) shown by curves 2 in Fig. 2.5.3 for different values of A_0. A typical feature of the curves for $h > 0$ is the disappearance of monotonicity in the decay of the potential and the appearance of oscillations, which implies the existence of particles which are untrapped by the wave (see §1.3).

The characteristic scale on which the transverse magnetic field acts on the beam is determined by equating the terms in the argument of the exponential of Eq. (2.5.7):

$$\xi_B \simeq \frac{\phi_0}{h} = \frac{ck\varphi_0}{vB_\perp}, \tag{2.5.11}$$

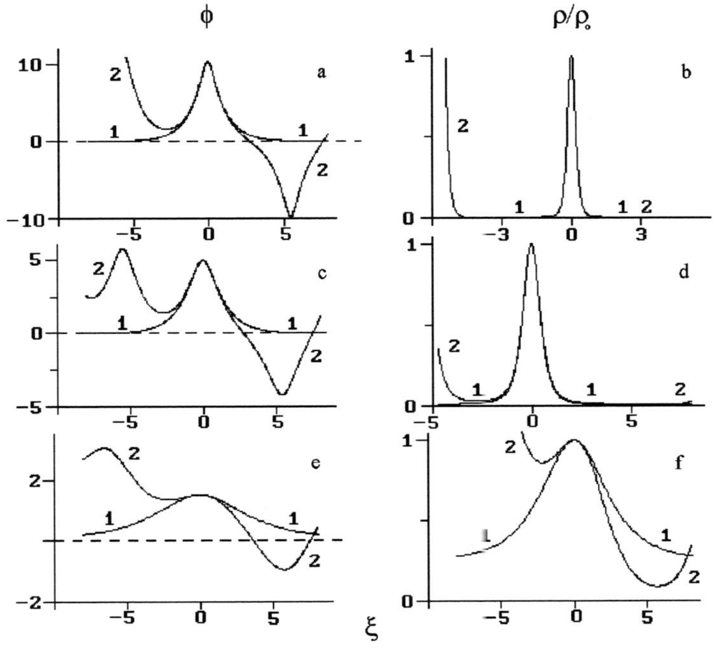

Fig. 2.5.3. Deformation of the potential ϕ and density ρ under the action of the transverse magnetic field for different values of $\phi_0 = 10.5$ (a,b), 5 (c,d) и 1.5 (e,f) .

i.e., this quantity equals the ratio of the focusing electric force to the Lorentz force. The action of the field on the bunch can be regarded as weak if this dimension exceeds the half-width ξ_0 of the equilibrium soliton. This quantity can be found in order of magnitude from equation (2.5.8):

$$\xi_B \geq \xi_0 \simeq \left(1 - \frac{1}{2A_0}\exp(-\phi_0)\right)^{-1/2}.$$

Passing to the asymptotic form (2.5.10), we find from Eqs. (2.5.9) and (2.5.11)

$$\frac{\omega_B}{\omega_p} \leq \frac{T}{mv^2}\begin{cases}(\phi_0/2)^2, & \phi_0 \ll 1,\\ \phi_0, & \phi_0 \gg 1.\end{cases} \tag{2.5.12}$$

Note that for a deep potential well $\phi_0 \gg 1$ the limitation on the maximum value of the transverse magnetic field is considerably less severe than in the case $\phi_0 \leq 1$. However we should include the condition that the equations for the plasma electrons be linear (see §1.4)

$$\phi_0 \leq (kR_0)^2 \frac{mv^2}{T}\sqrt{\delta} \simeq \frac{n_p\delta^{3/2}}{n_0}, \tag{2.5.13}$$

which was used in deriving Eq. (2.1.6) and which limits the maximum value of ϕ_o. The second inequality in (2.5.12) coincides with that found from the asymptotic equation (2.5.6) for $\phi_o \simeq mv^2\sqrt{\delta}/T$.

In the case of the laminated electron-ion beam (see §1.6) the additional term appears in the right-hand side of Eq. (2.5.7):

$$\frac{d^2\phi}{d\xi^2} - \phi = -\frac{1}{A}\left[I_1(\phi)\exp(-\phi_0 + h\xi) + QI_1(R\phi)\exp(-R\phi_0 - Rh\xi)\right], \tag{2.5.14}$$

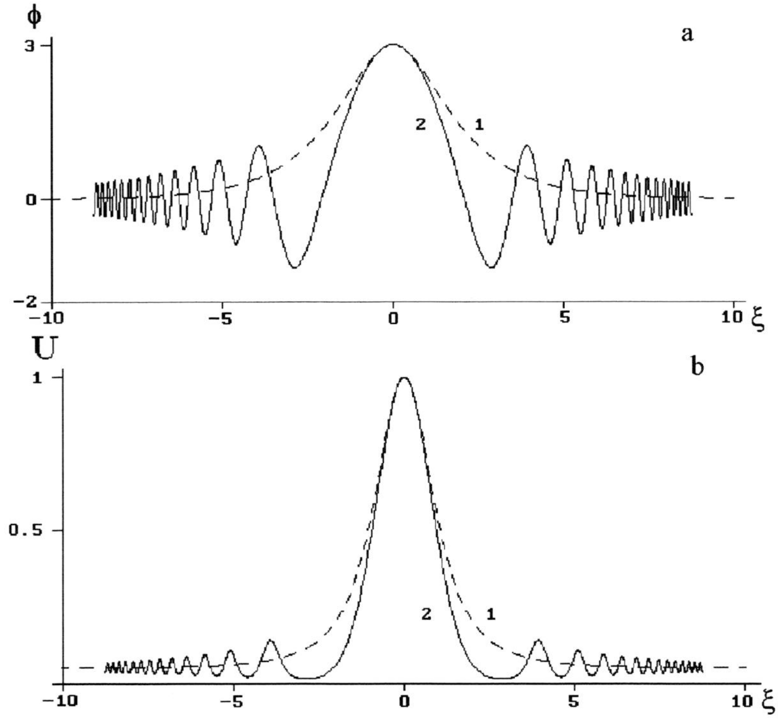

Fig. 2.5.4. Profiles of the potential $\phi(\xi)$ (a) and density of the bunches $U=n(\xi)/n_0$ (b) for $h=0$ (1) and $h=1$ (2).

where $n_{oe,i}$ and $T_{e,i}$ are the on-axis densities and temperatures of the electrons and ions, $Q = qn_{0i}/|e|n_{0e}$, and $R = qT_e/|e|T_i$.

It was shown by numerical integration of above equation that for the case of electrons $Q \ll 1$ or ions $Q \gg 1$ beams the additional terms in the right side exponents (caused by the transverse magnetic field $h > 0$) lead to the alteration of the solution character. There is no longer the monotonically decreasing solution due to the electron and ion runaway from the potential holes under the action of Lorenz force.

The numerical integration of the Eq. (2.5.14) shows that, as can be seen in Fig. 2.5.4, the situation drastically changes for the parameter domain $|Q-1| \ll 1$. In this case the densities of electron and ion bunches are close to each other, and the charge separation is accompanied by the generation of a strong electric field. A specific feature of the laminated electron-ion bunch radial equilibrium in the transverse magnetic field is the appearance of the potential oscillations. This oscillations correspond to the satellite electron and ion densities (see Fig. 2.5.4b).

§2.6. Self-focusing of rotating electron bunches.

In this section we consider the effect of a longitudinal magnetic field on the conditions for transport of a rotating bunches in a plasma. The rotation of the electron beam as a whole around its axis at the angular rate of Ω accompanies by the radial compression of a beam which interferes with electrostatic self-focusing [8].

In the presence of longitudinal magnetic field the electron distribution function satisfy the kinetic equation

$$\frac{\partial f}{\partial t}+v_z\frac{\partial f}{\partial z}+\left(-\frac{e}{m}\frac{\partial\varphi}{\partial r}+\omega_B v_\theta+\frac{v_\theta^2}{r}\right)\frac{\partial f}{\partial v_r}-\left(\omega_B v_\theta-\frac{v_\theta v_r}{r}\right)\frac{\partial f}{\partial v_\theta}-\frac{e}{m}\frac{\partial\varphi}{\partial z}\frac{\partial f}{\partial v_z}=0, \tag{2.6.1}$$

which in the most general form is an arbitrary function of the integrals of motion of the set of equations

$$\begin{gathered}\dot v_r=-\frac{e}{m}\frac{\partial\varphi}{\partial r}+\omega_B v_\theta+\frac{v_\theta^2}{r},\qquad \dot v_\theta=-\omega_B v_\theta+\frac{v_\theta v_r}{r},\\ \dot v_z=-\frac{e}{m}\frac{\partial\varphi}{\partial z},\qquad \dot r=v_r,\qquad r\dot\theta=v_\theta,\qquad \dot z=v_z.\end{gathered} \tag{2.6.2}$$

Carrying out the integration in Eqs. (2.6.2) , we find the energy integral and the momentum conservation law

$$E=\frac{mv^2}{2}+e\varphi,\qquad M_z=mr\left(v_\theta+\frac{\omega_B r}{2}\right),$$

which have to be used to define the electron distribution function.

To proceed, we consider the partial case of $f(\mathbf{r},\mathbf{v})\equiv f(W)$, where the combined integral

$$W=E-\Omega M_z=\frac{m\left[v_r^2+(v_\theta+\Omega r)^2+v_z^2\right]}{2}+\frac{m\Omega\left(\omega_B+\Omega\right)r^2}{2}+e\varphi, \tag{2.6.3}$$

takes into account a rotation of electron around the beam axis.

By analogy with §1.5 we write

$$f(\mathbf{v},\varphi)=\rho_0\left(\frac{m}{\pi e\varphi_m}\right)^2\left(-\frac{2W}{m}\right)^{1/2}\theta(-W), \tag{2.6.4}$$

where ρ_0 is the electron density at the bottom of the potential well, and $\varphi_m\,e<0$. The discontinuous function on the right-hand side of Eq. (2.6.4),

$$\theta(-W)=\begin{cases}1, & -W>0\\ 0, & -W<0\end{cases}$$

reflects the condition for capture of the electrons by the wave.

Integrating (2.6.4) over velocity, we find the density of trapped particles as a function of potential

$$\rho(\varphi)=\rho_0\left(\frac{\varphi}{\varphi_m}-\frac{r^2}{r_m^2}\right)^2\theta\left(\frac{\varphi}{\varphi_m}-\frac{r^2}{r_m^2}\right).$$

It follows from the above formula that the longitudinal magnetic field focuses an electron

$$r_m^2=\frac{2e\varphi_m}{m\Omega(\omega_B+\Omega)}>0 \tag{2.6.6}$$

over the frequency range $0<\Omega<|\omega_B|$ so that a minimal radius of a beam $r_m=\sqrt{8|e|\varphi_m/m\omega_B^2}$ corresponds to $\Omega=|\omega_B|/2$.

Substituting (2.6.5) in Eq. (2.1.6) and integrating we find an equation for the wave amplitude

$$(1+\eta)\xi^{-1}(\xi\Phi')'-\Phi=-q\Phi^2V\left(\xi^2/\xi_m^2\Phi\right), \tag{2.6.7}$$

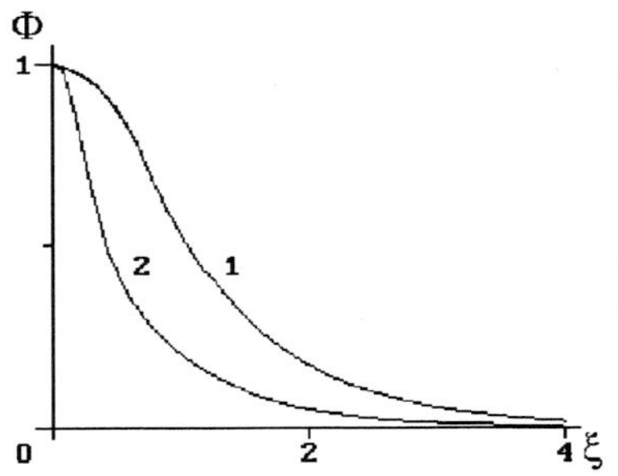

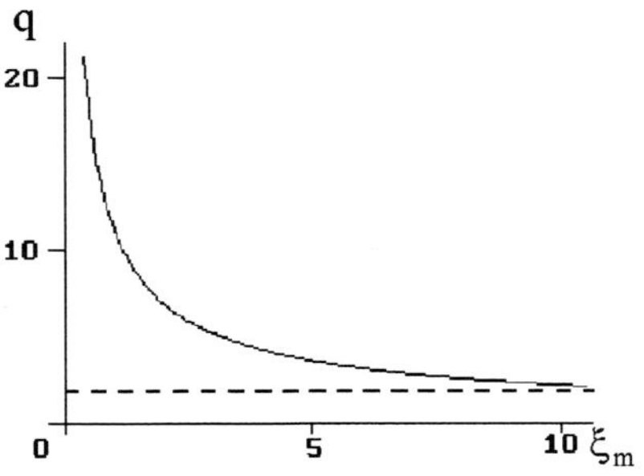

Fig. 2.6.1. Soliton solutions of Eq (2.6.7) for $q = 3.4$, $\xi_m = 5$ (1) and $q = 8.6$, $\xi_m = 0.5$ (2).

Fig. 2.6.2. Parameter q (inverse maximum soliton amplitude) as a function of ξ_m.

The above equation differs from Eqs. (1.5.5) and (2.4.1) by the presence of the parameter

$$V(x) = \begin{cases} (1 - x^2/2)\sqrt{1 - x^2} - 1.5x \arccos x, & x < 1, \\ 0, & x \geq 1, \end{cases}$$

at which $q = 2|e|\rho_0 l^2/3\pi^2\delta\varphi_m$, $\delta > 0$, $\xi_m = kr_m$, and the prime means the derivative with respect to the variable $\xi = 2\pi r/l$. In the limits $\Omega = 0$ and $\Omega = \omega_B$, the magnetic focusing does not occur (ξ_m=∞ and $V = 1$).

The density of rotating electron bunches (Ω >0) goes to zero at the point ξ_0 which satisfy the equality

$$\Phi(\xi_0) = \xi_0^2/\xi_m^2, \tag{2.6.8}$$

and an electron does not reaches the walls of the potential well because of the magnetic focusing occurs.

For $\xi > \xi_0$ the solution of Eq. (2.6.7) takes the form

$$\Phi = AK_0\left(\frac{\xi}{\sqrt{1+\eta}}\right), \tag{2.6.9}$$

where $K_0(x)$ is a Mcdonald function [5] and A is a constant of integration.

The continuity condition on the longitudinal and transverse components of the electrical fields at the point ξ_0

$$\Phi(\xi_0) = AK_0\left(\frac{\xi_0}{\sqrt{1+\eta}}\right), \quad \Phi'(\xi_0) = -\frac{A}{\sqrt{1+\eta}}K_1\left(\frac{\xi_0}{\sqrt{1+\eta}}\right) \tag{2.6.10}$$

determines the boundary surface of the beam $\xi = \xi_0$,

$$\sqrt{1+\eta}\,\frac{\Phi'(\xi_0)}{\Phi(\xi_0)} = -K_1\left(\frac{\xi_0}{\sqrt{1+\eta}}\right) \Big/ K_0\left(\frac{\xi_0}{\sqrt{1+\eta}}\right). \tag{2.6.11}$$

Numerical solution of Eq. (2.6.7), which corresponds to the boundary conditions at the beam axis $\Phi(0) = 1$, $\Phi'(0) = 0$, and the additional conditions (2.6.8) (2.6.11) is shown in Fig. 2.6.1. The function $q(\xi_m)$, which defines the existence of the soliton solutions, is presented in Fig. 2.6.2.

The relative force to influence of the electrostatic and magnetic fields on the bunch depends on the parameter

$$\xi_m^2 = \frac{r_m^2}{r_B{}^2} = \frac{\omega_B^2}{3\pi\Omega q\delta(1+\eta)(|\omega_B| - \Omega)(1+\eta)}. \tag{2.6.12}$$

For $\xi_m \gg 1$ the electrostatic self-focusing dominates, and contrastingly the magnetic compression occurs.

To estimate this ratio in order of magnitude, we consider that $\Omega = |\omega_B|/2$. In the limit $\rho_0 \gg n_p\delta^2$, which corresponds to a beam with a high density, the action of the longitudinal magnetic field accompanies by the radial defocusing of a beam $r_b \to r_m$ $(r_B/l \to \infty)$ if the following inequality holds

$$|\omega_B| \gg \omega_{Bc} = \sqrt{\omega_b\omega_p} \tag{2.6.13}$$

(here ω_p and ω_b are the Langmuir plasma frequency and beam frequency, respectively).

In the opposite case $\rho_0 \ll n_p\delta^2$, the action of the longitudinal magnetic field leads to the additional compression of a beam $(r_B/l \to 0)$. Therefore by analogy with (2.6.13), we find $\omega_{Bc} = \omega_b/\sqrt{2\delta}$.

Appendix 2.1

In conclusion, we consider a restricted electron bunches which longitudinal size a does not attain $L = \lambda_p(n_p/n_b)^{1/3}$ ($\lambda_p = 2\pi v_0/\omega_p$ is the plasma wavelength). In contrast with the opposite case $a \gg L$ [which have discussed above in §1.8; see also §3.8], the energy of plasma oscillations is not accumulated in the region occupied by the bunch so that the wave excited by the short bunch lag behind it forming a so-called wake. This channel of energy losses has a strong effect both on the development of perturbations in the bunch bulk and, as a whole, on the formation of the fine structure [9].

Fig. A2.2.1 shows the time dependence of the wake field amplitude corresponding to three bunches which sizes differ in their initial lengths. For small bunch $a << \lambda_p$, the amplitude of the excited wake field decreases gradually. In opposite case $a >> \lambda_p$, the initial amplitude of the excited wave is small, the macroscopic plasma dielectric permittivity is negative, and initially the bunch is compressed as a whole. However, the onset of the rapidly developing instability of the dissipative type leads to the periodically modulated bunch structure with the space period λ_p.

Further, the beam particle trapping by the field occurs; and the periodic order of the structure and, consequently, the coherent nature of generation are ruined (Fig. A2.2.1). Note that in the case of long bunches $a > \lambda_p$with a fixed number of particles the maximum value of the wakefield amplitude depends weakly on the initial longitudinal size of the bunch. This enables us to consider this value of amplitude as an ultimate. The formation of the structure enhances the energy flux to the dissipation channels. The maximum value of this flux corresponds to the peak amplitude of the structure. However, during the system relaxation, the structure may provide a minimum energy flux.

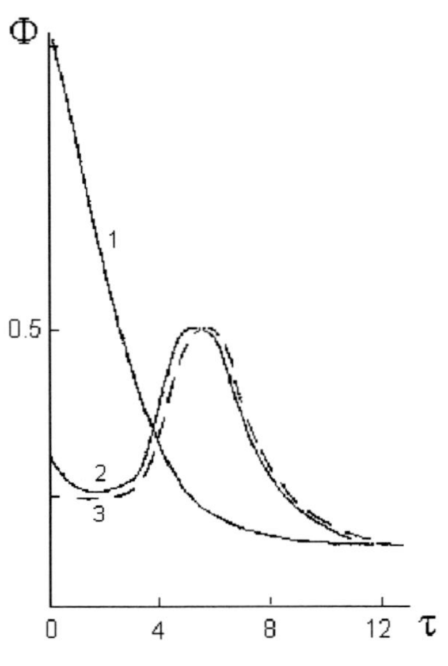

Fig. A2.2.1. Wakefield amplitude $\Phi = eE/mv_0\omega_p$ as a function of $\tau = \omega_p t$ for (1) a/λ_p=0.3, (2) 3 (3) 5.

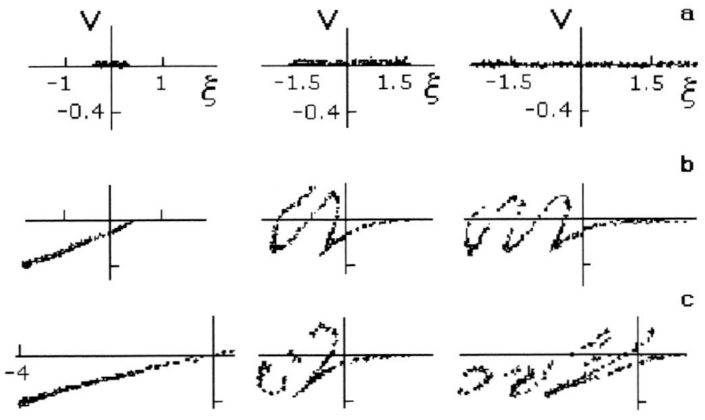

Fig. A2.2.2. Phase-plane (for the same bunches as in Fig. A2.1) at successive: times (a) τ=0, (b) τ=6, (c) τ=12 [$\xi=2\pi z/\lambda_p$ and $V=(v-v_0))/v_0$].

CHAPTER 3

SELF-FOCUSING OF RELATIVISTIC ELECTRON BUNCHES IN A PLASMA

In the present chapter the results obtained above for non-relativistic beams are generalized to the relativistic case when the radiation is non-potential and we must take into account the effect of the solenoidal electromagnetic field on the bunches [1,2]. We show that the system of Maxwell's equations for the field in a plasma with a relativistic beam may be reduced to an integral equation for the potential $\Psi = \varphi - vA_z/c$, that coincides in form with the analogous equation for the scalar potential φ in the non-relativistic limit (see Eq. 1.2.3).

At the same time the presence of an azimuthal magnetic field reduces the strength of the Coulomb repulsion of the electrons in the bunches (proportional to $\gamma^{-2} = 1 - v^2/c^2$), and the role of the polarization field, which focuses the beam for $\epsilon = 1 - \omega_p/\omega_m < 0$, grows. Because of the region of modulation frequencies corresponding to self-focusing expands since the focusing polarization field does not depend on relativistic factor γ. An analytic solution of the problem is derived in the nonresonant case $\omega_m < \omega_p$ for ultrarelativistic energies, $\gamma^2 \gg 1$, at which the nonlinear wave equation for the self-consistent potential simplifies, and variables can be separated. It thus becomes possible to treat the radial and longitudinal equilibria of the electron bunches independently.

In the region of relativistic energies, the growth rate of the satellite instability decreases in proportion to γ^{-1} and the length of beam transport in a plasma increases.

§3.1. Integro-differential equation.

We use the following system of equations to describe the radiation of an electron beam of charge density $\rho(t, \mathbf{r})$, propagating with a constant velocity $\mathbf{v}$ in a homogeneous dense plasma $n_p \gg \rho$:

$$\begin{aligned} \operatorname{curl}\mathbf{E} = -\frac{1}{c}\frac{\partial \mathbf{B}}{\partial t}, \quad \operatorname{curl}\mathbf{B} = \frac{1}{c}\frac{\partial \mathbf{E}}{\partial t} + \frac{4\pi e}{c}(n_p\mathbf{v}_p + \rho\mathbf{v}), \\ \frac{\partial \mathbf{v_p}}{\partial t} = \frac{e}{m}\mathbf{E}, \quad \frac{\partial n_p}{\partial t} + n_p \operatorname{div}\mathbf{v}_p = 0, \end{aligned} \tag{3.1.1}$$

where $n_p v_p$ is the plasma electron current density induced by the field of the beam.

Eliminating the magnetic field in Eq. (3.1.1), we find

$$\Delta\mathbf{E} - \operatorname{grad}\operatorname{div}\mathbf{E} - \frac{1}{c^2}\left(\frac{\partial^2}{\partial t^2} + \omega_p^2\right)\mathbf{E} = \frac{4\pi e\mathbf{v}}{c^2}\frac{\partial \rho}{\partial t}, \tag{3.1.2}$$

where $\omega_p = \sqrt{4\pi e^2 n_p/m}$ is the Langmuir plasma frequand Δ is the Laplacian operator.

It follows from Eq. (3.1.2) that the wave equations for the potentials

$$\left(\frac{\partial^2}{\partial t^2}+\omega_p^2\right)\left[\Delta-\frac{1}{c^2}\left(\frac{\partial^2}{\partial t^2}+\omega_p^2\right)\right]\varphi=-4\pi e\frac{\partial^2\rho}{\partial t^2}, \tag{3.1.3}$$

$$\left[\Delta-\frac{1}{c^2}\left(\frac{\partial^2}{\partial t^2}+\omega_p^2\right)\right]\mathbf{A}=-\frac{4\pi e}{c}\mathbf{v}\rho \tag{3.1.4}$$

satisfy the gage condition

$$\mathbf{E}=-\nabla\varphi-\tfrac{1}{c}\tfrac{\partial\mathbf{A}}{\partial t},\qquad \frac{\partial}{\partial t}\operatorname{div}\mathbf{A}+\frac{1}{c}\left(\frac{\partial^2}{\partial t^2}+\omega_p^2\right)\varphi=0.$$

The density of the beam, broken into bunches, is a periodic function of $z'=z-vt$ and, along with the scalar potential φ, may be represented as a Fourier series. Carrying out the summation on the right hand side of Eq. (3.1.3) (in analogy with §1.2), we find

$$\left(\Delta_r+\frac{1}{\gamma^2}\frac{\partial^2}{\partial z'^2}-\frac{\omega_p^2}{c^2}\right)\varphi=-4\pi e\rho_{\text{eff}}(r,z'),$$

$$\rho_{\text{eff}}(r,z')=\rho(r,z')-\frac{\alpha\pi}{\sin\alpha\pi}<\rho\cos\zeta>, \tag{3.1.5}$$

$$\zeta=2\pi\alpha\left(\frac{|\xi-z'|}{l}-\frac{1}{2}\right),\qquad <\ldots>=\frac{1}{l}\int_{-l/2}^{l/2}\ldots\,d\xi,$$

where $\alpha=\omega_p/\omega_m$, l is spatial period of the beam, $\omega_m=2\pi v/l$ is the modulation frequency,and Δ_r is the radial part of the Laplacian.

We direct the z axis of the cylindrical coordinate system along the beam velocity , and we consider an azimuthally symmetrical steady-state solution $A_r=A_\varphi=0$. It will be shown below that the equilibrium beam density depends on the potential $\psi=\varphi-\beta A_z$. Therefore we introduce an equation for ψ that follows from Eq. (3.1.4) and (3.1.5),

$$\left(\Delta_r+\frac{1}{\gamma^2}\frac{\partial^2}{\partial z'^2}-\frac{\omega_p^2}{c^2}\right)\psi=-4\pi e\tilde{\rho}_{\text{eff}}(r,z'),$$

$$\tilde{\rho}_{\text{eff}}(r,z')=\frac{\rho(r,z')}{\gamma^2}-\frac{\alpha\pi}{\sin\alpha\pi}<\rho\cos\zeta>, \tag{3.1.6}$$

where $\beta=v/c$ and $\gamma=\left(1-\beta^2\right)^{-1/2}$.

We note that the constant component of the potential $\langle\varphi\rangle=\langle\rho\rangle=0$ because of the screening of the static beam field by the cold plasma over the spatial period of the beam. At the same time, the current neutralization is incomplete, and $\langle A_z\rangle\neq 0$. Equation (3.1.6) differs from the corresponding equation (1.2.3) by the presence of the factor γ^{-2} and the additional term $\omega_p^2\psi/c^2$ due to additional radial screening of the field of a relativistic beam by the plasma [3,4].

§3.2. One-dimensional ultrarelativistic bunches.

We assume that the charge density of a beam depends on coordinates as:

$$\rho(r,z')=\begin{cases}\rho(r), & |z'|\le a/2,\\ 0, & a/2\le|z'|\le l/2.\end{cases} \tag{3.2.1}$$

Substituting (3.2.1) into (3.1.6) and evaluating the integral, we find

$$\frac{\tilde{\rho}_{\text{eff}}(r,z')}{\rho(r)} = \begin{cases} \dfrac{\sin[\alpha\pi(1-a/l)]}{\sin(\alpha\pi)}\cos kz' - \beta^2, & |z'| \le \dfrac{a}{2}, \\ -\dfrac{\sin(\alpha\pi a/l)}{\sin(\alpha\pi)}\cos\left[2\alpha\pi\left(\dfrac{|z'|}{l} - \dfrac{1}{2}\right)\right], & a/2 < |z'| \le \dfrac{l}{2}. \end{cases} \tag{3.2.2}$$

If the longitudinal size satisfies the condition

$$a_n = (1 - n\omega_m/\omega_p)l \equiv \delta_n l, \tag{3.2.3}$$

(n is a whole number) then each bunch for $\rho = \text{const}$ occurs in the force free equilibrium considered in §1.2 in the nonrelativistic case $\beta \ll 1$. According to Eq. (3.2.2), allowance for the solenoidal field does not destroy this equilibrium.

In the general actual case ($\rho \neq \text{const}$) the radial component of the electric field has a jump at the boundaries of bunches $|z'| = a_n/2$, and in the nonrelativistic case a three-dimensional analog of the solution of §1.2 does not exists. This jump is not eliminated in the relativistic case either, because it follows from Eq. (3.2.2) under condition (3.2.3) that

$$\tilde{\rho}_{\text{eff}}(r,z') = \rho(r)\begin{cases} \beta^2, & |z'| \le \dfrac{a_n}{2}, \\ \cos\left[k\alpha\left(|z'| - \dfrac{a_n}{2}\right)\right], & \dfrac{a_n}{2} < |z'| \le \dfrac{l}{2}. \end{cases} \tag{3.2.4}$$

However, in the ultrarelativistic limit $\beta^2 \to 1$ and the asymptotic solution exists, in which case $E_r - \beta B_\varphi = -\partial\psi/\partial r$ extends continuously through the boundary of a bunch.

Thus, under condition (3.2.3), ultrarelativistic electron bunches are in longitudinal equilibrium, but they experience a focusing magnetic force $F_r = -e\partial\psi/\partial r$ in the radial direction. If the initial beam radius is sufficiently large, $R_0 \gg c/\omega_p$, then it follows from (3.1.6) and (3.2.4) that

$$F_r = \frac{mv^2}{n_p}\frac{\partial\rho}{\partial r}. \tag{3.2.5}$$

The system of hydrodynamic equations

$$\frac{\partial v_r}{\partial t} + v_r\frac{\partial v_r}{\partial r} = \frac{F_r}{m\gamma}, \quad \frac{\partial\rho}{\partial t} + \frac{1}{r}\frac{\partial}{\partial r}(rv_r\rho) = 0,$$

describing the radial oscillations of bunches has a simple self-similar solution

$$v_r = \frac{r}{R}\frac{dR}{dt}, \quad \rho = \rho_0\begin{cases} (R_0/R)^2(1 - r^2/R^2), & r < R, \\ 0, & r > R, \end{cases} \tag{3.2.6}$$

where $R(t)$ is the bunch radius ρ_0 is the initial electron density on the beam axis($r = 0$).

The nonlinear equation following from Eq. (3.2.5),

$$\frac{d^2R}{dt^2} + \Omega_b^2\frac{R_0^4}{R^3} = 0, \quad \Omega_b^2 = \frac{2\rho_0 v^2}{n_p R_0^2\gamma} \tag{3.2.7}$$

describes a radial collapse of the beam

$$R(t) = R_0\sqrt{(1 - \Omega^2 t^2)} \tag{3.2.8}$$

over a time $t_b = \Omega_b^{-1}$. The physically correct solution, allowing for the presence of turning points in the region $R = 0$ due to the actually always significant kinetic pressure gradient, $-T\partial\rho/\partial r$, may be obtained by a periodic continuation of Eq. (3.2.8) into the region $t > t_b$.

Thus for the case of the plane ultrarelativistic electron bunches, $R_0 \gg c/\omega_p$ and $\gamma^2 \gg 1$, a physical mechanism of the longitudinal equilibrium, $a = \delta l$, is electrostatic (strong self-focusing). At the same time, the magnetic field of the wave leads to the radial compression of the bunches under the action of the Lorenz force (weak self-focusing).

The physical condition of beam equilibrium in a plasma is the absence of forces acting on the electrons in the bunches. In the hydrodynamic approximation the conditions of radial and longitudinal equilibrium are

$$e\rho(E_r - \beta B_\varphi) - T\frac{\partial\rho}{\partial r} = 0, \quad e\rho E_z - T\frac{\partial\rho}{\partial r} = 0, \tag{3.2.10}$$

where E_r, E_z, and B_φ are components of the electromagnetic field of the beam in the plasma; T is the electron temperature.

Transforming to potentials and taking into account that $\partial/\partial t = -v\partial/\partial z'$, we can reduce (3.2.9) to

$$\nabla'\left(\psi + \frac{T}{e}\ln\rho\right) = 0, \quad \nabla' = \left(\frac{\partial}{\partial r}, \frac{\partial}{\partial z'}\right),$$

which do not contain the component of the vector potential A_r. After integration we obtain the Boltzmann distribution

$$\rho(r, z') = \rho_0 \exp\left[-\frac{e\psi(r, z')}{T}\right],$$

which together with Eq. (3.1.6) determines the self-consistent potential in a plasma with a beam and the shape of equilibrium bunches.

The self-focusing occurs if the second term on the right-hand side of Eq. (3.1.6) exceeds the first one (the polarization field caused by the beam in the plasma exceeds the Coulomb field of the bunches). The appearance of the additional factor γ^{-2} decreases the role of the electron Coulomb repulsion, and the efficiency of the focusing mechanism grows because the condition that the parameter $\delta = 1 - \omega_m/\omega_p > 0$ be small is γ^2 times weaker than in a nonrelativistic beam,

$$\delta \ll \gamma^2 \frac{a}{l}. \tag{3.2.11}$$

The inequality (3.2.11) leads to the simplification of Eq. (3.1.6) to

$$\left(\Delta_r + \frac{1}{\gamma^2}\frac{\partial^2}{\partial z'^2} - \frac{\omega_p^2}{c^2}\right)\psi = \frac{4ke}{\delta}\cos kz' \int_0^{l/2} \rho(r, \xi)\cos k\xi \, d\xi. \tag{3.2.12}$$

We note that under resonance conditions $\delta \ll 1$, relativistic effects do not alter Eq. (3.2.12) for the potential, which has the same form as Eq. (1.3.1) (obtained for nonrelativistic beam), because the radiation field is determined mainly by the scalar potential $\varphi \simeq A_z/\delta \gg A_z$ and the contribution of the solenoidal field to the radiation of the beam in the plasma is small.

§3.3. Satellite instability.

The instability of a highly modulated electron beam in a plasma (satellite instability) was first considered in Ref. [5] for a sequence of charged electron planes. To estimate the growth rates we will use the dispersion relation obtained Ref. [6] for a nonrelativistic one-dimensional beam. In the beam frame of reference for the one-dimensional bunches of density (3.2.1) this dispersion relation is

$$\begin{aligned}\cos\left[\left(k' + \frac{\omega'}{v}\right) l'\right] &= \cos\frac{\omega'_p a'}{v\sqrt{\epsilon_b}} \cos\left[\frac{\omega'_p}{v}(l' - a')\right] \\ &= -\frac{1}{2}\left(\sqrt{\epsilon_b} + \frac{1}{\sqrt{\epsilon_b}}\right) \sin\frac{\omega'_p a'}{v\sqrt{\epsilon_b}} \cos\left[\frac{\omega'_p}{v}(l' - a')\right],\end{aligned} \tag{3.3.1}$$

where ω' and k' are the frequency and wave number of the perturbations, $\epsilon_b = 1 - \omega_b'^2/\omega'^2$, $\omega'_b = \sqrt{4\pi e^2 \rho'_0}$, $\omega'_p = \sqrt{4\pi e^2 n'_p/m\gamma^3}$, ρ'_0 and n'_p are the beam and plasma densities, l' is the distance between bunches, and a' is the thickness of bunches.

Carrying out the Lorentz transformation

$$\rho'_0 = \frac{\rho_0}{\gamma}, \quad n'_p = \gamma n_p, \quad l' = \gamma l, \quad a' = \gamma a,$$

we bring Eq. (3.3.1) to the plasma frame of reference:

$$\cos\frac{\omega l}{v} = \cos\frac{\omega_p a}{v\sqrt{\epsilon_b}} \cos\left[\frac{\omega_p}{v}(l - a)\right] - \frac{1}{2}\left(\sqrt{\epsilon_b} + \frac{1}{\sqrt{\epsilon_b}}\right) \sin\frac{\omega_p a}{v\sqrt{\epsilon_b}} \cos\left[\frac{\omega_p}{v}(l - a)\right], \tag{3.3.2}$$

where

$$\epsilon_b = 1 - \frac{\omega_b^2}{(\omega - kv)^2}, \quad \omega_b^2 = \frac{4\pi e^2 \rho_0}{m\gamma^3}, \quad \omega_p^2 = \frac{4\pi e^2 n_p}{m}.$$

For equilibrium bunches $a = \delta l$, it follows from Eq. (3.3.2) that

$$1 - \frac{a^2}{l^2}\frac{\omega_p^2}{(\omega - s\omega_m)^2} - \frac{\omega_b^2}{(\omega - kv)^2} = 0, \tag{3.3.3}$$

where $\omega_m = 2\pi v/l$ is the modulation frequency and s is a whole number.

In the case of sufficiently wide bunches

$$\nu = \frac{\langle\rho\rangle}{n_p\gamma^3}, \quad \langle\rho\rangle = \rho_0\frac{a}{l}, \quad \delta = \frac{a}{l} \gg \nu^{1/3},$$

the frequency $\mathrm{Re}\,\omega$ and maximum growth rate of instability $\mathrm{Im}\,\omega$ are

$$\begin{aligned}\omega_1 &= s\omega_m + \frac{a}{l}\omega_p + \Delta_1 = kv + \Delta_1, \\ \Delta_1 &= \frac{-1 \pm i\sqrt{3}}{2^{4/3}}\nu^{1/3}\omega_p.\end{aligned} \tag{3.3.4}$$

When the inequality opposite above is satisfied, $\delta \ll \nu^{1/3}$, we obtain

$$\begin{aligned}\omega_2 &= s\omega_m + \Delta_2 = kv \pm \omega_b + \Delta_2, \\ \Delta_2 &= \frac{-1 \pm i\sqrt{3}}{2^{4/3}}\nu^{1/6}\delta^{1/2}\omega_p.\end{aligned} \tag{3.3.5}$$

Combining Eq. (3.3.4) and Eq. (3.3.5), we find:

$$\mathrm{Im}\,\omega = \begin{cases}\nu^{1/3}, & a/l \gg \nu^{1/3}, \\ \nu^{1/6}(a/l)^{1/2}, & a/l \ll \nu^{1/3}.\end{cases} \tag{3.3.6}$$

The growth rate (3.3.6) coincides with the Cherenkov growth rate of a continuous beam with density $\langle\rho\rangle$. For thin bunches the growth rate is small in comparison with the Cherenkov growth rate, in the ratio $\sqrt{\delta\nu^{-1/3}} \ll 1$.

A nonlinear saturation at the amplitude of the satellite instability is reached when the bunch modulation depth $\tilde{\rho}_b$ reaches the value ρ_0,

$$\tilde{\rho}_b \simeq \frac{E_{NL} k \rho_0}{m\gamma^3(\omega - kv)^2} \simeq \rho_0.$$

Substituting $\omega - kv$ and k from Eq. (3.3.4) and Eq. (3.3.5), we find

$$E_{NL} \simeq \begin{cases} \nu^{2/3}, & \delta \gg \nu^{1/3}, \\ \nu/\delta, & \delta \ll \nu^{1/3}. \end{cases} \tag{3.3.7}$$

According to Eq. (3.1.7), the field of the stationary mode confining the bunches is $E_p \simeq \nu/\delta$, and when the first of inequalities (3.1.7) holds it is less than E_{NL}; consequently, the development of the satellite instability is accompanied by a destruction of bunches in a time $t_1 \simeq \nu^{-1/3}\omega_p^{-1}$.

In the second limiting case in (3.1.7) the field $E_{NL} \simeq E_p$ and the driving time of oscillations is $t_2 \simeq \left(\delta^{1/2}\nu^{1/6}\right)^{-1}$. Thus the possibility exists of transporting a relativistic electron beam in the form of thin bunches $a/l \ll \nu^{1/3}$ in a plasma over a distance $L \simeq ct_2$.

A relativistic electron beam divided into separate bunches in a dense plasma is in equilibrium with electromagnetic radiation if the modulation frequency is less than the plasma frequency, $\delta = 1 - \omega_m/\omega_p > 0$. Under conditions of resonance $\delta \ll 1$ the wave of radiation is close to being a potential wave, and self-focusing physical mechanism is analogous to the one considered in Chap. 7 for nonrelativistic beams with longitudinal bunch sizes $a \gg \delta l$. The volume potential well of depth $W \simeq mv^2\rho_0/n_p\delta$ confines electrons with a temperature $T \simeq W$, and the beam radius and bunch thickness are determined by the condition of compensation of the focusing collective force by the gradient of the kinetic pressure in the beam.

In the ultrarelativistic case $\gamma^2 \gg 1$ for a monoenergetic beam there exists an equilibrium in which the bunches have a thickness $a = \delta l$ and are found in a force-free longitudinal equilibrium, but radial focusing (collapse of electron disks) occurs due to Loretz force. Because in this case the potential well is anisotropic, $W_{||} \simeq mv^2\rho_0/n_p\delta$ and $W_\perp \simeq \delta W_{||}$, the presence in a beam with $\delta \ll 1$ of a small thermal spread stabilizes the radial collapse of bunches without changing their thickness $a = \delta l$. Thus the possibility exists of arranging a beam equilibrium with a small temperature $T \ll W_{||}$ and a radius $R_0 \simeq T(4\pi e^2\rho_0/m)^{-1/2}$, necessary for the transporting of large currents in a plasma in beams divided into separate bunches,

$$I_b \simeq \pi\left(\frac{c}{\omega_p}\right)^2 e < \rho > v \simeq I_a\gamma^2\nu \tag{3.3.8}$$

over long distances,

$$L \simeq \frac{c}{\omega_p\sqrt{\delta}}\left(\frac{I_a\gamma^2}{I_b}\right)^{1/6}, \quad \delta \ll \nu^{1/3} \simeq \left(\frac{I_b}{I_a\gamma^2}\right)^{1/3}. \tag{3.3.9}$$

Here $I_a = mc^3\beta\gamma/e$ is the Alfven current, and $\nu = \langle\rho\rangle/n_p\gamma^3$.

§3.4. Relativistic bunches in a heated plasma.

In this section the results obtained for a nonrelativistic beam in §2.2 are generalized to the relativistic case. Under resonant conditions the wave excited by the modulated beam in the plasma of finite temperature $T_p > 0$ is nearly electrostatic and thermal corrections to the cold-plasma dielectric function are proportional to the Debye radius $\lambda_d = V_T/\omega_p$, $V_T = \sqrt{2T_p/m}$ divided by the wavelength $\lambda/2\pi = v/\omega_p$. Hence in the equation for the self-consistent potential there is an additional parameter

$$\mu = \lambda_d^2/\lambda^2|\epsilon| = V_T^2/v^2|\epsilon|, \tag{3.4.1}$$

which accounts for the electron thermal motion and increases the order of the equation for $\mu > 0$.

In order to account for the thermal motion of the plasma electrons we include the gradient of the kinetic pressure in the equation of motion:

$$\frac{\partial \tilde{\mathbf{v}}_p}{\partial t} = \mathbf{E} - \frac{{V_T}^2}{n_p}\nabla \tilde{n}_p, \quad \frac{\partial \tilde{n}_p}{\partial t} = -n_p \mathrm{div}\tilde{\mathbf{v}}_p$$

where n_p is the background plasma density, and $\tilde{\mathbf{v}}_p$ and $\tilde{n}_p$ are the variable components of electron velocity and density with $\tilde{n}_p \ll n_p$.

Taking into account the above equations, one can modify the first equation (3.1.4) to the form:

$$\begin{aligned}\left(\frac{\partial^2}{\partial t^2} + \omega_p^2 - {V_T}^2\Delta\right)\left[\Delta - \frac{1}{c^2}\left(\frac{\partial^2}{\partial t^2} + \omega_p^2\right)\right]\varphi \\ = -4\pi e\left(\frac{\partial^2}{\partial t^2} - {V_T}^2\Delta\right)\rho.\end{aligned} \tag{3.4.2}$$

The second equation (3.1.4) and the gauge condition (3.1.3) do not change.

Transforming in Eqs. (3.1.4) and (3.4.2) to the function

$$\psi = \varphi - \beta A_z, \quad \beta = v/c$$

we find

$$\begin{aligned}\left(\frac{\partial^2}{\partial t^2} + \omega_p^2 - {V_T}^2\Delta\right)\left[\Delta - \frac{1}{c^2}\left(\frac{\partial^2}{\partial t^2} + \omega_p^2\right)\right]\psi \\ = -4\pi e\left[\frac{1}{\gamma^2}\left(\frac{\partial^2}{\partial t^2} - {V_T}^2\Delta\right) - \beta^2\omega_p^2\right]\rho.\end{aligned} \tag{3.4.3}$$

By analogy with §1.5, we determine the electron beam density as a function of the potential in the frame of reference of the beam by

$$\rho'(\varphi') = \rho_0'\left(\frac{\varphi'}{\varphi_m'}\right)^2\theta(\varphi'). \tag{3.4.4}$$

where ρ_0' is the electron density at the bottom of the potential well, φ_m'. The discontinuous function

$$\theta(\varphi') = \begin{cases} 1, & \varphi' > 0 \\ 0, & \varphi' < 0 \end{cases}$$

reflects the condition for capture of the electrons by the wave. A transformation to the frame of reference of the plasma is made with the help of a Lorentz transformation:

$$\varphi' = \gamma\psi, \quad \varphi'_m = \gamma\psi_m, \quad \rho_0 = \gamma\rho'_0,$$
$$\rho = \rho_0 \left(\frac{\psi}{\psi_m}\right)^2 \theta(\psi). \tag{3.4.5}$$

We consider the time-independent solution of Eq. (3.4.3),

$$\psi = \psi(\mathbf{r}_\perp, z'), \quad z' = z - vt$$

in the ultrarelativistic limit $\gamma^2 \gg 1$. Transforming to the variable z' and dropping terms of order $\gamma^{-2} \ll 1$, we find the following nonlinear partial differential equation:

$$\left(\frac{\partial^2}{\partial z'^2} + \frac{\omega_L^2}{c^2} - \frac{V_T{}^2}{c^2}\Delta_\perp\right)\left(\Delta_\perp - \frac{\omega_p^2}{c^2}\right)\psi = \frac{4\pi e\omega}{c^2}\rho(\psi), \tag{3.4.6}$$

where $\Delta_\perp$ is the transverse part of the Laplacian, $\omega_L^2 = \omega_p^2 + (kV_T)^2$.

For the ribbon beam the resonant solution has the form

$$\psi(\mathbf{r}, z') = \psi_m \Phi(x) \cos kz', \quad k = 2\pi/l. \tag{3.4.7}$$

Substituting (3.4.7) in (3.4.6) and averaging over the spatial period l, we find a nonlinear ordinary differential equation for the amplitude of the wave potential:

$$\mu\Phi^{(4)} - (1+\mu)\Phi'' + \Phi - q\Phi^2 = 0,$$
$$\mu = \frac{V_T{}^2}{2v^2\delta}, \quad q = \frac{2}{3\pi^2}\frac{|e|\rho_0 l}{\psi_m \delta}, \quad \delta = 1 - \frac{\omega_m}{\omega_L} \ll 1, \tag{3.4.8}$$

where the prime indicates differentiation with respect to $\xi = kx$.

The equation (3.4.8) does not differ from the nonrelativistic equation (2.2.2) with a precision of change of variable

$$\varphi \to \psi = \varphi - \beta A_z.$$

Because of this, the results of nonrelativistic theory can be used over a range of the relativistic energies.

§3.5. Space electron distribution of focused bunches.

In this section an analytic solution of the self-focusing problem is derived in the nonresonant case $\omega_m < \omega_p$ for ultrarelativistic energies, $\gamma^{-2} \ll 1$, at which the nonlinear wave equation for the self-consistent potential simplifies, and variables can be separated. It thus becomes possible to treat the radial and longitudinal equilibria of the electron beams independently [1].

We direct the z axis of the cylindrical coordinate system along the beam velocity , and we consider an azimuthally symmetrical steady-state solution $A_r = A_\varphi = 0$.

$$\psi = \varphi - \frac{v}{c}A_z, \; A_r = A_\vartheta = 0,$$
$$\psi = \varphi - \beta A_z, \quad z' = z - vt. \tag{3.5.1}$$

Switching to the function ψ in Eqs. (3.1.4), we find the equation

$$\left(\frac{\partial^2}{\partial z'^2}+\frac{\omega_p^2}{v^2}\right)\left(\Delta_r+\frac{1}{\gamma^2}\frac{\partial^2}{\partial z'^2}-\frac{\omega_p^2}{v^2}\right)\psi= -4\pi e\left(\frac{1}{\gamma^2}\frac{\partial^2}{\partial z'^2}-\frac{\omega_p^2}{c^2}\right)\rho, \tag{3.5.2}$$

where Δ_r is the radial part of the Laplacian, and $\beta = v/c$. The trapped-particle density $\rho(\psi)$ is determined by the formula (3.4.5).

Equation Eq. (3.5.2) simplifies substantially in the ultrarelativistic limit $\gamma^2 \gg 1$, at which terms of order $1/\gamma^2 \ll 1$ can be discarded:

$$\left(\frac{\partial^2}{\partial\zeta^2}+1\right)(\Delta_\xi-1)y=-qy^2\vartheta(y), \tag{3.5.3}$$

$$y=\frac{\psi}{\psi_m},\quad \xi=\frac{\omega_p r}{c},\quad \zeta=\frac{\omega_p z'}{c},\quad q=\frac{4\pi|e|\rho_0 c^2}{\psi_m\omega_p^2}.$$

The boundary conditions on Eq. (3.5.3), which reflect the conditions for self-focusing of the bunches, are

$$\begin{gathered}\frac{\partial y(0,\zeta)}{\partial\xi}=0,\quad y(\infty,\zeta)=0,\\ y(0,0)=1,\quad \frac{\partial y(0,0)}{\partial\zeta}=0.\end{gathered} \tag{3.5.4}$$

The point $\xi=\zeta=0$ corresponds to the center of a bunch, where the potential reaches its maximum with respect to both variables.

Separation of variables can be carried out in the nonlinear partial differential equation Eq. (3.5.3):

$$y(\xi,\zeta)=q^{-1}R(\xi)Z(\zeta). \tag{3.5.5}$$

In this case the radial potential profile is described by equation

$$\xi^{-1}(\xi R')'-R+R^2=0, \tag{3.5.6}$$

and the boundary conditions to Eq. (3.5.6):

$$R'(0)=0,\quad R(\infty)=0,$$

are a consequence of Eq. (3.5.4). The prime means the derivative with respect to the variable ξ. A solution of Eq. (3.5.6) in the form of a soliton with maximum at the beam axis corresponds to the radial self-focusing of a bunch. Numerical integration shows (see §1.5) that it is realized at $R(0)\simeq 2.392$. The results of calculation illustrates Fig. 1.5.2.

An equation for the function $Z(\xi)$ follows from (3.5.3) and (3.5.5) and describes the potential distribution along the beam:

$$\begin{gathered}Z''+Z-Z^2\theta(Z)=0,\\ Z'(0)=0,\quad Z(0)=q/R_0\quad R_0=R(0).\end{gathered} \tag{3.5.7}$$

Pursuing the mechanical analogy [7], we analyze the first integral of Eq. (3.5.7):

$$\begin{gathered}\frac{Z'^2}{2}+U(Z)=E,\quad U(Z)=\frac{Z^2}{2}-\frac{Z^3}{3}\theta(Z),\\ E=U(Z_0),\quad Z_0=Z(0).\end{gathered} \tag{3.5.8}$$

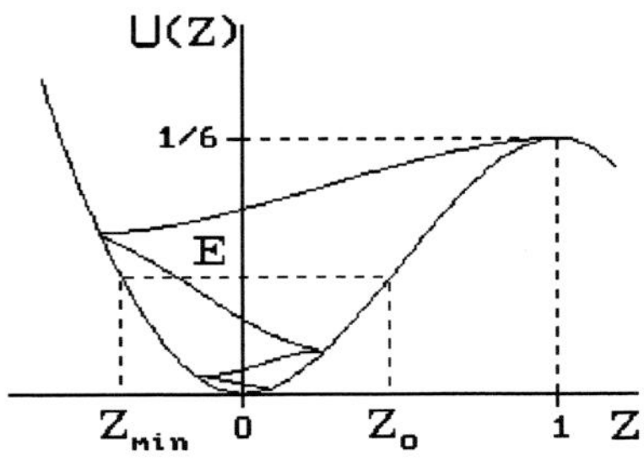

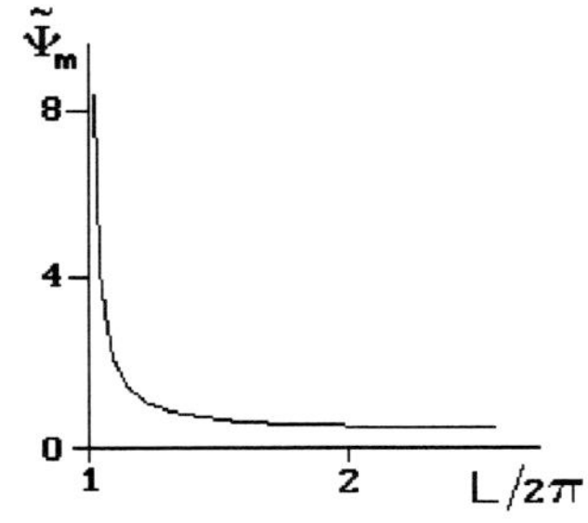

Fig. 3.5.1. Plot of the longitudinal potential well $U(Z)$. The dashed line shows the transformation of a soliton into shock wave in a collision plasma (see §3.7).
Fig. 3.5.2. The wave amplitude $\tilde{\psi}_{\rm m}=\psi_m/\psi_\infty$ versus the spatial period of the beam.

It follows from the plot of the of the function $U(Z)$ in Fig. 3.5.1 that the physically meaningful periodic solution exists in the region

$$Z_{\rm min} \leq Z \leq Z_0 < 1, \quad Z_{\rm min} = -\sqrt{2E}$$

and corresponds to motion of the particle between turning points $Z_{\rm min}$ and Z_0.

Integrating Eq. (3.5.8), we find

$$\zeta = \int_Z^{Z_0} \frac{dZ}{\sqrt{2(E-U)}}. \tag{3.5.9}$$

The period L is determined by Eq. (3.5.9): $L = 2\zeta(Z_{\rm min})$. Allowing for the discontinuous function $\theta(Z)$, we can write the period in the form

$$L = \pi + \sqrt{2}\int_0^{Z_0} \frac{dZ}{\sqrt{E - Z^2/2 + Z^3/3}}, \tag{3.5.10}$$

where the term π corresponds to the range of integration outside the bunch, and $Z_{\rm min} \leq Z \leq 0$.

The integral in Eq. (3.5.10) reduces to an incomplete elliptic integral of the first kind [8]

$$L = \pi + 2\kappa^{-1}F(\varphi, k), \tag{3.5.11}$$

$$2k^2 = 1 + \frac{2Z_0 - 1}{4\kappa^2}, \quad (2\kappa)^4 = (1+2Z_0)\left(1 - \frac{2Z_0}{3}\right),$$

$$\sin^2\varphi = 2\left[1 + \frac{3}{4\kappa^2}\left(1 - \frac{2Z_0}{3}\right)\right]^{-1}.$$

Since the arguments of the function $F(\varphi, k)$ depend on $Z_0 \sim \psi_M^{-1}$, Eq. (3.5.11) can be used to reconstruct the dependence of the depth of the potential well, ψ_M, on the dimensionless spatial period of the beam, $L = \omega_p l/c$. Fig. 3.5.2 shows a plot $\psi_M(L)$.

Similar manipulations lead to an expression for the function $Z(\zeta)$ in terms of elliptic functions [8]:

$$Z(\zeta) = Z_0 \begin{cases} \left[1 - \dfrac{\mathrm{sn}^2(\kappa\zeta, k)}{\sin^2\varphi}\right] \mathrm{dn}^{-2}(\kappa\zeta, k), & |\zeta - nL| \leq \zeta_0, \\ -\sqrt{1 - \dfrac{2Z_0}{3}} \cos\left(\dfrac{L}{2} - \zeta\right), & \zeta_0 \leq |\zeta - nL| \leq \dfrac{L}{2}, \end{cases} \tag{3.5.12}$$

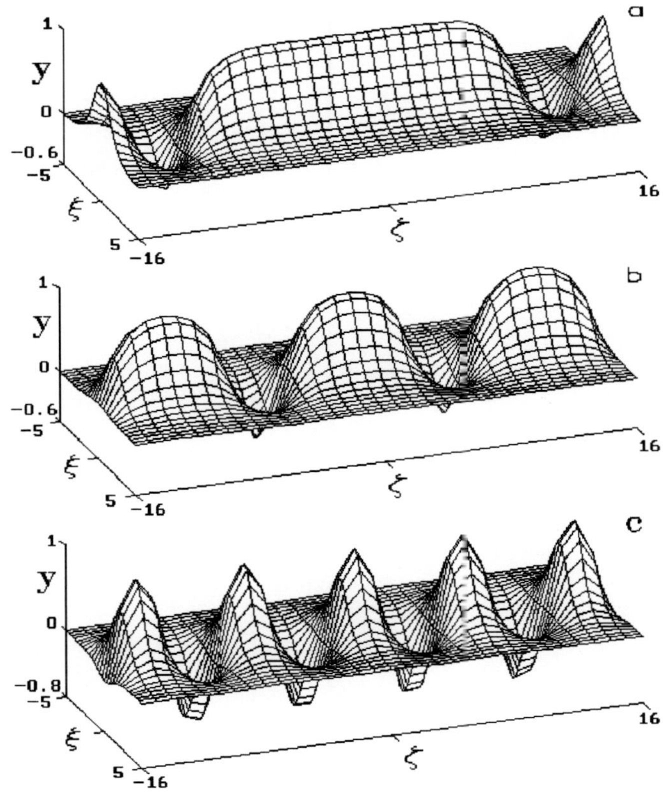

Рис.3.5.3. Spatial profile of the potential for various values of depth of the potential well (detunity δ): ψ_m/ψ_∞= 0.418 (a), 0.437(b) and 1.423 (c).

where $2\zeta_0 = L - \pi$, and the point $\zeta_n = nL$ corresponds to the center of the bunch of index n. Without any loss of generality, we can set $n = 0$ at this point; i.e., we can treat the case of a bunch which is centered at the origin of coordinates $\zeta = 0$. Fig. 3.5.3 shows the shape of the electron bunches for various values of the parameter ψ_M (**Appendix 3.1**).

The general expressions Eq. (3.5.11) and Eq. (3.5.12) simplify in limiting cases in which the asymptotic forms of the special functions can be used. In the limit $Z_0 \to 0$, we use power series in Z_0,

$$\begin{gathered} \mathrm{sn}(\kappa\zeta, k) \simeq \sin(\zeta/2), \quad \mathrm{dn}(\kappa\zeta, k) \simeq 1, \\ L \simeq 2\pi + \frac{4}{3}Z_0, \ k^2 \simeq \frac{4}{3}Z_0, \ 2\kappa \simeq 1 + \frac{2}{3}Z_0, \end{gathered} \tag{3.5.13}$$

finding

$$Z(\zeta) = Z_0 \cos\zeta, \tag{3.5.14}$$

where

$$\delta = \frac{L}{2\pi} - 1 \ll 1, \quad L = \frac{l\omega_p}{c},$$

$$\psi_m = \frac{2}{3\pi^2 R_0}\frac{|e|\rho_0 l^2}{\delta}.$$

In the case $v = c$ expression Eq. (3.5.14) is the same as that may be found from the integral equation Eq. (3.1.5) under resonant condition, such that the beam modulation frequency

$\omega_m = 2\pi c/l$ is close to ω_p, the Langmuir frequency of the plasma [1)].

Asymptotic expansions of Eq. (3.5.11) and Eq. (3.5.12) can also be derived in the case $\Delta = 1 - Z_0 \ll 1$:

$$k^2 = 1 - \frac{4\Delta}{3}, \quad 2\kappa = 1 + \frac{\Delta}{3}, \quad \varphi = \frac{\pi}{2} - \left(\frac{\Delta}{3}\right)^{1/2},$$

$$\mathrm{dn}(\kappa\zeta, k) = \left[1 + \frac{\Delta}{6}(\zeta + \sinh\zeta)\tanh\frac{\zeta}{2}\right]\mathrm{sech}\frac{\zeta}{2}. \tag{3.5.15}$$

Substituting Eq. (3.5.15) into Eq. (3.5.11), we find the period of the oscillations:

$$L = \pi + 2\ln\left[\frac{12}{\Delta}(2 - \sqrt{3})\right]. \tag{3.5.16}$$

In going from Eq. (3.5.11) to Eq. (3.5.16) we made use of the asymptotic expression $K(k) \cong \ln\left(4/\sqrt{1-k^2}\right)$ for the complete elliptic integral. As the parameter Δ decreases the modulation period L increases, and the longitudinal dimensions of the bunches also increase. The distance between bunches, on the other hand, remains constant, equal to half the wavelength, $\lambda = 2\pi c/\omega_p$, (Fig. 3.5.3).

In approximation Eq. (3.5.15) we find the following from expression Eq. (3.5.12):

$$Z(\zeta) = Z_0 \begin{cases} 1 + 2\Delta\left(1 - \mathrm{dn}^{-2}\dfrac{\zeta}{2}\right), & |\zeta| \le \zeta_0, \\ -3^{-1/2}\cos\left(\dfrac{L}{2} - \zeta\right), & \zeta_0 \le |\zeta| \le \dfrac{L}{2}. \end{cases} \tag{3.5.17}$$

Restricting the discussion to the first term in the asymptotic expression for the function $\mathrm{dn}(\kappa\zeta, k)$ in Eq. (3.5.15), $\mathrm{dn}(\kappa\zeta, k) \cong \mathrm{sech}(\zeta/2)$, we find the potential distribution near the origin from Eq. (3.5.17):

$$Z(\zeta) = Z_0\left(1 - 2\Delta\sinh^2\frac{\zeta}{2}\right), \quad |\zeta| \ll \zeta_0. \tag{3.5.18}$$

To simplify Eq. (3.5.17) at the boundary of the bunch, $\zeta \le \zeta_0$, we switch to the new variable $\zeta = \zeta_0 + \zeta_1$ and use the relations

$$\mathrm{dn}\frac{\zeta_0}{2} = (2\Delta)^{1/2}, \; \mathrm{cn}\frac{\zeta_0}{2} = \left(\frac{2\Delta}{3}\right)^{1/2}, \; \mathrm{sn}\frac{\zeta_0}{2} = 1.$$

The general formula from the theory of elliptic functions can that be put in the form

$$\mathrm{dn}\frac{\zeta_0 + \zeta_1}{2} = \sqrt{2\Delta}\left(\mathrm{dn}\frac{\zeta_1}{2} - \frac{1}{\sqrt{3}}\mathrm{sn}\frac{\zeta_1}{2}\mathrm{cn}\frac{\zeta_1}{2}\right)\mathrm{cn}^{-2}\frac{\zeta_1}{2},$$

$$\mathrm{dn}\frac{\zeta_1}{2} = \mathrm{cn}\frac{\zeta_1}{2} = \mathrm{sech}\frac{\zeta_1}{2}, \quad \mathrm{sn}\frac{\zeta_1}{2} = \tanh\frac{\zeta_1}{2}.$$

---

1) Note that the plasma electrons are linear $|\tilde{n}_p| \ll n_p$ in a case of small density beam when the inequality $\rho_0 \ll n_p\delta$ is fulfilled.

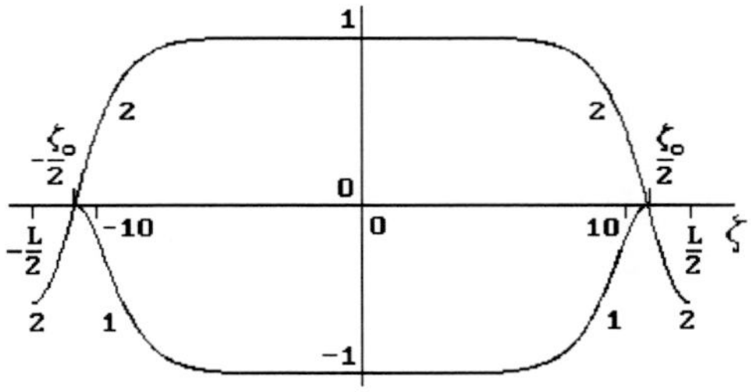

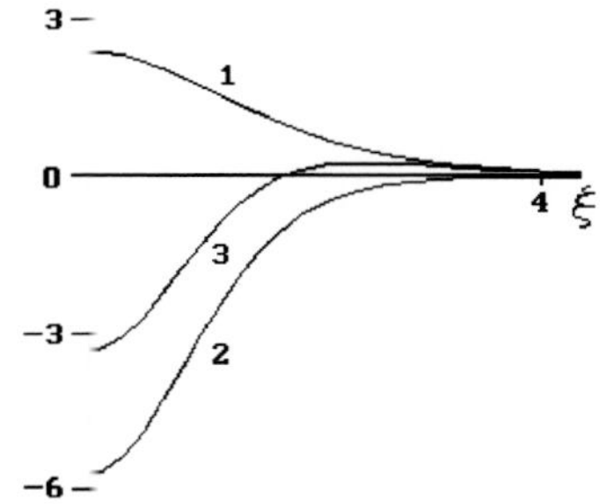

Fig. 3.6.1. Profiles of (1) the beam current I_b/I_0 and (2) the plasma current I_p/I_0 over a modulation period L for the case $\psi_m/\psi_\infty = 0.418$.

Fig. 3.6.2. Profiles of (1) the average current density of the plasma $< j >/j_0 < Z >$, (2) that of the beam, and (3) their sum along the radial variable ξ.

Substituting Eq. (3.5.19) into the first expression Eq. (3.5.17), we find the potential distribution near the boundary of a bunch:

$$Z(\zeta) = 1 - \frac{3}{2}\cosh^{-2}\left(\frac{\zeta_1}{2} - \operatorname{arccosh}\sqrt{\frac{3}{2}}\right). \qquad (3.5.20)$$

At the point $\zeta = \zeta_0(\zeta_1 = 0)$ function in Eq. (3.5.20) and its derivative are the same as the second expression in Eq. (3.5.17).

The maximum value of the potential which corresponds to the condition $Z_0 = 1$,

$$\psi_\infty = \frac{4\pi|e|\rho_0 c^2}{\omega_p^2 R_0} \qquad (3.5.21)$$

does not contain a resonant factor. It is an asymptotic function of $\psi_m(L)$ in the limit $L \to \infty$ (Fig. 3.5.2).

As the deviation from resonance increases, for $L \gg \pi$, the potential well becomes shallower, and the temperature of the captured electrons does not exceed

$$T \cong |e|\psi_\infty \cong mc^2\rho_0/n_p, \quad L \gg 2\pi. \qquad (3.5.22)$$

Thus it is shown above that electromagnetic radiation emitted by a relativistic beam in a plasma simultaneously creates longitudinal and radial potential wells for the electron bunches if the spatial period of the beam l is greater than the wavelength $\lambda_p = 2\pi c/\omega_p$.

The length of bunch varies from $\lambda_p/2$ to $(L - \pi)c/\omega_p$ as the modulation frequency of the beam is reduced in the region $\omega_m < \omega_p$. However, the beam radius $r_0 \cong \lambda_p$ and the distance between bunches, $\delta z' = \lambda_p/2$, are not changed.

§3.6. Induced charge and current of the plasma.

To determine the electron current density $\mathbf{j}_p$ induced in the plasma by the beam, we use equations which follow from Eq. (3.1.4) for the scalar potential φ and vector potential A_z:

$$\varphi = -\frac{\partial^2\psi}{\partial\zeta^2}, \quad A_z = -\frac{\partial^2\psi}{\partial\zeta^2} - \psi, \qquad (3.6.1)$$

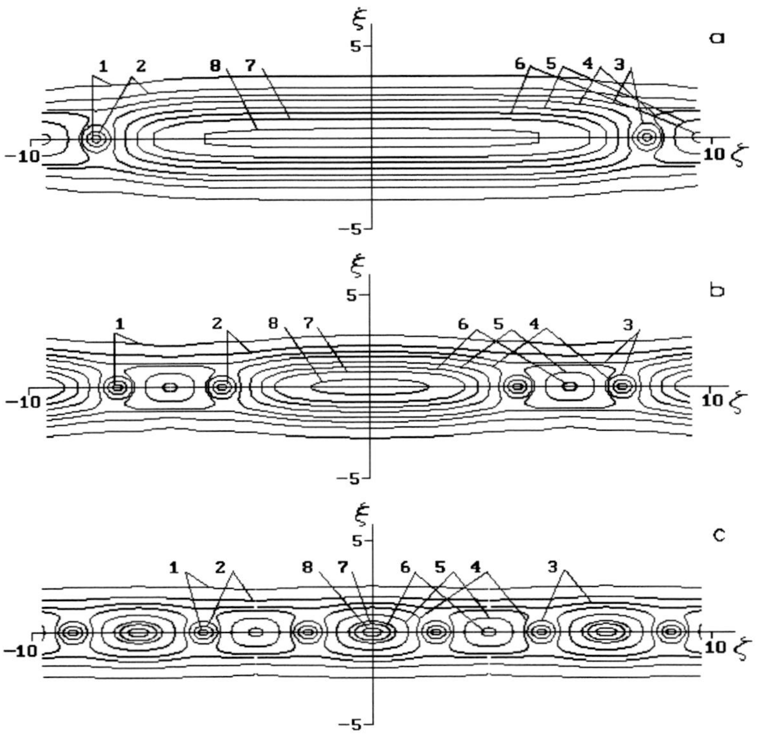

Fig. 3.6.3. Contours of constant current density in the plasma for various values of $j = N j_0$, $N = 0.3$ (1), 0.6 (2), 0.9 (3); $j = j_1 + N(j_2 - j_1)$, $N = 0.15$ (4), 0.55 (5), 0.95 (6); $j = j_2 + N(1 - j_2)$, $N = 0.3$ (7), 0.8 (8) ($\psi_m/\psi_\infty = 0.418$ (*a*), 0.437 (*b*), 1.123 (*c*)).

Substituting Eq. (3.6.1) into Maxwell's equations, we find

$$j_{pz} = \frac{\omega_p^2}{4\pi c}\psi, \quad j_{pr} = -\frac{\partial^2 j_{pz}}{\partial\zeta\partial\xi}. \tag{3.6.2}$$

where j_{pz} and j_{pr} are longitudinal and radial projections of the vector $\mathbf{j}_p$.

To compare j_{pz} with the beam current density $j_b = e\rho v$, we switch to the dimensionless functions

$$j_{pz} = j_0 RZ, \quad j_b = -j_0 R^2 Z^2 \theta(Z), \quad j_0 = \frac{|e| c\rho_0}{q^2} \tag{3.6.3}$$

and we use the relations

$$\begin{gathered}\overline{R} = 2\pi\int_0^\infty R\ \xi d\xi = 2\pi\int_0^\infty R^2\ \xi d\xi, \\ \langle Z\rangle = \frac{1}{L}\int_{-L/2}^{L/2} Z\ d\zeta = \frac{1}{L}\int_{-L/2}^{L/2} Z^2\theta(Z)\ d\zeta,\end{gathered} \tag{3.6.4}$$

which follows from Eq. (3.5.6) and Eq. (3.5.7).

Integrating Eqs. (3.6.3) over the cross section, we find the distributions of the plasma current and the beam current along the length of a bunch:

$$I_{pz} = I_0 Z, \ I_b = -I_0 Z^2\theta(Z), \ I_0 = 2\pi j_0 c^2 \overline{R}\omega_p^{-2} \tag{3.6.5}$$

(Fig. 3.6.1). The constant $\overline{R} = 9.9$ is found by numerical integration.

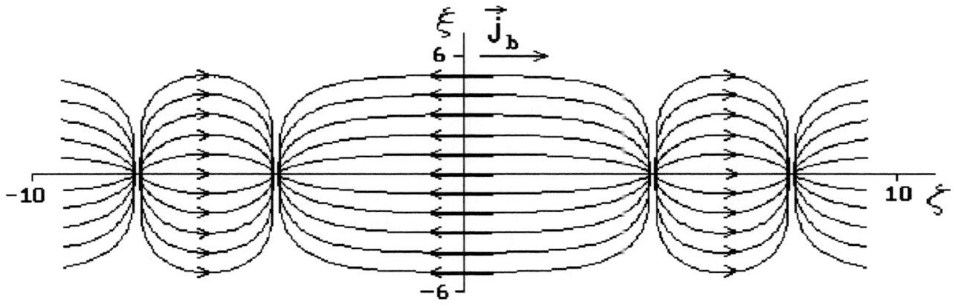

Fig. 3.6.4. Profiles of the force line of the electron current density for various values of $r(\zeta_0)$=1 (1), 2 (2), 3 (3), 4 (4), 5 (5) (ψ_m/ψ_∞= 0.437).

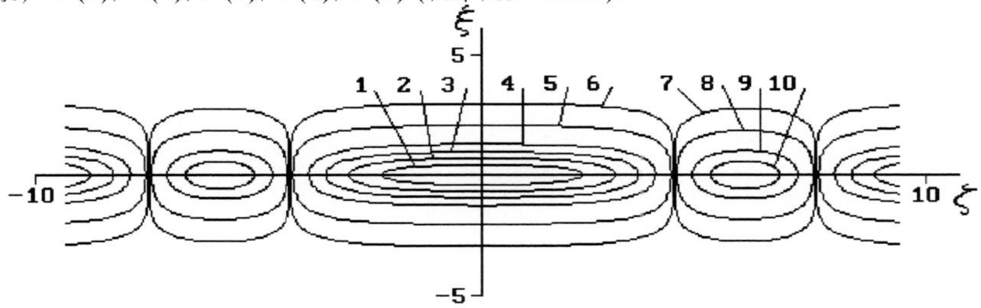

Fig. 3.6.5. Contours of constant charge density in the plasma for various values of ρ_p/ρ_0=-0.9 (1), -0.7 (2), -0.5 (3), -0.3 (4), -0.1 (5), -0.01 (6), 0.01 (7), 0.1 (8), 0.3 (9), 0.5 (10) (ψ_m/ψ_∞=0.437).

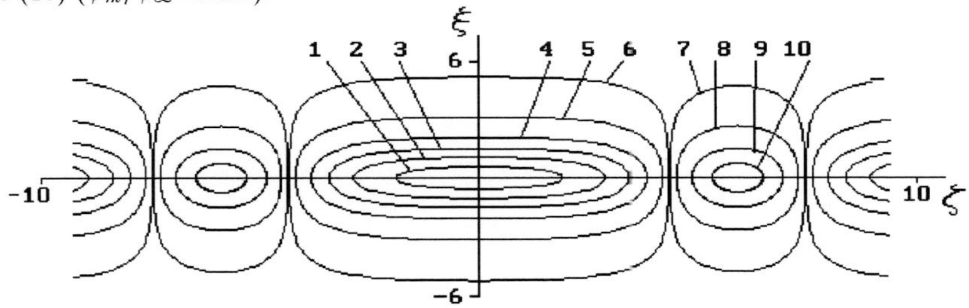

Fig. 3.6.6. Profiles of the equipotential surfaces of the plasma and beam for various values of y= 0.9 (1), 0.7 (2), 0.5 (3), 0.3 (4), 0.1 (5), 0.01 (6), -0.01 (7), -0.1 (8), -0.3 (9), -0.5 (10) (ψ_m/ψ_∞=0.437).

The radial profile of the current density,

$$j_{pz} = j_0 R\langle Z\rangle, \quad \langle j_b\rangle = -j_0 R^2 < Z > \tag{3.6.6}$$

is shown in Fig. 3.6.2. The constant $\langle Z\rangle$ (the area under the curve over the interval $|\zeta| \leq L/2$) reduces to incomplete elliptic integrals [8]:

$$\langle Z\rangle = \frac{2}{L}\Big[Z_0 \left(\frac{12h^2}{g^2}\right)^{1/4} \frac{g+\sqrt{h}}{gh+\sqrt{3}(Z_0-1/2)}\, F(\varphi,k)$$

$$- 3^{3/4}\sqrt{2hg}\; E(\varphi,k) - \sqrt{\frac{3}{2}}\,(\sqrt{3}g-h) - \sqrt{\frac{2}{3}}\,Z_0 h\Big], \tag{3.6.7}$$

$$g = \sqrt{Z_0+1/2}, \quad h = \sqrt{3/2-Z_0}.$$

The spatial distribution of a beam current density is shown in Fig. 3.6.3 and Fig. 3.6.4.

From Eq. (3.6.2) and the continuity equation for the plasma electrons, we find the induced-charge density to be

$$\rho_p = \frac{1}{4\pi|e|}\left(\Delta_\xi - \frac{\omega_p^2}{2}\right)\psi = -\frac{\rho_0}{q^2}ZR^2, \tag{3.6.8}$$

(Fig. 3.6.5); where Δ_ξ is the radial part of the Laplacian. From Eqs. (3.6.3), (3.6.8), and (3.6.4) we find

$$\langle I_{pz}\rangle = -\langle I_b\rangle, \quad \langle \rho_p\rangle = -\langle \rho_b\rangle \tag{3.6.9}$$

so the plasma with the beam has charge neutrality and also current neutrality on the average over a period [2)].

The results of the numerical calculation which reflect the spatial distribution of density of plasma current, Figs. 3.6.3 and 3.6.4, and of density of plasma charge, Figs. 3.6.5, over a modulation period are shown correspondingly. The equipotential surfaces of the plasma and beam are shown in Fig. 3.6.6. The numerical algorithm is described in details in **Appendix 3.1**.

§3.7. Shock wave with trapped electrons in a collisional plasma.

In a real dense plasma, a perturbation caused by the beam is damped by Coulomb collisions, and the lifetime of a focused beam is limited by the time scale of this effect. For this reason, the one dimensional periodic solution found in [10] is time-varying. It describes a slowing of the nonrelativistic beam as whole. If the solution in a collisional plasma is written formally as a monochromatic plane wave, the dielectric constant of the plasma acquires an imaginary increment $\epsilon_c = \epsilon + i\nu_c/\omega_p$ (ν_c is the collision rate). This increment implies the appearance of a phase difference between the polarization wave and the charge density wave of the beam. A numerical integration of the equation for the radial equilibrium of a bunch under the condition $\nu_c > 0$ shows that there is no steady-state periodic solution in a collisional plasma with a modulated electron beam §1.6.

A steady-state solution in the form of an electromagnetic shock wave exists in a collisional plasma with relativistic electron bunches [11]. This solution differs in a fundamental way from this of §3.5, in that there is no periodicity in terms of the variable $z' = z - vt$, because of the temporal damping of the waves in the plasma. Since the longitudinal dimensions of the bunches increase with decreasing strength of the focused field, an observer in the frame of reference of the plasma sees bunches moving at a constant velocity and becoming longer as time elapses. There is a corresponding increase in the amount of Coulomb charge which they must have in order to offset the energy loss and to maintain a steady-state wave in the plasma.

Allowance for collisions in the form of a friction force $-m\nu_c \mathbf{v}_p$ in the equations of motion of the plasma electrons leads to additional terms in the equation for the effective potential Eq. (3.1.4):

$$\left(\frac{\partial^2}{\partial t\partial\tau} + \omega_p^2\right)\left[\frac{\partial}{\partial\tau}\left(\Delta - \frac{1}{c^2}\frac{\partial^2}{\partial t^2}\right) - \frac{\omega_p^2}{c^2}\frac{\partial}{\partial t}\right]\psi = -4\pi e\frac{\partial}{\partial\tau}\left(\frac{1}{\gamma^2}\frac{\partial^2}{\partial t\partial\tau} - \frac{v^2}{c^2}\omega_p^2\right)\rho, \tag{3.7.1}$$

[2)] Contrary to paper [3] there are charge neutrality and current neutrality on the average over a period of the beam so that the dimension of bunch is consistent with the wave.

where $\partial/\partial\tau = \partial/\partial t + \nu_c$, $\psi = \varphi - vA_z/c$, ρ and v are the density and velocity of the beam, $\gamma = \left(1 - v^2/c^2\right)^{-1/2}$ is the relativistic factor, and φ and A_z are the scalar and vector potentials. These potentials satisfy the gauge condition

$$\frac{\partial^2 A_z}{\partial t \partial z} + \frac{1}{c}\left(\frac{\partial^2}{\partial t \partial \tau} + \omega_p^2\right)\varphi = 0.$$

Since the mean free path $l_c = v/\nu_c$, of an electron in the plasma, is considerably larger than the characteristic size $\lambda = 2\pi v/\omega_p$, of a bunch, $\nu_c\partial/\partial t \ll \omega_p^2$ is the most important term near the plasma resonance. Incorporating this term only in the first operator on the left side of Eq. (3.7.1), discarding the terms on the order of $\gamma^{-2} \ll 1$, and switching to the variable $z' = z - vt$, we find the equation

$$\left(\frac{\partial^2}{\partial\zeta^2} - \nu\frac{\partial}{\partial\zeta} + 1\right)\left[\frac{1}{\xi}\frac{\partial}{\partial\xi}\left(\xi\frac{\partial\psi}{\partial\xi}\right) - \psi\right] = -\psi_0\frac{\psi^2}{\psi_m^2}\theta(\psi),$$
$$\xi = \frac{\omega_p r}{c}, \quad \zeta = \frac{\omega_p z'}{c}, \quad \psi_0 = \frac{4\pi|e|\rho_0 c^2}{\omega_p^2}, \quad \nu = \frac{\nu_c}{\omega_p}, \tag{3.7.2}$$

where we have retained the notations of Eq. (3.5.3).

Variables can be separated in Eq. (3.7.2)

$$\psi(\xi,\zeta) = \psi_0 R(\xi) W(\zeta). \tag{3.7.3}$$

The ν-independent equation for the radial function,

$$\xi^{-1}(\xi R')' - R + R^2 = 0,$$
$$R'(0) = 0, \quad R(\infty) = 0$$

is the same as that in Eq. (3.5.6). It has at $R_0 = R(0) = 2.39$ the soliton solution which is necessary for the occurrence of a self-focusing.

The longitudinal equilibrium of the bunches is described by the equation of a nonlinear oscillator with a small negative friction:

$$W'' - \nu W' + W - \left(\frac{W}{R_0 W_m}\right)^2 \theta(W) = 0,$$
$$W'(0) = 0, \quad \psi_m = \psi_0 R_0 W_m. \tag{3.7.4}$$

In the collisionless plasma the change of variables $W = (R_0 W_m)^2 Z$ make it possible to eliminate the constant factor W_m from the equation for the auxiliary function $Z(\zeta)$. When collisions are taken into account ($\nu > 0$), on the other hand, the wave amplitude $W_m(\zeta)$ varies slowly along the beam,

$$W' \simeq \nu W_m \ll W_m$$

and an addition term (a nonlinear friction) appears in the equation:

$$Z'' - \left(\nu + 4\frac{Z_0'}{Z_0}\right) + Z - Z^2\theta(Z) = 0 \tag{3.7.5}$$

(the point $\zeta = 0$ corresponds to the center of the bunch). In going from Eq. (3.7.4) to Eq. (3.7.5) we made use of the obvious relation

$$W_m = (R_0^2 Z_0)^{-1},$$

where $Z_0(\zeta)$ is maximum value of the function $Z(\zeta)$ at the centers of the bunches.

Equation Eq. (3.7.5) can be used for a qualitative analysis, by analogy with problems in mechanics [7]. In the absence of dissipation ($\nu = 0$), the integral of Eq. (3.7.5) is

$$\frac{Z'^2}{2} + U(Z) = E, \; U(Z) = \frac{Z^2}{2} - \frac{Z^3}{3}\theta(Z), \; E = U(Z_0). \tag{3.7.6}$$

From the plot of the potential energy $U(Z)$ in Fig.3.5.1 we see that in the region

$$Z_{\min} \le Z \le Z_0 < 1, \; Z_{\min} = -\sqrt{2E}$$

the motion of a particle is periodic, with turning points $Z_{\min}$ and Z_0. As $Z_0 \to 1$, the solution transforms into soliton.

The solution soliton $Z_0 = 1$ is a separatrix in the phase plane. It describes two semi-infinite electrons beam, separated by gap $\lambda/2 = \pi c/\omega_p$. The physical meaning of this solution becomes apparent when collisions are taken into account, since under the condition

$$\nu_{\text{eff}} = \nu + \frac{4Z_0'}{Z_0} < 0,$$

Eq. (3.7.5) describes slightly damped nonlinear oscillations corresponding to a slow descent of the particle from the point $Z_0 = 1$ to the bottom of the potential well, $Z_0 = 0$ (Fig. 3.5.1), at which point the solution transforms into a shock wave.

The small parameter $\nu \ll 1$ in Eq. (3.7.5) can be utilized to find an analytic solution in the adiabatic approximation. Since this solution describes weakly damped oscillations, we can take the unperturbed solution to the periodic solution found for a collisionless plasma ($\nu = 0$) in Eq. (3.5.12):

$$Z(\zeta) = Z_0 \begin{cases} \left[1 - \dfrac{\operatorname{sn}^2(\kappa\zeta, k)}{\sin^2\varphi}\right] \operatorname{dn}^{-2}(\kappa\zeta, k), & |\zeta - nL| \le \zeta_0, \\[2ex] -\sqrt{1 - \dfrac{2Z_0}{3}} \cos\left(\dfrac{L}{2} - \zeta\right), & \zeta_0 \le |\zeta - nL| \le \dfrac{L}{2}. \end{cases} \tag{3.7.7}$$

The nonlinear spatial period of the beam is expressed in terms of an incomplete elliptic integral of the first kind in Eq. (3.5.11):

$$\begin{gathered} L = \pi + 2\kappa^{-1}F(\varphi, k), \\ 2k^2 = 1 + \frac{2Z_0 - 1}{4\kappa^2}, \quad (2\kappa)^4 = (1 + 2Z_0)\left(1 - \frac{2Z_0}{3}\right), \\ \sin^2\varphi = 2\left[1 + \left(\frac{3}{4\kappa^2}\right)\left(1 - \frac{2Z_0}{3}\right)\right]^{-1}. \end{gathered} \tag{3.7.8}$$

It increases without bound in accordance with $L \cong \ln(1 - Z_0)^{-1}$ as the periodic solution transforms into a soliton.

We now move on the next approximation in the small parameter ν. Working from Eq. (3.7.5) and Eq. (3.7.7), and taking an average over the fast oscillations, we find an equation for the slowing varying function $E(\zeta)$:

$$\frac{dE}{d\zeta} = \left(\nu + 4\frac{Z_0'}{Z_0}\right)\frac{1}{L}\int_0^L \left[\frac{dZ(\zeta', E)}{d\zeta'}\right]^2 d\zeta'. \tag{3.7.9}$$

The integral on the right side of Eq. (3.7.9) is the following adiabatic invariant from mechanics [12]:

$$I = 2^{3/2} \int_{Z_{\min}}^{Z_0} \sqrt{E(\zeta) - U(Z)}\, dZ. \tag{3.7.10}$$

Using the relation $I'(E) = L$, and integrating Eq. (3.7.9), we find

$$Z_0^{-4} I(Z_0) = I_\infty \exp[\nu(\zeta - \zeta_\infty)], \tag{3.7.11}$$

where ζ_∞ corresponds to the soliton solution $Z_0 = 1,\ E = U(Z_0)$.

The integral in Eq. (3.7.10) can be expressed in terms of incomplete elliptic integrals of the first and second kinds (**Appendix 3.2**):

$$I(Z_0) = \frac{\pi Z_0^2}{2}\left(1 - \frac{2}{3}Z_0\right) + \frac{3}{5}\Big\{(2\kappa)^5[2(1 - k^2 + k^4)E(\varphi, k)$$

$$- (2 - k^2)(1 - k^2)F(\varphi, k)] - 2Z_0\left(Z_+^{-1} - \frac{1}{3}\right)\left(1 - \frac{2}{3}Z_0\right)^{1/2}\Big\}, \tag{3.7.12}$$

$$Z_+ = \frac{3}{4}\left\{1 - \frac{2}{3}Z_0 + \left[\left(1 - \frac{2}{3}Z_0\right)(1 + 2Z_0)\right]^{1/2}\right\}.$$

For highly compressed bunches with $Z_0 \ll 1$, and with a nearly monochromatic wave, Eq. (3.7.8) and Eq. (3.5.12) simplify:

$$L = 2\pi + \frac{4}{3}Z_0 + \frac{Z_0^2}{3}\left(\frac{5\pi}{4} - \frac{4}{3}\right), \quad I = \pi Z_0^2. \tag{3.7.13}$$

For near-soliton a soliton solution (with $\Delta = 1 - Z_0 \ll 1$), we find the following results from Eq. (3.7.8) and Eq. (3.5.12):

$$I = I_\infty - \frac{\Delta^2}{2}(1 + L), \quad I_\infty = \frac{\pi}{6} + \frac{6}{5}\left(1 - \frac{2}{3^{3/2}}\right),$$
$$L = \pi + 2\ln\left[\frac{12}{\Delta}(2 - \sqrt{3})\right]. \tag{3.7.14}$$

Equations (3.7.11), (3.7.12), and (3.7.14) determine function $Z_0(\nu\zeta)$ and thus the slowly varying coefficients $k(\nu\zeta)$, $\kappa(\nu\zeta)$, and $\varphi(\nu\zeta)$ in Eqs. (3.5.7) (3.5.8). The results of the adiabatic theory is shown in Fig. 3.7.1, where line 1 is a plot of $Z_0(\nu\zeta)$. Lines 2 and 3 show the slow spatial evolution of the beam modulation period and of the function $W_m = (R_0^2 Z_0)^{-1}$. These curves are in qualitative agreement with the asymptotic expressions Eq. (3.7.13) and Eq. (3.7.14). The two-scale solution $Z(\nu\zeta, \zeta)$, which follows from Eq. (3.7.7), Eq. (3.7.11), and Eq. (3.7.12), is shown in Fig. 3.7.2.

The auxiliary function $Z(\nu\zeta, \zeta)$, which has the form of a classical shock wave, determines the profile of the potential along the longitudinal coordinate:

$$\psi(\xi, \zeta) = \psi_0 \frac{R(\xi) Z(\nu\zeta, \zeta)}{[R_0 Z_0(\nu\zeta)]^2}.$$

A simple physical explanation can be found by introducing the time $T = -\zeta$. The time $T \to \infty$ corresponds to the creation of a infinitely deep potential well for the bunches whose longitudinal dimension is equal to half the wavelength , $\pi c/\omega_p$. At later times, Coulomb

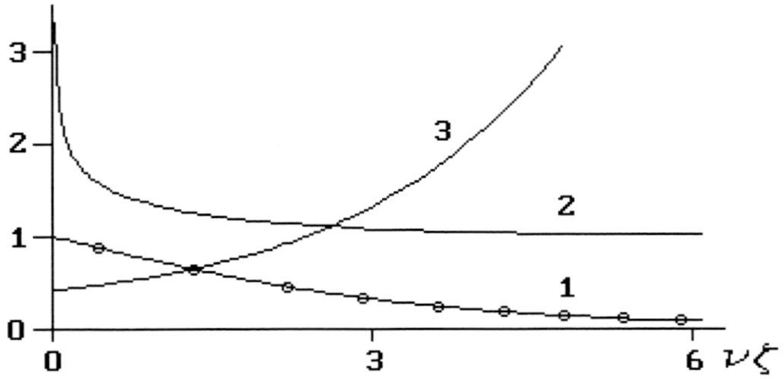

Fig. 3.7.1. The slowly varying functions (1) Z_0, (2) $L/2\pi$, and (3) W_m. The points on line 1 correspond to the maxima of the function $Z(\nu\zeta,\zeta)$.

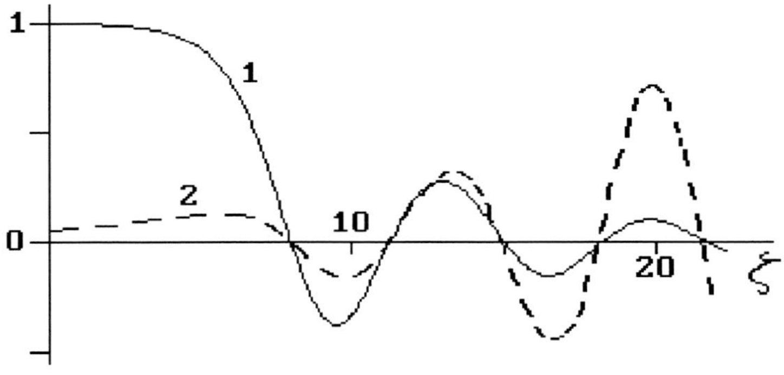

Fig. 3.7.2. Shock wave at ν=0.3. 1-The function $Z(\nu\zeta,\zeta)$; 2-potential at the beam axis, $\psi(0,\zeta)/20\psi_\infty$.

collisions in the plasma reduce the field amplitude. As $T \to \infty$, this amplitude tends toward the asymptotic value $\psi_\infty = \psi_0/R_0$ (see §3.5). Fig. 3.7.2 shows the evolution of the potential at the beam axis.

In the adiabatic approximation ($\nu \to 0$), descends infinitely slowly to the bottom of the potential well Fig. 3.5.1, and the shock wave is an infinite set of periodic solutions. At nonzero values of the parameter ν, discreteness appears in this system. However, since the period varies only slightly over the length of an individual bunch, the constant-density surfaces

$$\frac{\rho(\xi,\zeta)}{\rho_0} = \left[\frac{R(\xi)}{R_0}\frac{Z(\nu\zeta,\zeta)}{Z_0(\nu\zeta)}\right]^2 = \mathrm{C} \qquad (3.7.15)$$

are approximately collisionless. Fig. 3.7.3 shows the results of some numerical calculations based on Eq. (3.7.3) and Eq. (3.7.5) and those of some calculations based on Eqs. (3.7.11) and (3.7.12). The curves are correspond to the intersection of the surfaces in Eqs. (3.7.15) with a plane passing through the beam axis. As the parameter ν decreases, the spatial evolution of the longitudinal dimensions of the bunches becomes more noticeable Fig. 3.7.3b.

This study was a continuation and development of the study reported in Sec. §3.5 for a collisional plasma. Consequently, the physical mechanism for the self-focusing which was discussed in the collisionless approximation remains in force when collisions are infrequent, i.e. under the condition $\nu \ll 1$. At the same time, the absence of periodic solutions in a collisional plasma means that the beam modulation regime must be corrected if the steady-state wave required for transporting relativistic electron bunches over large distances is to

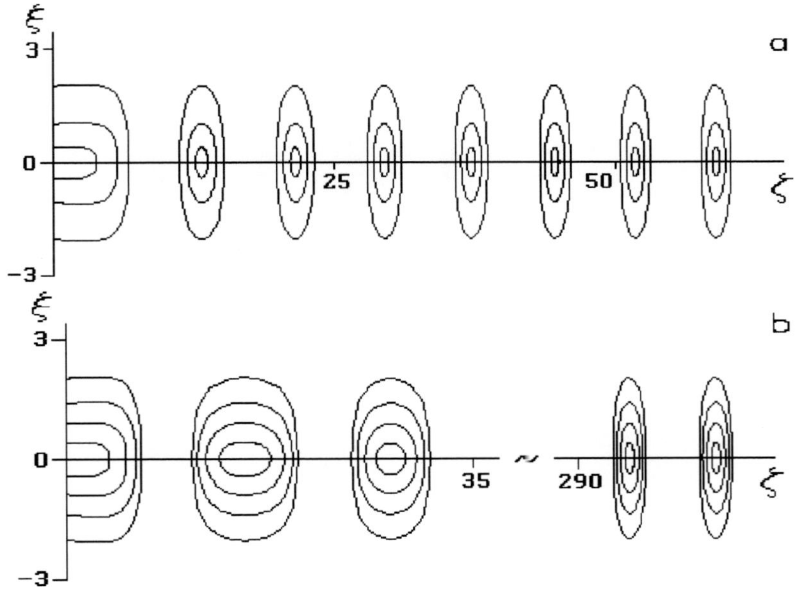

Fig. 3.7.3. Curves of constant electron density in the bunches. a: $\nu= 0.1$, $C= 0.9, 0.5, 0.1$ and b: $\nu= 0.01$, $C= 0.9, 0.6, 0.3, 0.1$.

be maintained in the plasma.

§3.8. Dynamics of electron bunches in a plasma.

The mathematical difficulties arising in the search for analytical solutions of nonlinear partial differential equations allow only the conditions under which beam equilibrium can occur in the plasma and the shape of the electron bunches to be determined. The process by which bunches form and decay, including the stage in which potential wells develop, the compression of bunches to high density in the wave field as it reaches large amplitude, and the approach to the steady state of beam transport in the plasma can only be analyzed numerically [13].

Particles is used to simulate the beam and plasma systems separately [14,15], modified so as to describe the transport of the beam in the plasma in the form of separate relativistic bunches [16]. The results of numerical calculations described below shows that the preliminary bunching of the beam substantially alters the nature of the beam-plasma instability in comparison with the case of a continuous beam [17-19]. The beam electrons settle to the bottoms of the already existing potential wells whose depths are increasing, remaining at fixed phases of the field. In contrast to a continuous beam, where the initial perturbation is small and beam bunching occurs as the instability develops and the wave amplitude grows, the randomization of the beam which gives rise to scattering of the electron clumps is significantly reduced.

1. Periodic system.

The self-consistent system of equations for the problem includes Vlasov equations for beam and plasma particles and the Poisson equation for the electric $\mathbf{E}(t, \mathrm{r})$ field:

$$\frac{\partial f_\alpha}{\partial t} + \mathbf{v}\frac{\partial f_\alpha}{\partial \mathbf{r}} + e_\alpha \mathbf{E}\frac{\partial f_\alpha}{\partial \mathbf{p}} = 0, \quad \mathrm{div}\mathbf{E} = 4\pi \sum_\alpha e_\alpha \int f_\alpha \, d\mathbf{p}, \tag{3.8.1}$$

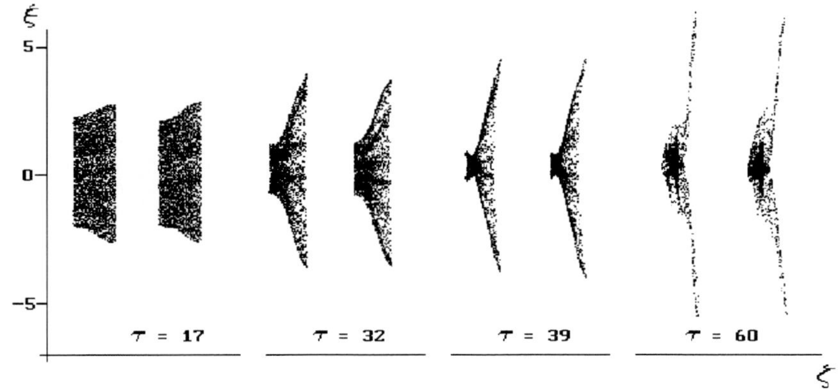

Fig. 3.8.1. Spatial distributions of beam particles at successive times τ=17, 32, 39, 60.

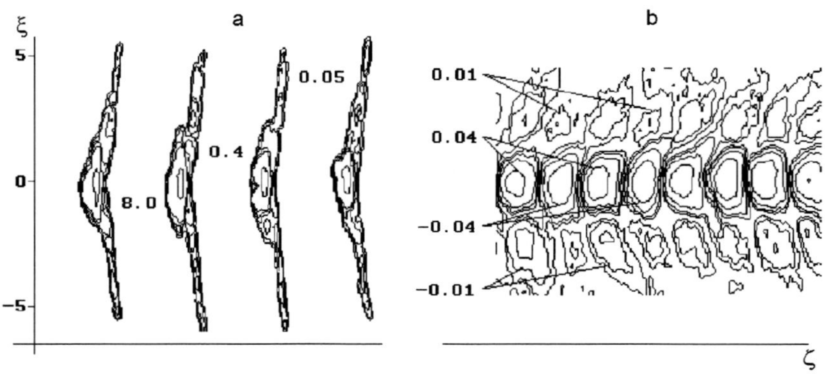

Fig. 3.8.2. Contours of constant density in the beam (a) and equipotentials (b) at time τ=60.

where the summing is carried out over all particle species.

We use the particle-pushing technique to simulate the self-consistent dynamics of the beam-plasma system. Accordingly, the continuous distribution function $f(t, \mathbf{r}, \mathbf{p})$ defined in the six-dimensional space $(t, \mathbf{r}, \mathbf{p}) = (t, x, z, p_x, p_y, p_z)$ is replaced by a discretized function:

$$f(t, \mathbf{r}, \mathbf{p}) \to \sum_{\alpha} \sum_{i\alpha=1}^{N_\alpha} G(|\mathbf{r} - \mathbf{r}_{i\alpha}(t)|)\delta(|\mathbf{p} - \mathbf{p}_{i\alpha}(t)|), \tag{3.8.2}$$

where N_α is the number of particles of species α, G is the particle form factor, and the coordinates $\mathbf{r}_{i\alpha}(t)$ and momenta $\mathbf{p}_{i\alpha}(t)$ satisfy the equations of motion:

$$\begin{gathered} \dot{\mathbf{p}}_{i\alpha}(t) = e_\alpha \int \mathbf{E}(t, \mathbf{r}) G(|\mathbf{r} - \mathbf{r}_{i\alpha}(t)|)\, d\mathbf{r}, \\ \dot{\mathbf{r}}_{i\alpha}(t) = \mathbf{v}_{i\alpha}, \quad \mathbf{v}_{i\alpha} = \mathbf{p}_{i\alpha}/m\gamma_{i\alpha}. \end{gathered} \tag{3.8.3}$$

The dynamics of the interaction between the modulated relativistic electron beam and the initially cold plasma have been studied in periodic geometry, simulating the propagation in a cold plasma of an infinite train of electron clumps. The initial parameters for the main case are following: length and width of the clumps are $\lambda_p/2$ and $3\lambda_p$ respectively,

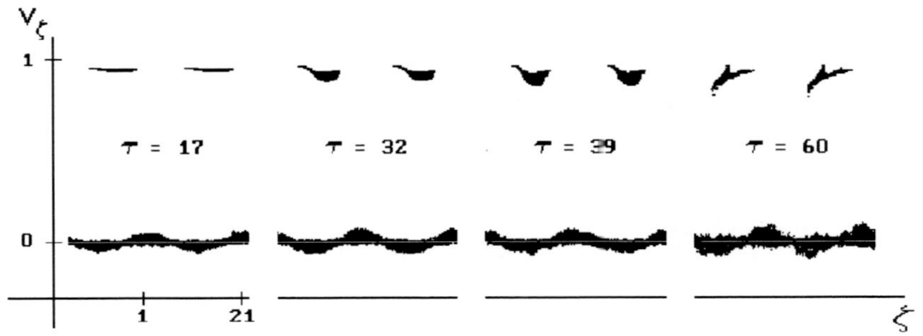

Fig. 3.8.3. Projection of the distribution of the centers of the beam electrons (above) and of the plasma on longitudinal phase plane (z, v_z) at times τ= 17, 32, 39, 60.

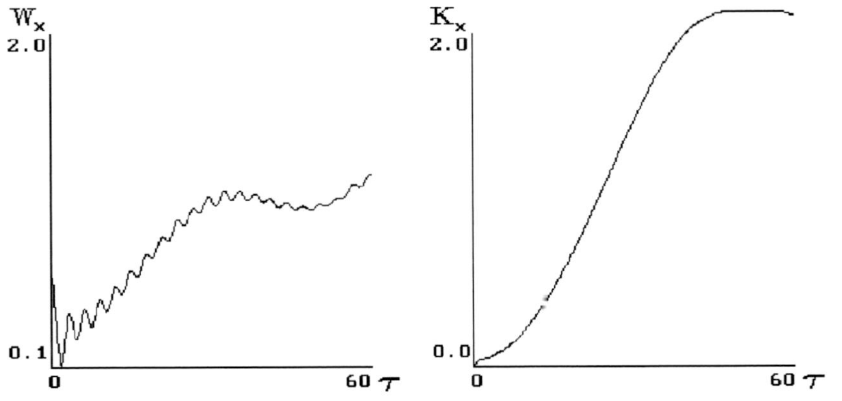

Fig. 3.8.4. Time dependence of the mean density of the transverse energy W_x (scale factor 10^{-3}) of the self-consistent field and of the transverse beam kinetic energy K_x (scale factor 10^{-2}).

$(\lambda_p = 2\pi v/\omega_p)$, relativistic factor $\gamma = 3$, mismatch between the modulation frequency and plasma frequency $\delta = 0.1$, ratio of beam density to plasma density $\nu = \rho/n_p = 10^{-2}$, plasma ion mass equal to 1840 electron mass, total number of beam and plasma particles $N = 5 \cdot 10^5$, and dimensionless time and coordinates are $\tau = \omega_p t$, $\xi = \omega_p x/c$ and $\zeta = \omega_p z/c$ respectively. The initial distribution of beam and plasma particles satisfies the condition of charge neutrality: i.e., the charge of the plasma and beam electrons is completely neutralized by the charge of the plasma ions.

Fig. 3.8.1 illustrates the transverse and longitudinal self-compression of the clumps, which are initially in the form of square waves in (x, z) space. As time passes the particle density at the centers of the clumps increases and by the time $\tau = 60$ it has increased by a factor of ten over the initial value Fig. 3.8.2. The equipotential surfaces of the plasma and beam at this time are shown in Fig. 3.8.2b.

The evolution of the clumps and the heating of the plasma electrons and ions in the beam propagation region can be seen in the (z, v_z) phase plain. The bulk of the beam electrons slow down, but at the leading edge of the clumps electrons are accelerated Fig. 3.8.3.

The change in the dimensionless energy densities of the transverse field $W_x = \langle E_x^2/8\pi n_p mc^2 \rangle$ and transverse energy of the beam $K_x = \langle f_b m v_x^2/2n_p mc^2 \rangle$ are shown versus time by the traces of Fig. 3.8.4. The wave amplitudes increase during the stages in which beam neutralization occurs and the clumps collapse under the action of the polarization

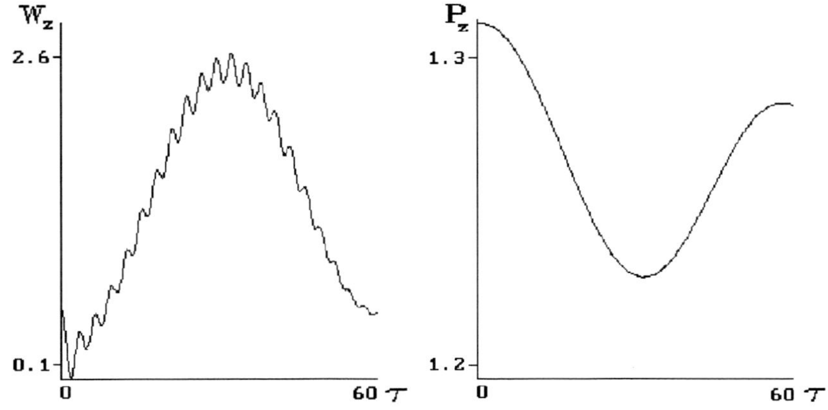

Fig. 3.8.5. Time dependence of the mean density of the longitudinal energy W_z (scale factor 10^{-3}) of the self-consistent field and of the total longitudinal electron momentum P_z (scale factor 10^{-2}).

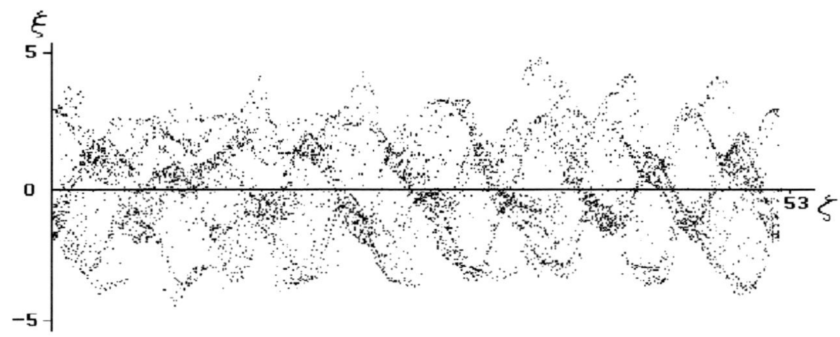

Fig. 3.8.6. Spatial electron distribution of a continuous beam at time τ=47 .

field in the plasma. Then over a distance of a dozen clumps the wave energy saturates and the beam reaches a quasisteady state.

The dimensionless energy of the longitudinal component $W_z = \langle E_z^2/8\pi n_p mc^2 \rangle$ of the field and the longitudinal beam electron momentum $P_z = \langle f_b p_z/mc \rangle$ change in time with a fairly long period Fig. 3.8.5, which corresponds to pumping of beam energy into the field and back due to acceleration of the beam of preformed clumps as a whole [19,20].

For comparison a picture of the growth of the beam instability of a continuous beam with the same parameters is shown in Fig. 3.8.6. As the field amplitude rises electron phase mixing occurs, accompanied by randomization of the beam. In contrast with the one-dimensional problem [17,18], there is considerable transverse defocusing at the beam under the action of the instability field. Note that preferential excitation of transverse modes in a plasma with an relativistic electron beam occurs due to the anisotropy of the longitudinal and transverse mass [21].

The calculations reveal that the beam instability also exerts a significant effect on the dynamics of low-density ($\nu = 10^{-3}$) clumps. However, in that case phase mixing is inhibited because of self-focusing of the clumps. When the beam density reaches $\nu = 10^{-2}$ the instability is essentially fully suppressed.

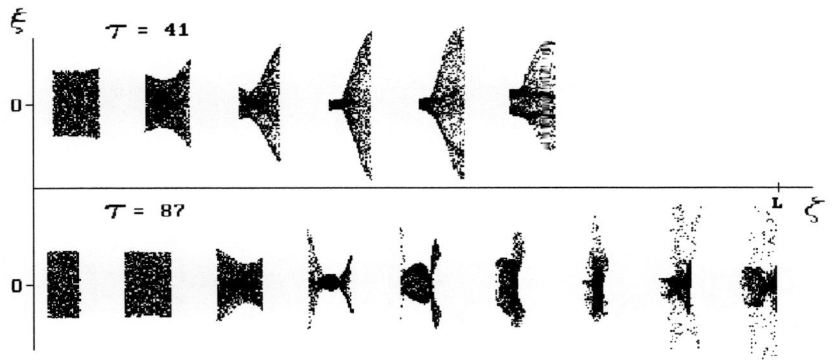

Fig. 3.8.7. Spatial distribution of beam particles in an open system at successive times τ=41 и τ=87.

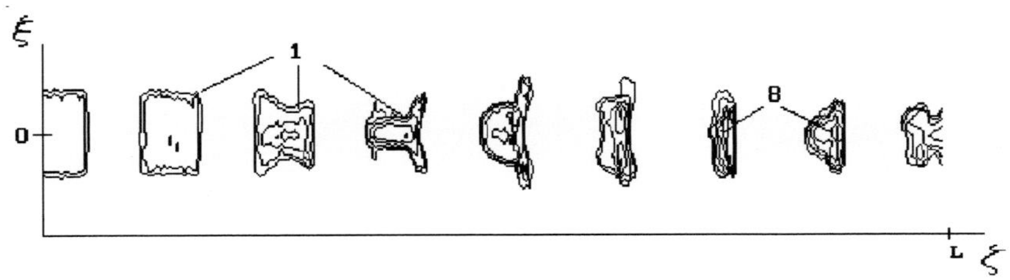

Fig. 3.8.8. Contours of constant density in the beam at time τ = 87 in an open system.

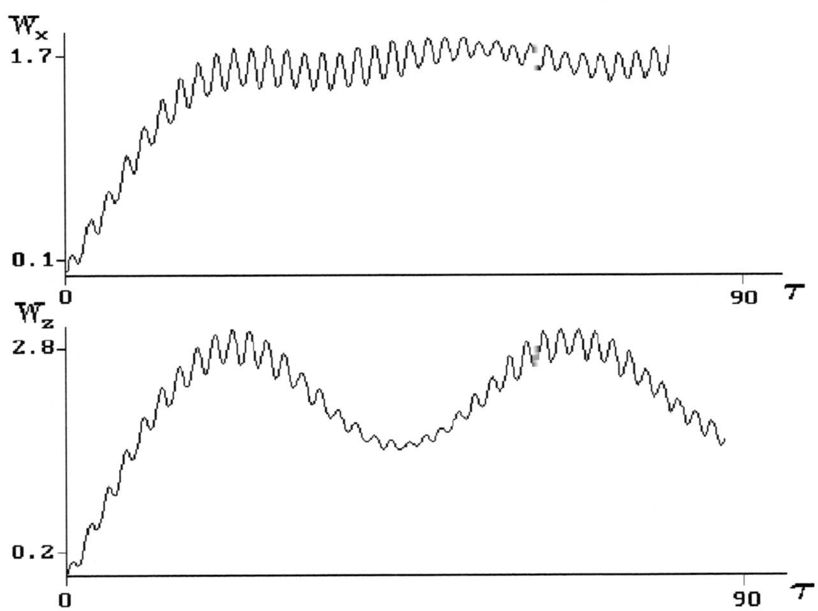

Fig. 3.8.9. Time dependence of the mean density of the transverse energy W_x (scale factor 10^{-2}) and the longitudinal energy W_z (scale factor 10^{-2}) of the self-consistent field in an open system.

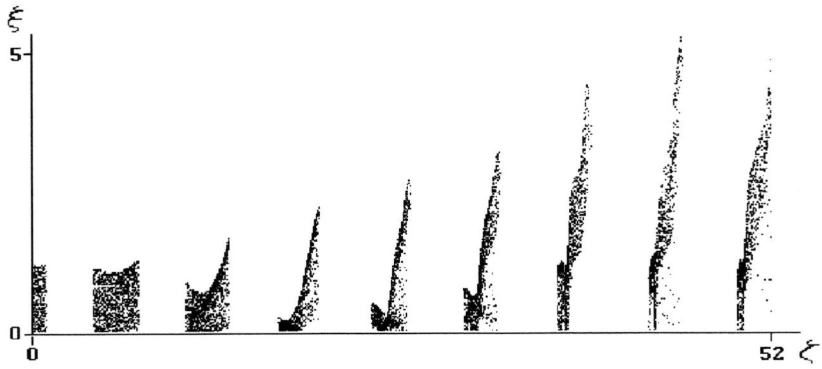

Fig. 3.8.10. Spatial distributions of beam particles at time $\tau = 71$ in a cylindrical beam.

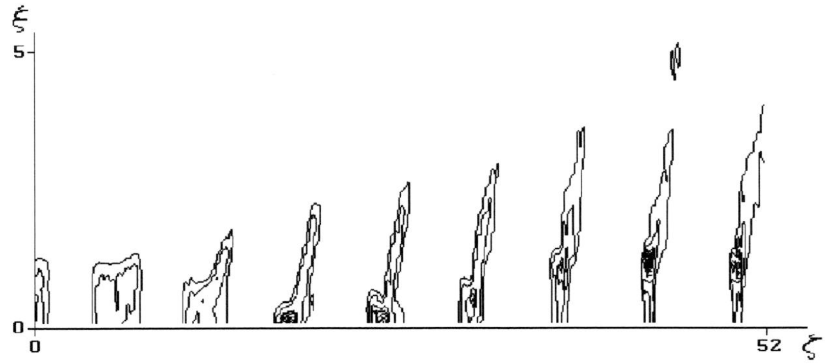

Fig. 3.8.11. Contours of constant density in the beam at time $\tau = 71$ in a cylindrical beam.

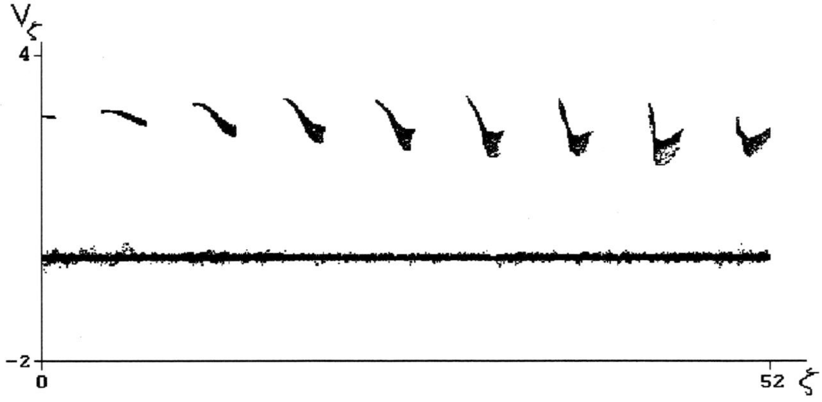

Fig. 3.8.12. Longitudinal phase plane (z, v_z) at time $\tau = 71$ in case of cylindrical beam.

2. Open system.

An advantage of the periodic model described above is the possibility of studying the nonlinear process by which relativistic electron clumps interact with the self-consistent field without distortions with the presence of the plasma boundary. However, a model that takes into account the presence of injector in the $z = 0$ plane is closer to the experimental formulation of the problem. It is also distinguished by the fact that it takes into account the process of charge neutralization of the clumps continuously injected into a previously unperturbed finite-temperature plasma. Moreover, the clumps are found in the different situations, depending on when they were injected into the plasma, which enables us to study in particular the processes at the leading edge of the beam. Note that the presence of an external source of relativistic electrons (with an infinite supply of energy) causes the wave amplitude in the beam-plasma system to increase relative to that in the periodic case.

The results of the calculations for an open system with clumps having the same parameters ($\gamma = 3$, $\delta = 0.1$, $\nu = 10^{-2}$) are shown in Figs. 3.8.7 - 3.8.9.

Fig. 3.8.7 illustrate the spatial distribution of electrons at successive times. The leading clump which propagates in the unperturbed plasma and excites a plasma wave in it, appears distinctive. The subsequent clumps, which move in the field radiated by the first, exhibit a self-focusing effect qualitatively similar to that discussed above in the periodic model. A clump undergoes the maximum compression over a distance of order three wavelengths from the injection plane. At that point the density of the clump has increased by factor of ten Fig. 3.8.8.

The global properties of the system (the longitudinal and transverse field energies) are shown in Fig. 3.8.9. In contrast to periodic model Figs. 3.8.4 and 3.8.5 the beam-plasma system now passes to a quasisteady state.

3. Principal results.

The particle simulation technique has been used to analyse the self-consistent dynamics of the beam-plasma system for continuous beams and beams broken up into separate relativistic clumps. The continuous beam is unstable against multidimensional electrostatic waves whose field gives rise to phase mixing of the electrons and randomization of the beam.

The beam instability has a considerable effect also on the dynamics of the clumps with low ($\nu = 10^{-3}$) density. However, when the density is raised to $\nu = 10^{-2}$, self-compression of the clumps substantially suppresses the instability.

For clumps with initial square-wave cross sections the compression is both transverse and longitudinal. Initially the electrons undergo transverse compression at the trailing edge of the clumps; then the bulk of the particles slow down in the longitudinal direction and gather at the focal points of the trailing edge. As shown by the evolution of the system in (z, v_z) phase space, the longitudinal velocity is adjusted nonuniformly over the length of a clump, and electrons at the leading edge are accelerated.

The plasma with a modulated beam of fairly high density $\nu = 10^{-2}$ goes over to the quasisteady regime at time $\tau = 10$, which for $\gamma = 3$ agrees in order of magnitude with the time $\tau_b \simeq (\gamma^3/\nu)^{1/3}$ for the beam instability to develop.

The self-focusing effect also takes the place for an open system, which resembles the experiment much more closely. It takes into account continuous injection of relativistic clumps into plasma, which allows the self-focusing dynamics of the leading edge of the beam to be analyzed. We have established that in this case the focusing effect of the beam goes over to quasisteady state corresponding to the formation of equilibrium clumps at a distance from the injection plane equal to several wavelengths.

The results of the calculations for a periodic system in case of a cylindrical beam is shown in Figs. 3.8.10 and 3.8.12.

Appendix 3.1

The sections of the equipotential surfaces $y(\xi, \zeta) = \text{const}$ by the plane passing through the axis of the beam are considered as a family of integral curves for the equation

$$d\,[y(\xi, \zeta)] = 0, \tag{A3.1.1}$$

written in the polar coordinate system (r, η):

$$\zeta = \zeta_0 + r \cos \eta, \quad \xi = r \sin \eta,$$

whose center $\zeta_0 = nL/2$ (L is the beam modulation period) coincides with the center of the bunch ($n = 0, \mp 2, \cdots \; y > 0$) or with the center of a gap ($n = \mp 1, \mp 3 \cdots$ and $y < 0$).

Then (A3.1.1) takes the following form in these coordinates

$$\frac{1}{r}\frac{dr}{d\eta} = \frac{B \sin \eta - A \cos \eta}{A \sin \eta + B \cos \eta}, \tag{A3.1.2}$$

where $A = R'Z$, $B = Z'R$, and the derivatives R' and Z' are taken, respectively, with respect to ξ and ζ. The initial conditions for (A3.1.2) are taken as the points on the axis $\xi = 0$ at which the potential $y(0, \zeta)$ takes present values. Figure 3.6.6 shows these equipotential surfaces for a series of values of q.

As q increases (reduction in the potential ψ_M for fixed values of ρ_0 and n_p), the electron bunches are stretched in the longitudinal direction with unaltered transverse dimensions and distances between them of $\lambda/2$ ($\lambda = 2\pi c/\omega_p$ is the wavelength), which coincides with the conclusion from the analytic theory.

Equation (A3.1.2) is the general one for numerical tabulation and enables one to recover the spatial distribution of the charge density and current induced in the plasma by the sequence of relativistic electron bunches.

The lines of constant induced-charge density (3.6.8) are defined as the solutions to (A3.1.2) for $A = 2R'Z$, $B = Z'R$ and by the initial conditions at the $\xi = 0$ axis, where the density has a preset value (Fig. 3.6.5).

The lines of constant current density in the plasma are defined in Eq. (3.6.3). Correspondingly, one needs to put in (A3.1.2) that

$$A = R'(RZ^2 + R''Z'^2), \quad B = Z'(R^2Z + R'^2Z''). \tag{A3.1.3}$$

However, in that case in the spatial period L there will be three differing intervals of current values separated by critical points:

$$j_2 = \left(1 - \frac{2q}{3R_0}\right)^{1/2} > j_1 = \frac{R_0'}{R_0} j_2 \tag{A3.1.4}$$

(here $R_0{}' = 1,08$ is the derivative at the point ξ_0 where $R''(\xi_0) = 0$).

The constant current density curves in the interval $0 < j < j_1$ (Fig. 3.6.3) are unconnected. They have closed components surrounding the zeros of $Z(\zeta)$ (the points $j = 0$) and unclosed ones that lie increasingly far from the $\xi = 0$ axis as $j = j_p q^2/ec\rho_0$ decreases.

The curves are also unconnected for currents $j_1 < j < j_2$ but the two components are closed. One of them surrounds the center of a bunch and the other the center of a gap.

Finally, for currents in the range $j_2 < j < 1$, all the curves are singly connected and are closed around the center of a bunch. They contract towards that center as j increases.

The current lines in Fig .3.6.4 are described by

$$\frac{j_{rp}}{j_{zp}} = \frac{d\xi}{d\zeta} = -\frac{R'Z'}{RZ} \tag{A3.1.5}$$

and at each point in space coincide with the direction of the plasma electron current. The numerical integration is based on (A3.1.2), in which one puts $A = RZ$, $B = R'Z'$, while the initial conditions are selected as arbitrarily specified values of r on the axes $\zeta = \zeta_0$ passing through the centers of the bunches and gaps.

Appendix 3.2

Taking into account (3.7.6), we rewrite integral (3.1.10) in the form

$$I = \sqrt{\frac{8}{3}} \int_{Z_{\min}}^{Z_0} \sqrt{(Z_+ - Z)(Z_0 - Z)(Z - Z_-)}\, dZ, \tag{A3.2.1}$$

$$Z_\pm = \frac{3}{4}\left\{1 - \frac{2}{3}Z_0 \pm \left[\left(1 - \frac{2}{3}Z_0\right)(1 + 2Z_0)\right]^{1/2}\right\}.$$

Substituting

$$(Z_0 - Z)/(Z_+ - Z) = k^2 \sin^2\varphi/\Delta^2(\varphi),$$

$$\Delta^2(\varphi) = 1 - k^2\sin^2\varphi,\ k^2 = (Z_0 - Z_-)/(Z_+ - Z_-)$$

to the above formula, and carrying out the integration by parts in (A3.2.1), we obtain:

$$\begin{aligned} I = \pi E + \frac{3}{5}(2\kappa)^5 \Bigg\{ & \frac{3k^2(1-k^2)^2\sin^2\varphi}{2\Delta^5(\varphi)} + 2(1 - k^2 + k^4)\Big[E(k,\varphi) \\ & - \frac{k^2\sin 2\varphi}{2\Delta(\varphi)}\Big] - (2-k^2)(1-k^2)\left[F(k,\varphi) + \frac{k^2\sin 2\varphi}{2\Delta^3(\varphi)}\right]\Bigg\}. \end{aligned} \tag{A3.2.2}$$

In going from (A3.2.2) to (3.1.12), we made use of the relations

$$(1-k^2)/\Delta^2(\varphi) = 2Z_+/6\kappa^2,\ 2 - k^2 = (2Z_+ - 1)/4\kappa^2,$$

$$k^2\sin 2\varphi/2\Delta(\varphi) = (Z_+/2\kappa Z_0)\sqrt{(1 - 2Z_0/3)}.$$

We use power series in $Z_0 \ll 1$,

$$\begin{aligned} F(\varphi,k) &= \varphi + k^2(2\varphi - 1)/8 + 3\left(k^2/8\right)^2(3\varphi - 2), \\ E(\varphi,k) &= \varphi - k^2(2\varphi - 1)/8 - \left(k^2/8\right)^2(3\varphi - 2), \\ 2\kappa &= 1 + Z_0/3 - Z_0^2/2,\ k^2 = 4Z_0(1 - Z_0)/3, \\ 2\varphi &= \pi/2 + Z_0(1 - Z_0/3) \end{aligned} \tag{A3.2.3}$$

to calculate (3.1.13), and in going from (3.1.12) to (3.1.14) we made use of the asymptotic expressions

$$\begin{aligned} F(\varphi,k) &= \frac{R}{2} + \frac{\Delta}{6}(R-2), \quad R = \ln\left[\frac{12}{\Delta}(2-\sqrt{3})\right], \\ E(\varphi,k) &= 1 + \frac{\Delta}{3}(R - 1 - \sqrt{3}) - \frac{\Delta^2}{6}(R + 1/6 - 2\sqrt{3}), \\ \varphi &= \frac{\pi}{2} - \sqrt{\frac{2\Delta}{3}}\left(1 - \frac{2\Delta}{9}\right),\quad k^2 = 1 - \frac{4\Delta}{3}(1-\Delta), \\ 2\kappa &= 1 + \Delta/3 - \Delta^2/2,\quad \Delta = 1 - Z_0 \ll 1. \end{aligned} \tag{A3.2.4}$$

In the limit case of $\Delta \ll 1$ the adiabatic invariant I and the nonlinear spatial period L are connected by the condition

$$I = I_\infty - \int_0^{1/6} L \, dE.$$

CHAPTER 4

STATIONARY SPACE-CHARGE WAVES IN A PLASMA WITH RESONANT PARTICLES

In the theory of the interaction of the electron beam with a highly turbulent plasma it is assumed that intense plasma waves are exited in the plasma at the time of the beam injection, and these plasma waves convert into a gas of solitons in the course of modulational instability. Energy exchange between the beam and the plasma is accompanied by the evolution of this solitons gas and by a relaxation of the beam [1]. The necessary condition for this approximation, in which the effect of the beam on the shape of the pulses can be ignored, and the field in the plasma can be described as a superposition of solitons, holds if the energy density of the external field in the plasma is comparable to the beam energy density.

Steady-state nonlinear waves with the carrier frequency and the carrier wave number satisfying the conditions for plasma and Cherenkov resonance exist in a plasma if the external field is only a small perturbation which is amplified by the beam [2]. Their amplitude and nonlinear period are determined by a plasma-beam instability and by the dispersion and self-contraction of Langmuir wave packets (a modulational instability). It is possible in principle for a steady-state solitary Langmuir wave to exist in a plasma with an intense electron beam [3,4]. From the physical standpoint, this circumstance corresponds to a zero value for the work performed by the wave field on an electron which passes through the wave. A bell-shape solution founded in Ref. [5] corresponds to capture of an electron beam by the field of solitary Langmuir wave.

The situation qualitatively parallels that in a fully ionized plasma occurs in a plasma with an admixture of resonant atoms [6]. At resonance, when the frequency of the atomic transitions is close to the plasma frequency, a weak plasma nonlinearity and pronounced nonlinearity of the atoms appear simultaneously. An important point is that even a small admixture of resonant atoms is accompanied by the slowing of the wave in group velocity by comparison with the group velocity of the wave in a plasma.

The energy of monochromatic Langmuir wave in a kinetic plasma gets transformed into the energy of resonance electrons in a time on the order of γ_L^{-1} (γ_L is Landau damping [7]). However, the representation of perturbations in the form of a plane wave with time-dependent amplitude is possible only for quite large wave packets, when the transit time of the resonance electron through the pulse $\Delta t \simeq \Lambda/v_{\rm ph}$ is sufficiently large $\Delta t \gg \gamma_L^{-1}$ (Λ is the width of the pulse and $v_{\rm ph}$ is the phase velocity). When this inequality is not satisfied, the electron gets appreciably displaced along the wave packet and there arises a need to take the form of the pulse into consideration [8].

§4.1. Solitary pulse in a plasma with an electron beam.

As a self-consistent system of equations which our starting point we will use the nonlinear hydrodynamic equations of a plasma, taking into account the plasma pressure, and the Poisson equation for the field

$$\frac{\partial}{\partial x}\left(\frac{\partial^2}{\partial t^2} - v_T^2 \frac{\partial^2}{\partial x^2} + \omega_e^2\right) E = 4\pi e \left(\frac{\partial^2}{\partial t^2} - v_T^2 \frac{\partial^2}{\partial x^2}\right) \tilde{n}_b, \tag{4.1.1}$$

where E is the electric field, $v_T^2 = T_e/m$, T_e is the temperature of the plasma electrons, $\omega_e^2 = 4\pi e^2 n_e/m$, $n_e = n_p + \delta n_e$, n_p is the equilibrium plasma density, and δn_e is the nonlinear change in the electron density.

The linearized hydrodynamic equations of a monoenergetic electron beam with density $n_b \ll n_p$ have the form

$$\left(\frac{\partial}{\partial t} + v_0 \frac{\partial}{\partial x}\right) \tilde{v}_b = \frac{e}{m} E, \quad \left(\frac{\partial}{\partial t} + v_0 \frac{\partial}{\partial x}\right) \tilde{n}_b = -n_b \frac{\partial \tilde{v}_b}{\partial x}, \tag{4.1.2}$$

where $\tilde{v}_b$ and $\tilde{n}_b$ are the small depth of a modulation of the beam density and velocity, respectively.

Let us assume that the electric field of stationary nonlinear wave packet has the form

$$E(t,x) = \operatorname{Re} E(\xi) \exp[i(\omega_0 t - k_0 x), \quad \xi = x - ut,$$

where the amplitude slowly varying with time and coordinate,

$$E'_t \ll \omega_0 E, \quad E'_x \ll k_0 E.$$

It follows from the equations (4.1.1) and (4.1.2) that the function $E(\xi)$ satisfies the equation

$$\begin{gathered}(u^2 - v_T^2)E'' + 2i\omega(u - v_g)E' \\ - \left[\omega^2 - k_0^2 v_T^2 - \omega_p^2 \left(1 - \frac{\delta n_e}{n_p}\right)\right] E = -i4\pi e \omega_p v_0 \tilde{n}_b, \\ (v_0 - u)^2 \tilde{n}_b'' + 2i(v_0 - u)\Delta\omega \tilde{n}_b' - \Delta\omega^2 \, \tilde{n}_b = \frac{e n_b \omega_p}{m v_0} E,\end{gathered} \tag{4.1.3}$$

where

$$|\Delta\omega| = |\omega_0 - k_0 v_0| \ll \omega_0, \quad |\omega - \omega_0| \ll \omega_0,$$
$$|u - v_g| \ll v_g, \quad v_g = k_0 v_T^2/\omega_0,$$

and v_g is the group velocity of the wave. The prime means the derivative with respect to the variable ξ.

The propagating of a wave packet in a plasma is accompanied by the separation of electrons and ions under the action of the ponderomotive force (2.3.1) so that the electrostatic field $E_s = -d\varphi/d\xi$ appears. It follows from the equation of motion of the plasma electrons that the nonlinear equilibrium condition has the form [9]

$$\frac{m}{2} < \tilde{v}_e^2 > + e\varphi = -T_e \frac{\delta n_e}{n_p}, \tag{4.1.4}$$

where averaging is fulfilled over the plasma wavelength, $\lambda_p = 2\pi v_0/\omega_p$.

According to the equation of continuity and the equation of motion of plasma ions, the density $n_i(\xi)$ and velocity $v_i(\xi)$ are determined by the equalities,

$$(v_i - u)n_i = -un_p,$$
$$M\left(-uv_i + \frac{v_i^2}{2}\right) = e\varphi - T_i \ln\frac{n_i}{n_p}.$$

Substituting $\delta n_i = n_i - n_p \ll n_p$ into the above expressions, we find

$$\frac{\delta n_i}{n_p} = \frac{e\varphi}{T_i - Mu^2}. \tag{4.1.5}$$

It follows from the equations (4.1.4) and (4.1.5):

$$\frac{\delta n_e}{n_p} = -\frac{|E|^2}{E_T^2}, \quad E_T^2 = 16\pi n_p(T_e + T_i)\left(1 - \frac{u^2}{c_s^2}\right), \tag{4.1.6}$$

if the charge neutrality condition is fulfilled, $\delta n_e = \delta n_i$. Since these plasma waves are subsonic, $u < c_s = \sqrt{(T_e + T_i)/M}$, and $\delta n_e < 0$ (modulation instability [10]), we have omitted in Eq. (4.1.3) a correction to the coefficients which is of order $u/v_T \ll 1$.

The relationships $v_{\text{ph}} = \omega_0/k_0 = v_0$ and $u = v_g$ enable us to reduce the set of equations given by Eq. (4.1.3) to a forth-order equation. In the general case the detunity of the group velocity can be compensated by the displacement of the phase velocity. Substituting

$$E = A\exp(-i\beta\xi), \quad \tilde{n}_b = -i\rho\exp(-i\beta\xi),$$

it is possible to represent the set of equations as

$$\begin{aligned} A'' + 2i\left(\frac{\omega u_g}{v_T^2} - \beta\right)A' + \Bigg[\beta\left(\frac{2\omega u_g}{v^2} - \beta\right) \\ + \frac{\omega^2}{v_T^2}\left(\epsilon + \frac{\delta n_e}{n_p}\right)\Bigg]A = \frac{4\pi e v_0 \omega_p}{v_T^2}\rho, \\ \rho'' + 2i\left(\frac{\Delta\omega}{v_0} - \beta\right)\rho' - \left(\frac{\Delta\omega}{v_0} - \beta\right)^2\rho = -\frac{en_b\omega_p}{mv_0^3}E, \end{aligned} \tag{4.1.7}$$

where

$$u_g = u - v_g, \quad \epsilon = 1 - \frac{\omega_L^2}{\omega_0^2}, \quad \omega_L^2 = \omega_p^2 + k_0^2 v_T^2.$$

Assuming that the phase of the wave is equal to

$$\beta = \frac{\omega u_g}{v_T^2} = \frac{\Delta\omega}{v_0} \ll k_0. \tag{4.1.8}$$

we reduce the Eqs. (4.1.7) to the form

$$\begin{aligned} A'' + \frac{\omega^2}{v_T^2}\left(\epsilon_g + \frac{\delta n_e}{n_p}\right)A = \frac{4\pi e v_0\omega_p}{v_T^2}\rho, \\ \rho'' = -\frac{en_b\omega_p}{mv_0^3}E, \end{aligned} \tag{4.1.9}$$

where $\epsilon_g = \epsilon - (u_g/v_T)^2$ and δn_e are defined by Eq. (4.1.6).

The equations for the beam electrons remain linear under the condition $|\tilde{n}_b| \ll n_b$, which by means of Eqs. (4.1.2), can be represented as

$$\frac{eE_m\omega_p L^2}{mv_0^3} \ll 1, \tag{4.1.10}$$

where E_m and L are the amplitude and the width of pulse, respectively [4]. This condition is satisfied if the time required for the electron to cross the pulse, L/v_0, is short in comparison with the period of the nonlinear phase oscillations, $\sqrt{mv_0/eE_m\omega_p}$, and there is no enough time for the wave field to trap the beam electrons.

The set of nonlinear equations in (4.1.9), which reflects the presence of the beam in the plasma, has no soliton solution. In the limit $n_b \to 0$ and $\epsilon_g = -|\epsilon| < 0$, however, when the beam density is low an asymptotic soliton solution does exist [1,11]:

$$E(\xi) = \sqrt{2|\epsilon|}\, E_T \cosh^{-1}(k_s\xi), \quad k_s = |\epsilon|^{1/2}\lambda_d^{-1}, \tag{4.1.11}$$

where $\lambda_d = v_T/\omega_p$ is the Debye length.

The change in the velocity of resonance electron ($v_0 = v_{\text{ph}}$) under the action of the wave field is equal to

$$\dot{v} = \frac{e}{m}E(k_s\xi)\cos\left[k_0(x - v_0t)\right], \tag{4.1.12}$$

$$\xi = x - v_gt, \quad x = v_0t + x_0,$$

where k_s^{-1} determines a spatial scale of the amplitude. Carrying out the integration in Eq. (4.1.12), we find

$$v - v_0 = \frac{e\cos\psi_0}{mk_s(v_0 - v_g)}S(\eta, \eta_0), \tag{4.1.13}$$

$$S(\eta_0, \eta) = \int_{\eta_0}^{\eta} E(\eta)\, d\eta, \ \eta = k_s(v_0 - v_g)t, \ \psi_0 = k_0x_0,$$

where η_0 is an arbitrary function of the initial time t_0. Because of it a passage of electron through a soliton pulse

$$\eta_0 = -\infty, \quad S(-\infty, \infty) = \pi$$

leads to the energy exchange between a beam and a plasma. In the case of a low beam density, $n_b \to 0$, this effect is accompanied by the evolution of a soliton gas and by the relaxation of the beam [1]. In other words, the soliton which does not take to account a beam contribution to the shape of Langmuir wave packets is not stationary.

The contribution of beam term in the equations (4.1.9) is determined by the parameter

$$\mu = \sqrt{\frac{n_bv_g}{n_pv_0}\frac{1}{|\epsilon|^2}} \tag{4.1.14}$$

which is equal to the ratio of the spatial growth rate of the plasma-beam instability $k_b = \omega_b/v_0\sqrt{|\epsilon|}$ to the inverse width of the Langmuir soliton $k_s = \sqrt{|\epsilon|}/\lambda_d$. In the special case of $\mu > 0$ a solution of the solitary-wave type must satisfy the condition [3)]

$$S(-\infty, \infty) = \int_{-\infty}^{\infty} E(\xi)\, d\xi = 0 \tag{4.1.15}$$

[3)] It is noted by L.I.Rudakov that a solution of the solitary-wave type in a plasma with an electron beam (μ>0) does not coincide with the Langmuir soliton.

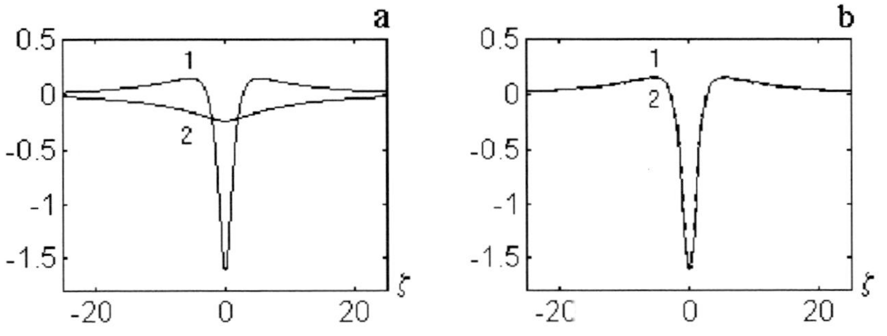

Fig. 4.1.1a. The field y (1) and the beam density ρ (2) versus the coordinate ζ for μ=0.1; Fig. 4.1.1b. Numerical soliton $y(\zeta)$ (1) and approximating function $y_a(\zeta)$ (2) which is closed to (1).

which follows from Eqs. (4.1.9). Substituting $u_g = 0$ and introducing the dimensionless variables

$$\zeta = k_s\xi, \quad y = E/E_T\sqrt{|\epsilon|}, \tag{4.1.16}$$

we represent the system of equations (4.1.9) in the form

$$y'' - y + y^3 = \rho, \quad \rho'' = -\mu^2 y, \tag{4.1.17}$$

where ρ is the amplitude of the modulation of the beam density, expressed in units of $n_p v_T|\epsilon|^{3/2}/v_0$ [4].

The result of numerical integration of Eq. (4.1.17) for $\mu = 0.1$ is represented in Fig. 4.1.1a. A function $y(\zeta)$ is well approximated by the following expression

$$y_a(\zeta) = \frac{y(0)}{2-\mu\pi}\left[\frac{2}{\cosh\zeta} - \mu\pi\exp(-\mu|\zeta|)\right] \tag{4.1.18}$$

(see Fig. 4.1.1b, curve 3). In the absence of the beam the function $y_a(\zeta)$ corresponds to Langmuir soliton in a plasma. But a small term $\mu \ll 1$ is a cause for the existence of a slow-varying "tails" of an opposite sign which allow to satisfy the condition (4.1.15).

According to the numerical solutions present above (Figs. 4.1.2), a growth of the field amplitude and a spatial compression of the pulse width are observed if the beam density increases. The arising of the depth of the density modulation, $\rho(\zeta)$, occurs for $\mu > 1$ when the spatial growth rate of the plasma-beam instability k_b^{-1} is less than the inverse width of the Langmuir soliton k_s^{-1}.

Thus the shape of the wave packets envelope is characterized by two nonlinear scales. The first of its is determined by the parameter k_b^{-1} and second one is of order k_s^{-1}. In the limit case of $\mu \ll 1$, we find that $L \simeq k_s^{-1}$ and $E_m \simeq E_T\sqrt{|\epsilon|}$. Therefore, the inequality (4.1.10), which can be represented as

$$n_b v_0 \gg |\epsilon|^{3/2} n_p v_T, \tag{4.1.19}$$

corresponds to a dense beam [12].

§4.2. Multi-soliton Langmuir pulse.

The numerical integration of Eqs. (4.1.9) shows that the alternative multi-soliton solution exists for $\mu = 0.1$ (see Fig. 4.2.1 and Fig. 4.2.2). It differs from the similar one-soliton solution presented in Fig. 4.1.1 by an increase of the depth of beam modulation

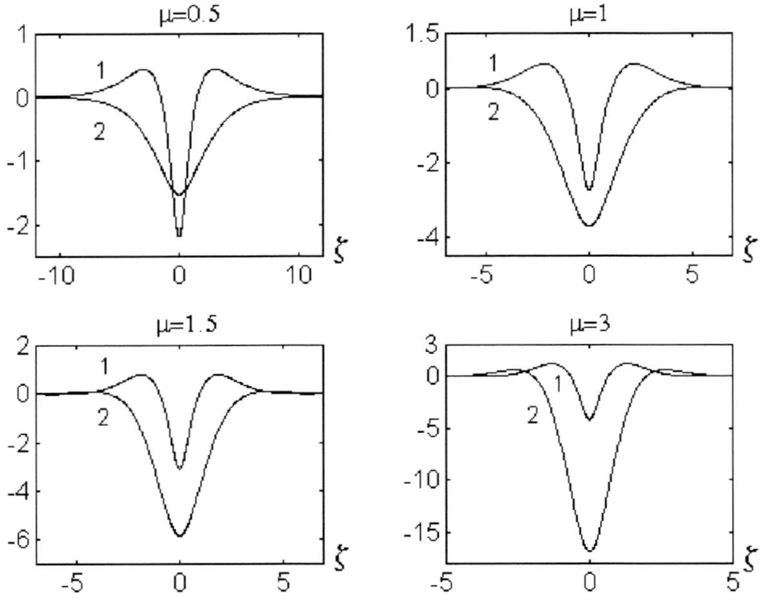

Fig. 4.1.2. Evolution of solitary wave shape $y(\zeta)$ (1) and of the beam density $\rho(\zeta)$ (2) versus the parameter μ.

$\rho(\zeta)$ and by the positive beam part of the electric field $y_+(\zeta) > 0$. Since an area of this curve, $S_+ = \int y_+(\zeta)\, d\zeta$, exceeds the area of Langmuir soliton $S(-\infty, \infty) = \pi$, the condition (4.1.15) holds for the negative part of electric field, $S_- = \int y_-(\zeta)\, d\zeta$, $y_-(\zeta) < 0$, containing the several Langmuir solitons [4].

1. Singular point of nonlinear equation.

Let us put $\epsilon_g = \epsilon$ and $u = v_g$ in Eqs. (4.1.9). Noting that some information about the nonlinear solution is incorporated in its linear asymptotic expressions, we write the corresponding soliton:

$$E \sim \exp(\pm k_g \xi), \quad k_g^2 = k_s^2 \left[\frac{1}{2} \pm \left(\frac{1}{4} - \mu^2 \right)^{1/2} \right]. \tag{4.2.1}$$

In the limit $|\epsilon| \gg n_b/n_p$, we find from (4.2.1)

$$k_s = \frac{\sqrt{|\epsilon|}}{\lambda_d}, \quad k_b = \frac{\omega_b}{v_0\sqrt{|\epsilon|}}, \quad \mu = \frac{k_b}{k_s} \ll 1. \tag{4.2.2}$$

The physical meaning of these parameters is quite obvious. The quantity k_s^{-1} determines the width of the Langmuir soliton, while k_b^{-1} is the same as the inverse growth rate of the plasma-beam instability if the frequency ω is not too close to the plasma frequency ω_p (Ref. [12]).

Introducing the slow variables

$$\zeta_b = k_b \xi, \quad y = E/E_T\sqrt{|\epsilon|}, \tag{4.2.3}$$

we rewrite Eqs. (4.1.7) as

$$\mu^2 y^{(4)} - (y - y^3)'' + y = 0. \tag{4.2.4}$$

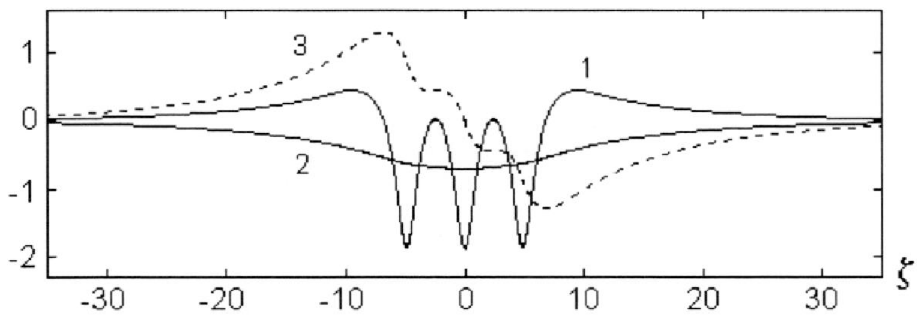

Fig. 4.2.1. Long-soliton Langmuir pulse for μ=0.1. The electric field $y(\zeta)$ (1), the depth of beam modulation $\rho(\zeta)$ (2) and the curve area $S(\zeta)/4$ (3).

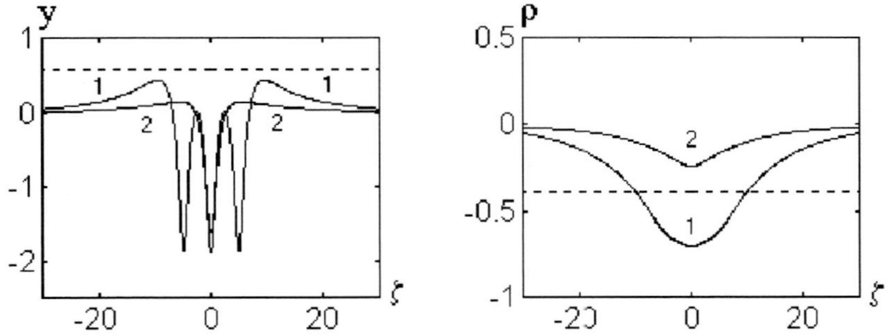

Fig. 4.2.2. The numerical solutions of Eqs. (4.1.20) for μ=0.1: $y_1(0)\simeq-1.87$ (1) and $y_2(0)\simeq-1.61$ (2). Dashed lines correspond to a singular point of Eq. (4.2.6).

We can make use of the small parameter $\mu \ll 1$ to study Eq. (4.2.4) in analytic form [4].

An integral of (4.2.4) which satisfies homogeneous conditions on the function and its derivatives is

$$[(\mu^2 y'' - y \pm y^3)']^2 = -\mu^2(2yy'' - y'^2) + y^2 - 3y^4/2. \qquad (4.2.5)$$

Making use of the small parameter, we omit the corresponding terms from Eq. (4.2.5):

$$y'^2 = y^2 \frac{1-(3/2)y^2}{(1-3y^2)^2}. \qquad (4.2.6)$$

It is easy to see that this equation has linear asymptotic expressions, which are the same as (4.2.2). As y increases, however, the derivative y' acquires a singularity at $y_c = \pm 1/\sqrt{3}$, so that in the region $y^2 > y_c^2$ the solution is not defined, and it becomes necessary to consider the higher-order derivatives (terms of order μ^2) which have been omitted.

The transition from a monotonic variation of the function to an oscillatory variation occurs near critical points y_c, and the minimum value of the function is on the order of -$\sqrt{3}$. The appearance of oscillations in the function $y(\zeta_b)$ can be explained qualitatively by drawing from the analogy with the motion of a particle in a potential well, as is done in the study of second-order nonlinear differential equations [11].

Noting that Eq. (4.2.4) is equivalent to the system of equations

$$\mu^2 y'' - y + y^3 = \rho, \quad \rho'' = -y, \qquad (4.2.7)$$

we rewrite (4.2.5) as

$$\frac{\rho'^2}{2} = \frac{\mu^2 y'^2}{2} + U(y,\rho), \quad U(y,\rho) = \frac{y^4}{4} - \frac{y^2}{2} - \rho y. \qquad (4.2.8)$$

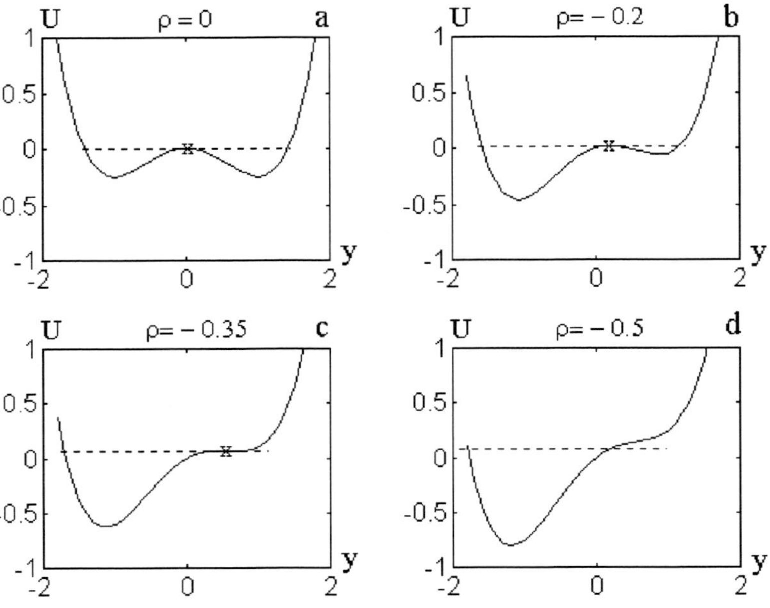

Fig. 4.2.3. The potential energy $U(y,\rho)$ versus the coordinate y. a) $\xi_b \to -\infty$; b) $\xi_b=-1.2$; c) $\xi_b=-0.7$; d) $\xi_b=-0.5$. The points of an unstable equilibrium state are determined by "x".

We assume that $E = \rho'^2/2$ is the energy of a particle with a mass μ^2 which is moving in a potential well $U(y,\rho)$. Since the functions ρ and ρ' vary only slightly over the oscillation period, the functions E and U remain constant over a distance on the order of $\zeta_b \simeq \mu$.

The slow evolution of the potential well is illustrated by Fig. 4.2.3. The symmetrical curve (Fig. 4.2.3a) corresponds to the asymptotic form of the solution $\zeta_b \to -\infty$, for which the line $E = 0$ runs tangent to the curve $U(y,0)$ at the point of a local maximum $y = 0$. Fig. 4.2.3b corresponds $\rho = -0.2$, in which case the potential well has already been deformed, but the curve U still has a maximum, and the particle is still in the unstable equilibrium state. Fig. 4.2.3c corresponds to the critical value $y_c = 1/\sqrt{3}$, at which the point of the maximum disappears, and the particle slides into the potential well. Beginning at this point (i.e., at the transition through the singularity) the monotonic motion of the particle gives way to an oscillatory motion (Fig. 4.2.3d). Once the function ρ reaches its maximum, the process repeats itself, in the opposite order.

Let us examine Eq. (4.2.4) analytically in a qualitative fashion, allowing for the presence of the singular point y_c. Equation (4.2.6) describes a monotonic change in the function y in the region $0 < |y| < y_c$, i.e., over distances $\zeta_b \simeq 1$. As the singular point is approached, $|y| \to y_c$, however, the scale changes, and the width of the transition layer between the monotonic change in y to the oscillatory change is on the order $\mu^{4/5}$. Setting (4.2.4) $y = y_c + w$, $|w| \ll 1$, (4.2.4), we find the Painleve equation [12]

$$\frac{d^2u}{d\zeta_1^2} + \sqrt{3}u^2 = \sqrt{6}\zeta_1, \qquad u = \mu^{-2/3}w, \quad \zeta_1 = \mu^{-4/5}(\zeta_b - \zeta_0), \tag{4.2.9}$$

where ζ_0 is the coordinate of the singular point. In the region $\zeta_1 < 0$ we have

$$u_- = -(-2^{-1/2}3^{-1}\zeta_1)^{1/2}, \quad -\zeta_1 \gg 1 \tag{4.2.10}$$

and the solution becomes (4.2.6). To the right of the singular point, the curve falls sharply:

$$u_- = -2\sqrt{3}(\zeta_1 - \zeta_+)^{-2}, \quad \zeta_1 > 0, \tag{4.2.11}$$

where the numerical value of ζ_+ is determined by asymptotic expression (4.2.10) (see Fig. 6.3.6).

2. Adiabatic approximation.

According to the numerical solutions present above (Fig. 4.2.2), the nature of the solution changes when the singular point is crossed, $|\rho| > \rho_c = 2/3^{3/2}$. Allowing for the appearance of oscillations, we introduce a fast variable ζ and we rewrite (4.2.7) and (4.2.8) as

$$\begin{gathered}\frac{\rho'^2}{2\mu^2} = \frac{y'^2}{2} + \frac{y^4}{4} - \frac{y^2}{2} - \rho y,\\ \rho'' = -\mu^2 y,\\ \rho' = \frac{d\rho}{d\zeta}, \quad \zeta = k_s\xi = \frac{\zeta_b}{\mu}, \quad \mu^2 \ll 1.\end{gathered} \tag{4.2.12}$$

According to the second of Eqs. (4.2.12), the functions ρ and ρ' are the integrals of motion at the small scales $\zeta \simeq 1$, while the first of Eqs. (4.2.12) can be integrated at constant ρ and ρ'. Its solution leads to an incomplete elliptic integral and can be written [9]

$$F(\varphi, k) = \pm\kappa(\zeta - \zeta_0), \tag{4.2.13}$$

$$\begin{gathered}\varphi = 2\mathrm{arcctg}[q(\alpha - y)/p(y - \beta)]^{1/2},\\ k^2 = 1/2 - 3t/2\kappa^2,\\ p^2 = 4\{t + 1/3 + R - [(t + 1/3)(2/3 - t + 2R)]^{1/2}\},\\ q^2 = 4\{t + 1/3 + R + [(t + 1/3)(2/3 - t + 2R)]^{1/2}\},\\ \alpha = -(t + 1/3)^{1/2} + (2/3 - t - 2R)^{1/2},\\ \beta = -(t + 1/3)^{1/2} - (2/3 - t - 2R)^{1/2},\\ R = (t^2 - t/3 + 6u + 1/9)^{1/2},\\ \kappa = [12(t^2 + 2u)]^{1/4},\\ t = \{v - u + [8u^3 + (v - u)^2]^{1/2}\}^{1/2} +\\ + \{v - u - [8u^3 + (v - u)^2]^{1/2}\}^{1/2},\\ v = (\rho - 4/27)/8, \quad u = (\rho'^2/\mu^2 - 1/6)/12.\end{gathered}$$

Expression (4.2.13) describes two-period oscillations of the function $y(\zeta)$, since the period of the fast oscillations $T = 4K(k)/\kappa$, is a function of the slow parameters ρ and ρ'/μ with a region of variation $\zeta \simeq 1/\mu$ [$K(k)$ is the complete elliptic integral of the first kind]. This result explains the appearance of slow envelope in Fig. 4.2.1 and Fig. 4.2.2.

Introducing the elliptic functions [13], we rewrite (4.2.13) in the form [4]

$$\begin{gathered}y = -\sqrt{t + 1/3} + \sqrt{2/3 - t + 2R}\,\frac{a - \mathrm{cn}(\alpha_0, k)}{1 - a\,\mathrm{cn}(\alpha_0, k)},\\ \alpha_0 = \kappa(\zeta - \zeta_0), \quad a = \sqrt{\frac{t + 1/3 - R - \kappa^2/2}{t + 1/3 - R + \kappa^2/2}}.\end{gathered} \tag{4.2.14}$$

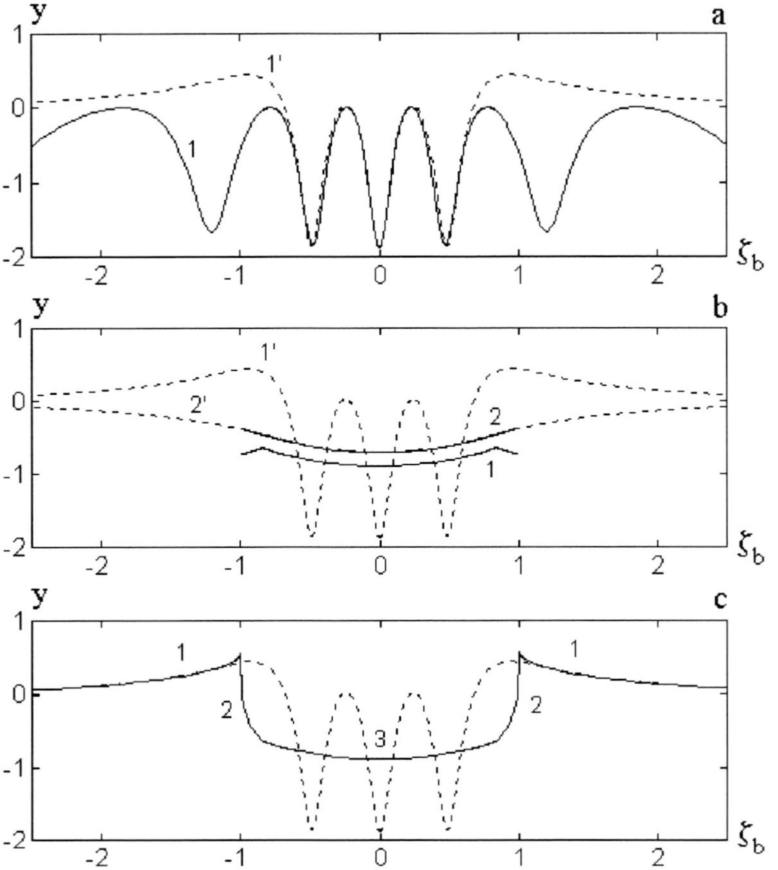

Fig. 4.2.4a. The analytical (1) and the numerical (1') solutions of Eq. (4.2.4) for μ=0.1.
Fig. 4.2.4b. The solution of the average Eq. (4.2.16) (curves 1 and 2) and the initial Eq. (4.2.4) (curves $1'$ and $2'$).
Fig. 4.2.4c. The solitary pulse for $\mu \to 0$. The solution of Eq.(4.2.6) (1), the asymptotics for transition layer (2), and the function $<y>$ (3).

As the critical points

$$y_c^2 = 1/3, \quad \rho_c^2 = 4/27, \quad (\rho'/\mu)_c^2 = 1/6$$

are approached, the oscillations period $T \sim (\rho - \rho_c)^{-1/4}$ increases. This increase corresponds to a transition from the scale μ to the scale $\mu^{4/5}$ (see Eq. (4.2.9)). In this case the asymptotic form of expression (4.2.14) is

$$y = \frac{1}{\sqrt{3}} \frac{2(\zeta - \tilde{\zeta}_+)^2/3 - 3}{2(\zeta - \tilde{\zeta}_+)^2/3 + 1}, \quad \tilde{\zeta}_+ = \frac{\zeta_+}{\mu} \tag{4.2.15}$$

and is the same as the asymptotic expression in (4.2.11) in the transition region near the singularity $y_c = 1/\sqrt{3}$.

A curve 1 in Fig. 4.2.4a corresponds to the solution of equation (4.2.14) found in an adiabatic approximation. The numerical solution of equation (4.2.14) (curve 1') is used to determine the slow functions ρ and ρ'.

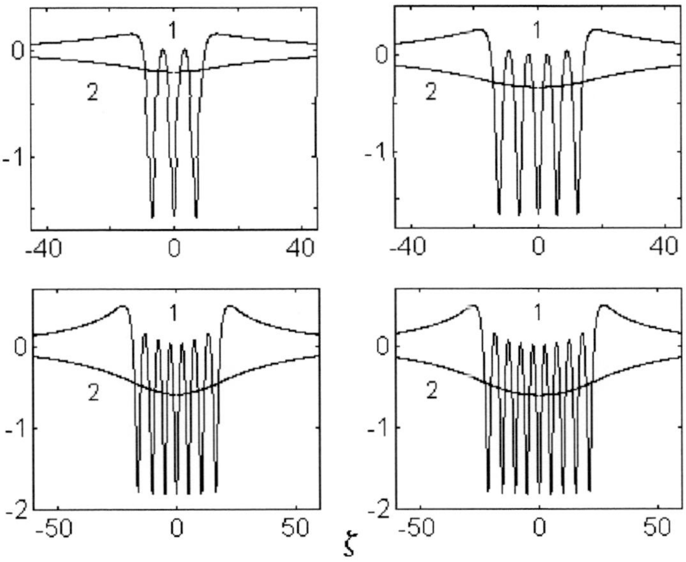

Fig. 4.2.5. A solitary pulse for μ=0.03. (1) The electric field $y(\zeta)$ and (2) the beam density $\rho(\zeta)$. A shape of pulse depends on the beam density amplitude $\rho(0)$.

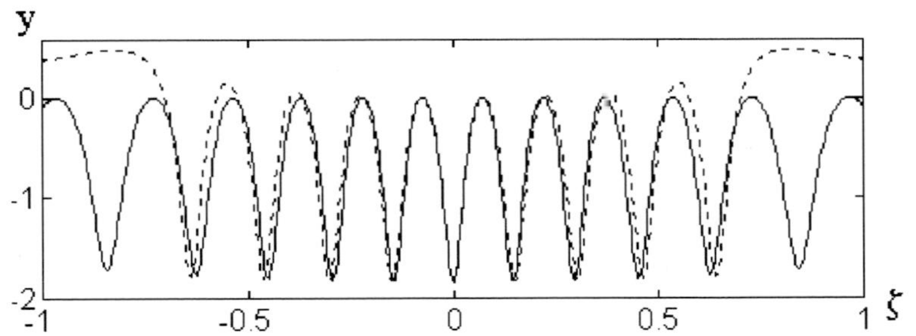

Fig. 4.2.6. The electric field for μ=0.03. The numerical solution (dashed curve) and the analytical solution in an adiabatic approximation (4.2.14).

An equation for the slowly varying function ρ can be found by the averaging the second of Eqs. (4.2.12) over the period of the nonlinear oscillations:

$$\frac{d^2\rho}{d\zeta_b^2} = -<y>, \tag{4.2.16}$$

$$\langle y \rangle = (t+1/3)^{1/2}\left[1 - 2(1-n)\Pi(n,k)/K(k)\right],$$

$$n = \frac{t + 1/3 + R - \kappa^2/2}{2(t+1/3)},$$

$\Pi(n,k)$ is the complete elliptic integral of the third kind, ζ_b is the slow variable (4.2.3). The expressions (4.2.13) and (4.2.16) are derived more details in **Appendix 4.1**.

The solution $\rho(\zeta_b)$ of Eq. (4.2.16), which satisfies initial condition, $\rho = -0.7094$, $\rho' = 0$, is practically the same as the exact solution of Eq. (4.2.4) (see Fig. 4.2.4b).

The function ρ is given by the averaging over the nonlinear period an electric field $\langle y \rangle$ (curve 1), which can be used to define a shape of the solitary wave in the limit case of $\mu = 0$.

Since this solution has the singularity at $y(\zeta_c) = 1/\sqrt{3}$, the function $y(\zeta_b)$ can be found by joining of the asymptotics y_- [Eq. (4.2.6)] and y_+ [Eq. (4.2.16)] in the singular points ζ_c (see Fig. 4.2.4c). The asymptotic form of Eq. (4.2.16) near the singular point is

$$y_+ = \frac{1}{\sqrt{3}}\left[1 - \left(\frac{\zeta_b - \zeta_c}{\Delta\zeta}\right)^{1/4}\right], \tag{4.2.17}$$
$$|\zeta_c| \simeq 1, \quad \Delta\zeta = 2^{1/2}3^{-1}[K(2^{-1/2})/\pi]^4.$$

It follows that the size of the transition layer, *langley*$\langle$, is $\Delta\zeta \simeq 0.002$.

3. Let us summarize the results.

A solitary Langmuir wave exists in a nonlinear plasma with a fairly strong electron beam with phase and group velocities which are the same as the beam velocity and the group velocity of the plasma.

A distinctive feature of this solution is that the area under the pulse is zero; this result corresponds to a situation in which the wave performs no work on an electron which passes through it.

A decrease in the parameter μ corresponds by the appearance of small-scale oscillation (with a period of order μ) for width of order $\zeta_b \simeq 1$. The number of peaks is determined by the relation between these scales and depends on the field amplitude at the center of the pulse. The solutions of Eq. (4.2.4) found for $\mu = 0.03$ are represented in Figs. 4.2.5 and 4.2.6.

Thus, the above investigation shows that the shape of envelope of a solution of the solitary-wave type is characterized by two nonlinear scales. The first of the other is determined by the inverse growth rate of the beam-plasma instability, and the second one is equal to the width of the Langmuir soliton. At a qualitative level, an arise of two length scales can be interpreted as a chain of Langmuir solitons whose phases are correlated by an electron beam.

§4.3. Solitary wave with an arbitrary group velocity.

Let us examine the equations (4.1.3) in the general case when a group velocity of the wave u does not coincide with the plasma group velocity v_g, $u_g = u - v_g \neq 0$. We assume that the carrier frequency and the carrier wave number of the soliton-like wave satisfy the conditions for plasma and Cherenkov resonances, $\omega_0/k_0 = v_0$, and we rewrite the above set of equations can in the form

$$\frac{d^4E}{d\xi^4} + 2i\frac{u_g\omega_p}{v_T^2}\frac{d^3E}{d\xi^3} + \frac{\omega_p^2}{v_T^2}\frac{d^2}{d\xi^2}\left(\epsilon + \frac{|E|^2}{E_T^2}\right)E + \frac{\omega_b^2\omega_p^2}{v_0^2v_T^2}E = 0, \tag{4.3.1}$$

where ω_b is the Langmuir frequency of the beam.

To analyze Eq. (4.3.1) for a range of $\mu \simeq 1$, we represent it in terms of dimensionless variables

$$\eta = \Delta k\xi, \quad Y = \frac{1}{\Delta k\lambda_d}\frac{E}{E_T}, \quad v = \frac{2}{\Delta k\lambda_d}\frac{u_g}{v_T},$$
$$\Delta k = \left(\frac{\omega_p\omega_b}{v_Tv_0}\right)^{1/2}, \quad \kappa = \frac{\epsilon}{(\Delta k\lambda_d)^2}, \tag{4.3.2}$$

which are connected with the analogous values of §4.1 by means of expressions

$$\kappa = \mathrm{sign}(\epsilon)/\mu, \quad \Delta k\lambda_d = \sqrt{\mu|\epsilon|}, \quad \eta = \sqrt{\mu}\,\zeta.$$

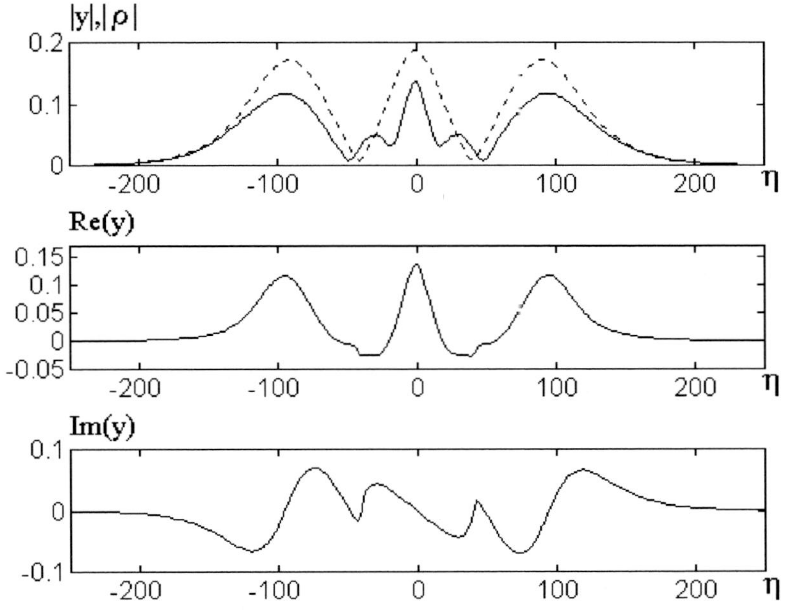

Fig. 4.3.1. The electric field $y(\zeta)$ and the beam density $\rho(\zeta)$ (dashed curve) for $\kappa \simeq 0.67$ and $v \simeq 0.14$.

Using Eqs. (4.3.1) and (4.3.2), we obtain the following equation for field amplitude

$$Y^{(4)} + ivY^{(3)} + [(\kappa + |Y|^2)Y]'' + Y = 0. \tag{4.3.3}$$

To compute the energy integral and the momentum integral, we rewrite Eq. (4.3.3) as a system of equations:

$$Y'' + ivY' + (\kappa + |Y|^2)Y = R, \tag{4.3.4}$$

$$R'' = -Y. \tag{4.3.5}$$

Multiplying Eq. (4.3.4) by Y'^*; Eq. (4.3.5) by R'^*, and putting these results together with the complex-conjugate expressions, we find

$$[|Y'|^2 + \kappa|Y|^2 + |Y|^4/2]' = 2\mathrm{Re}(Y'R^*),$$

$$(|R'|^2)' = -2\mathrm{Re}(Y^*R').$$

Subtracting the above expressions and carrying out a simple integration, we find the energy conservation law

$$|Y'|^2 + |Y|^2(\kappa + |Y|^2/2) - |R'|^2 - 2\mathrm{Re}(YR^*) = C_1. \tag{4.3.6}$$

Multiplying Eq. (4.3.4) by Y'^*; Eq. (4.3.5) by R'^*, and subtracting these results from the complex-conjugate expressions, we find

$$Y^*Y'' - Y(Y^*)'' - iv(|Y|^2)' = Y^*R - YR^*,$$

$$R^*R'' - R(R^*)'' = Y^*R - YR^*.$$

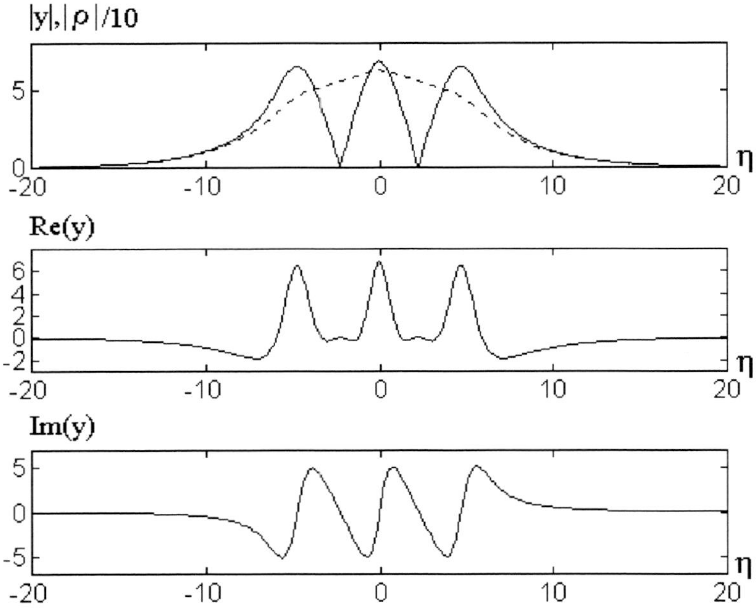

Fig. 4.3.2. The electric field $y(\zeta)$ and the beam density $\rho(\zeta)$ (dashed curve) for κ=−7.6669 and v=14.0624.

Subtracting above expressions and carrying out the integration, we find the momentum conservation law

$$2\mathrm{Re}\,(iY^*Y' - iR^*R') + v|Y|^2 = C_2. \qquad (4.3.7)$$

The soliton-like solution of Eq. (4.3.3) has been obtained by G.V.Fomin in 1998. The results of numerical calculations taking into account a change of the amplitude and the phase of the wave are shown in Fig. 4.3.1 and Fig. 4.3.2. The accuracy of calculations was monitored on the basis of integrals (4.3.6) and (4.3.7).

In the region of parameters $E_m \simeq \Delta k \lambda_d E_T$ and $L \simeq 1/\Delta k$ the condition (4.1.10) at which the beam still remains linear is restricted by the condition

$$(n_b/n_p)^{1/4} \gg (v_T/v_0)^{5/2}. \qquad (4.3.8)$$

This inequality is the same as (4.1.19) in the case of $\epsilon \simeq (\Delta k \lambda_d)^2$ in the latter one.

If the group velocity displacement is large, $|v| \gg 1$, the fourth derivative in Eq. (4.3.2) can be omitted and the equation reduces to the form

$$ivY^{(3)} + [(\kappa + |Y|^2)Y]'' + Y = 0. \qquad (4.3.9)$$

It is easy to see that by a change of scale one can represent Eq. (4.3.9) as one-parameter equation which has a discrete spectrum of the soliton solutions found in §1.3.

§4.4. Bell-shape wave in a plasma.

The hydrodynamic solution obtained above are valid only for a sufficiently dense and fast beam, so that the dominant nonlinear effect is the dependence of plasma density on field amplitude, while the beam remains linear if the inequality (4.1.10) holds. Thus, effects related to beam particle capture by the wave remain insignificant for $\Delta t \ll \tau_{\text{ph}}$, where $\Delta t = v_0/L$ is the transit time of the resonance electron through the pulse, and $\tau_{ph} = (mv_0/eE_m\omega_p)^{1/2}$ is the period of nonlinear phase oscillations of the trapped electron.

The goal of the present study is to consider effects of nonlinear interaction of a beam with both wave and plasma nonlinearity [5]. To obtain an analytical solution of this problem, the fine electron bunching model is used, which provides a qualitatively correct description of the intensely nonlinear stage of plasma beam instability accompanied by bunching of the beam in the field of the main wave. We consider times shorter than the characteristic time for development of satellite instability (see §3.3), so that the single-mode approximation remains valid.

Let us analyze a possibility of the existence of solitary Langmuir wave in a hydrodynamic nonlinear plasma with a kinetic electron beam. We rewrite the equation (4.1.1) in the form

$$\frac{\partial}{\partial t}\left[\frac{\partial^2}{\partial t^2} - v_T^2\frac{\partial^2}{\partial x^2} + \omega_p^2(E)\right] E = -4\pi e\left(\frac{\partial^2}{\partial t^2} - v_T^2\frac{\partial^2}{\partial x^2}\right) j_b, \tag{4.4.1}$$

where $\omega_p(E)$ is the plasma frequency taking into account the ponderomotive plasma nonlinearity [see (4.1.6)], and j_b is the beam current density.

We consider a stationary envelope wave

$$E(t,x) = \mathrm{Re}\, E(\xi)\exp[i(\omega_0 t - k_0 x)], \tag{4.4.2}$$

where $\xi = x - ut$, u is the group velocity, and ω_0 and k_0 satisfy the conditions of plasma and Cherenkov resonance (see §4.1).

In the group velocity reference frame which depends only on coordinate, it will be convenient to use the traveling-wave tube theory method for solution, writing the beam current density in the form

$$j_b(t,x) = \frac{j_0 T}{S}\sum_{s=-\infty}^{\infty}\delta[t - t_s(\xi)], \tag{4.4.3}$$

where the functions $t_s(x)$ satisfy the equation of motion

$$\frac{d^2 t_s}{d\xi^2} = -\frac{e}{m}\left(\frac{dt_s}{d\xi}\right)^3 E\left[\xi, t_s(\xi)\right], \tag{4.4.4}$$

$T = 2\pi/(\omega_0 - k_0 u)$ is the bunch repetition period in the fixed plane $\xi = \mathrm{const}$,

$$j_0 = \frac{1}{T}\int_0^T j_b\, dt$$

is the mean beam current density, and S is a whole number.

Let us substitute Eq. (4.4.2) in Eq. (4.4.1). After averaging over the time period, we obtain the equation for the slowly varying complex amplitude

$$\begin{aligned}(u^2 - v_T^2)E'' + 2i\omega_0(v_g - u)E' + [-\omega^2 + \omega_p^2(1 - \delta n_e/n_p) + k_0^2 v_T^2]E \\ = -i8\pi e j_0\omega_p S^{-1}\sum_{s=1}^{S}\exp i[k_0\xi - (\omega_0 - k_0 u)t_s(\xi)],\end{aligned} \tag{4.4.5}$$

where $\omega_0 = k_0 v_0$, and $v_g = k_0 v_T^2/\omega_0$ is the plasma group velocity.

Taking into account that the term $\xi/(v_0 - u)$ corresponds to uniform motion of electron bunches, and switching to the variable

$$t_s(\xi) = \xi/(v_0 - u) + s\Delta t + \tau_s(\xi), \quad \psi_s(\xi) = 2\pi s/S + k_0(v_0 - u)\tau_s(\xi),$$

we rewrite Eq. (4.4.4) and (4.4.5) in the form

$$\begin{gathered} \frac{v_T^2}{\omega_p^2} E'' + 2i\frac{u_g}{\omega_p} E' + \left(\epsilon + \frac{|E|^2}{E_T^2}\right) E = i\frac{8\pi e j_0}{\omega_p S} \sum_{s=1}^{S} \exp(-i\psi_s), \\ \psi_s'' = -\frac{ek_0}{mv_0^2}\left(1 + \frac{\psi_s'}{k_0}\right)^3 \operatorname{Re} E \exp(i\psi_s), \end{gathered} \tag{4.4.6}$$

where $u_g = u - v_g$, $\epsilon = 1 - \omega_L^2/\omega_0^2$, ω_L is the Langmuir plasma frequency, and δn_e is determined by Eq. (4.1.9) (in case of $u < c_s = \sqrt{T_e/M}$). Terms of order $u^2/v_T^2 \ll 1$ and $u^2/v_0^2 \ll 1$ are omitted in Eqs. (4.4.6).

The self-consistent set of equations (4.4.6) corresponds to the nonlinear hydrodynamic plasma and the kinetic electron beam. Integrating (4.4.6) yields the conservation of the energy and momentum represented in **Appendix 4.2.**

The transition to the hydrodynamic (linear) description of the beam [see Eqs. (4.1.7)] can be introduced by the expanding in a series exponentials in Eqs. (4.4.6) in the small deviations of the electron trajectories from the initial trajectories

$$\tilde{\psi}_s = \psi_{so} - \psi_{so}, \quad \psi_{so} = 2\pi s/S, \quad |\tilde{\psi}_s| \ll \psi_{so}$$

if the inequality (4.1.10) holds (see **Appendix 4.2**).

In the limiting case of $S = 1$, when only one electron bunch is injected in a plasma over the period of beam modulation, an analytical solution of Eq. (4.4.6) can be found [5]. By substituting the variables

$$E = A(\xi)\exp\left(-i\frac{u_g\xi}{v_T\lambda_d}\right), \quad \lambda_d = \frac{v_T}{\omega_p}, \tag{4.4.7}$$

we reduce Eq. (4.4.6) to the form

$$\begin{gathered} \lambda_d^2\frac{d^2A}{d\xi^2} - \left(\epsilon_g - \frac{|A|^2}{E_T^2}\right) E = i\frac{8\pi e j_0}{\omega_p}\exp(-i\vartheta), \\ \vartheta'' = -\frac{ek_0}{mv_0^2}\left[1 + \frac{1}{k_0}\left(\vartheta' + \frac{u_g}{v_T\lambda_d}\right)\right]^3 \operatorname{Re} A\exp(i\vartheta), \end{gathered} \tag{4.4.8}$$

where

$$\vartheta = \psi - \frac{u_g\xi}{v_T\lambda_d}, \quad \epsilon_g = |\epsilon| - \frac{u_g^2}{v_T^2} > 0.$$

It can easily be seen that at $\vartheta = \pi/2$ energy exchange between bunch electrons and beam is absent and the equation of motion is satisfied automatically. After differentiation this equality with respect to time, we find the velocity of trapped bunch

$$V = u + \frac{v_g}{u}(v_{ph} - u), \tag{4.4.9}$$

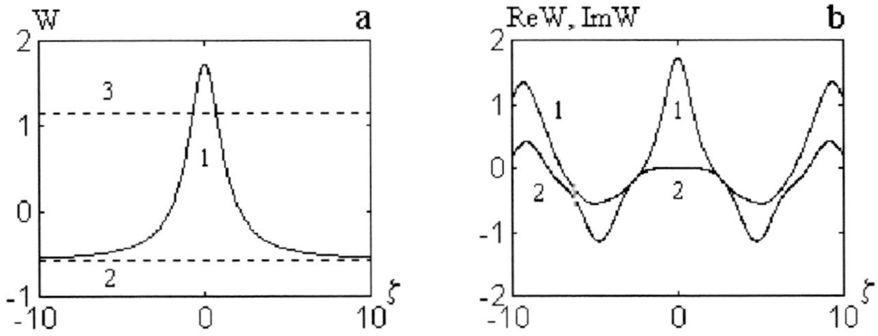

Fig. 4.4.1a. The solutions of Eq. (4.4.10): bell-shape (1), $W_1 = -1/\sqrt{3}$ (2) and $W_{2,3} = 2/\sqrt{3}$ (3).
Fig. 4.4.1b. The solution of Eq. (4.4.8) for $\vartheta_0 - \pi/2 = 10^{-2}$. Real part (1) and imaginary part (2) of the complex field amplitude.

while the equation for the field (4.4.8) becomes nonimaginary:

$$W'' - (1 - W^2)W = J_0,$$

$$W = \frac{A}{E_T\sqrt{\epsilon_g}}, \quad \zeta = \frac{\sqrt{\epsilon_g}}{\lambda_d}\xi, \quad J_0 = \frac{8\pi e j_0}{\omega_p E_T \epsilon_g^{3/2}}. \tag{4.4.10}$$

The stationary points of Eq. (4.4.10) are determined by the roots of the following algebraic equation

$$W^3 - W - J_0 = 0. \tag{4.4.11}$$

Let us seek the solution of Eq. (4.4.10) in the case of $J_0 = 2/3^{3/2}$ [14] when the roots of above cubic equation are

$$W_1 = 2/\sqrt{3}, \quad W_2 = W_3 = -1/\sqrt{3}. \tag{4.4.12}$$

We will integrate Eq.(4.4.10), assuming the constant of integration is $C = 1/12$. The corresponding first-order equation has the form

$$W' = \sqrt{\frac{1}{2}(\sqrt{3} - W)\left(W + \frac{1}{\sqrt{3}}\right)^3}. \tag{4.4.13}$$

The solution of Eq. (4.4.13) can be represented by elementary functions:

$$W = \frac{1}{\sqrt{3}}\left(-1 + \frac{12}{3 + 2\zeta^2}\right). \tag{4.4.14}$$

This expression describes a bell-shaped pulse and corresponds to capture of an infinite beam having infinite energy as a whole [5]. The solutions of Eqs. (4.4.8) are represented in Fig. 4.4.1. Curve 1 in Fig. 4.4.1a corresponds to $\vartheta = \pi/2$ [bell-shape pulse (4.4.14)]; and the dashed lines 2 and 3 are the roots of cubic equation (4.4.12). Numerical integration in case of $\vartheta \neq \pi/2$ shows that a small phase perturbation accompanies by the arising of nonlinear phase oscillation (see Fig. 4.4.1b).

Combining (4.4.7) and (4.4.14), we obtain the expression

$$\widetilde{W} = \frac{1}{\sqrt{3}}\left(-1 + \frac{12}{3 + 2\zeta^2}\right)\exp(-i\beta\zeta), \quad \beta = \frac{u_g}{v_T \epsilon_g}, \tag{4.4.15}$$

determining a complex amplitude of the solitary wave in the presence of the trapped electrons. An increase of the group velocity difference β ($\epsilon_g \to 0$) is accompanied by the spreading of the wave packet. Besides, a solitary wave transforms in a periodic structure if the new scale, $\zeta_g \simeq \beta^{-1}$ ($\xi_g \simeq v_T \lambda_d / u_g$), dominates in Eq. (4.4.15).

Note that in the special case of $u_g = 0$ and $\epsilon_g = |\epsilon|$, the amplitude, $E_m \simeq E_T \sqrt{|\epsilon|}$, and the width, $\Delta\xi \simeq \lambda_d/\sqrt{|\epsilon|}$, of the bell-shape pulse agree in order of magnitude with the analogous parameters for the Langmuir soliton (4.1.13).

§4.5. Langmuir solitons in a plasma with an admixture of resonant atoms.

In the present section it is assumed that active atoms are present in the plasma and interact with plasma waves. At resonance, when the frequency of the atomic transitions, Ω, is closed to the plasma frequency, ω_p, a week plasma nonlinearity and a pronounced nonlinearity of the atoms appear simultaneously. An important point is that even a small admixture of resonant atoms is accompanied by the slowing of the wave in group velocity by comparison with the group velocity of a wave in a plasma. A distinctive feature of the soliton solutions found below is an intense exchange of energy between the wave and the medium, accompanied by the transition of the resonant atoms to a state with a population inversion among the energy levels [6].

The electric field of the Langmuir wave in a plasma-gas medium satisfy the Poisson equation:

$$\frac{\partial}{\partial x}(E + 4\pi P) = 4\pi e \tilde{n}_p, \tag{4.5.1}$$

where the polarization vector P of the gas satisfy the equation (see **Appendix 6.1**)

$$\left(\frac{\partial^2}{\partial t^2} + \Omega^2\right) P = \frac{2d}{\hbar}\left[(Nd\Omega)^2 - \Omega^2 P^2 - \left(\frac{\partial P}{\partial t}\right)^2\right]^{1/2} E, \tag{4.5.2}$$

N and d are the density and dipole moment of the two-level atoms [15,16].

The plasma density $\tilde{n}_p$ can be excluded from the equation (4.5.1) by means of the hydrodynamic equations. As result, we obtain the equation:

$$\left(\frac{\partial^2}{\partial t^2} - v_T^2 \frac{\partial^2}{\partial x^2}\right)(E + 4\pi P) + \omega_p^2\left(1 - \frac{|E|^2}{E_T^2}\right) E = 0. \tag{4.5.3}$$

In the absence of a gas $P = 0$ it determines nonlinear plasma wave of plasma Eq. (4.1.1). A going from (4.1.1) to (4.5.3) can be realized by means of formal substitution $e\tilde{n}_p = -\partial P/\partial x$.

For small perturbations $\exp(i\omega t - ikx)$ we find from the self-consistent system of equations (4.5.2) and (4.5.3), the dispersion relation

$$1 - \frac{\omega_p^2}{\omega^2 - k^2 v_T^2} + \frac{\omega_g^2}{\Omega^2 - \omega^2} = 0, \quad \omega_g^2 = \frac{8\pi N d^2 \Omega}{\hbar}, \tag{4.5.4}$$

which determines the frequency ω as a function of the wave number k.

Assuming that the wave packet is narrow,

$$k = k_0 + \Delta k, \quad \omega = \omega_0 + \Delta\omega, \quad \Delta\omega = u\Delta k,$$
$$|\Delta\omega| \ll \omega_0, |\Delta k| \ll k_0,$$

and expanding (4.5.4) in a power series in Δk, we find

$$\epsilon + \frac{2}{\omega_p}(u - v_g)\Delta k - (\frac{\Delta k v_T}{\omega_p})^2 + \frac{\omega_g^2}{\Omega^2 - \omega_0^2 - 2\omega_0 u \Delta k} = 0, \tag{4.5.5}$$

where $v_g = k_0 v_T^2/\omega_0$ is the group velocity of a wave in the plasma in the absence of atoms, $\omega_L^2 = \omega_p^2 + k_0^2 v_T^2$, $\epsilon = 1 - \omega_L^2/\omega_0^2$ is the permittivity of a plasma. In going from (4.5.4) to (4.5.5) we made use of the obvious relations $\omega_0 \gg k_0 v_T$ and $v_T \gg$u.

The "tails" of a solution are described by a pair of a complex-conjugate roots of the algebraic equation (4.5.5). It is easy to see that the required solution exists if the density of atoms is low,

$$\begin{gathered} \Delta k_1 = \pm v_T^{-1}(\omega_L^2 - \omega_0^2)^{1/2}, \\ |\Delta k_1|^3 \gg \omega_g^2 \omega_p / v_g v_T^2, \quad |\Delta k_1| \gg |u - v_g|\omega / v_T^2 \end{gathered} \tag{4.5.6}$$

and its corresponds to a Langmuir soliton in a fully ionized plasma (see §4.1). Furthermore, under the opposite in equalities there is the asymptotic behavior

$$\Delta k_2 = \frac{\Omega^2 - \omega_0^2}{4u\Omega} \pm i \left[\frac{\omega_g^2 \omega_p}{4\Omega(v_g - u)u} - \left(\frac{\Omega^2 - \omega_0^2}{4u\Omega} \right)^2 \right]^{1/2}, \tag{4.5.7}$$

which determines the tails of a dynamic Langmuir soliton analogous to an optical soliton. The quantities $|\Delta k_1|^{-1}$ and $|\mathrm{Im}(\Delta k_2)|^{-1}$ determined typical spatial dimensions of the solitons.

We seek a nonlinear solution of system (4.5.2) and (4.5.3) in the form of a plane wave with a slowly varying complex amplitude:

$$\begin{gathered} E(t, x) = \mathrm{Re}\, E(\xi) \exp(i\Phi), \quad P(t, x) = \mathrm{Re}\, E(\xi) \exp(i\Phi), \\ \Phi = \omega_0 t - k_0 x, \quad \xi = x - ut. \end{gathered} \tag{4.5.8}$$

In terms of the dimensionless variables

$$\zeta = \frac{\omega_p}{v_T}(x - ut), \quad \beta = \frac{u}{v_T}, \quad \beta_g = \frac{u_g}{v_T}, \quad \epsilon_T = \frac{E_T}{Nd},$$

$$\delta = \frac{\Omega^2 - \omega_0^2}{\omega_p^2}, \quad Q = \frac{\omega_L^2 - \omega_0^2}{\omega_p^2}, \quad q^2 = \frac{\omega_g^2}{\omega_p^2}$$

we find from (4.5.2) and (4.5.3)

$$\begin{gathered} \epsilon'' - 2i(\beta_g - \beta)\epsilon' - \left(Q - \frac{|\epsilon|^2}{\epsilon_T^2} \right) \epsilon = -4\pi a, \\ -2i\beta a' + \delta a = \frac{q^2}{4\pi}\sqrt{1 - |\epsilon|^2}\, \epsilon, \end{gathered} \tag{4.5.9}$$

where the prime denotes the derivative with respect to the variable ζ; $|\delta| \ll 1$ and $|Q| \ll 1$. In deriving (4.5.9) we have omitted the second derivative of the field amplitude with respect to the time, which is small in comparison with the spatial derivative, by a factor on the order of the parameter $k_0 v_T/\omega_0 \ll 1$. Furthermore, in the equation for the polarization we ignored the second derivative of the polarization amplitude, as we are justified in doing when the gas density is low, $q \ll 1$.

In the absence of atoms, with $q^2 = a = 0$ [see inequalities (4.5.6)], the system (4.5.9) becomes a nonlinear parabolic equation and has a soliton solution:

$$\begin{gathered}\epsilon_1 = \epsilon_{m1} \cosh^{-1}(\xi/\Lambda_1), \\ Q_1 = \epsilon_{m1}^2/2\epsilon_T^2, \quad \Lambda_1 = \sqrt{2}\epsilon_T/\epsilon_{m1}, \quad \beta = \beta_g.\end{gathered} \tag{4.5.10}$$

In the linear plasma without dispersion [in the approximation (4.5.7)] there is also a soliton solution:

$$\begin{gathered}\epsilon_2 = \epsilon_{m2} \cosh^{-1}(\xi/\Lambda_2), \\ Q_2 = \frac{q^2}{\delta}\left(1+\mu^2\right)^{-1}, \quad \beta = \beta_g\left[1 + \frac{q^2}{\delta}\left(1+\mu^2\right)^{-1}\right]^{-1}, \\ \Lambda_2 = \frac{16\pi\beta}{q^2\epsilon_{m2}}, \quad \mu = \frac{q^2\epsilon_{m2}}{8\pi\delta},\end{gathered} \tag{4.5.11}$$

which is analogous to an optical soliton [16].

The asymptotic nonlinear solutions (4.5.10) and (4.5.11) can be used to seek solution of (4.5.9) in the more general case in which the plasma simultaneously is nonlinear and contains atoms. For this purpose we set $Q = Q_1 + Q_2$, and represent (4.5.9) formally as two nonlinear second-order equations:

$$\epsilon_1'' - \left(Q_1 - \frac{|\epsilon_1|}{\epsilon_T^2}\right)\epsilon_1 = 0, \tag{4.5.12}$$

$$\left\{\begin{aligned} & 2i(\beta-\beta_g)\epsilon_2' + Q_2\epsilon_2 = 4\pi a, \\ & -2i\beta a' + \delta a = \frac{q^2}{4\pi}\sqrt{1-|a|^2}\,\epsilon_2 \end{aligned}\right. \tag{4.5.13}$$

each of which has a soliton solution. Clearly, our original system of equation, (4.5.9), has a soliton solution under the conditions $\epsilon_{m1} = \epsilon_{m2} = \epsilon_m$ and $\Lambda_1 = \Lambda_2 = \Lambda$. Expressing Q_1 and Q_2 in terms of the wave amplitude ϵ_m, we find from (4.5.10) and (4.5.11) a nonlinear dispersion law,

$$Q = \frac{\epsilon_m^2}{2\epsilon_T^2} + \frac{q^2}{\delta}\left(1+\mu^2\right)^{-1}, \tag{4.5.14}$$

which is in the linear approximation ($\epsilon_m \to 0$) the same as (4.5.4). In contrast (4.5.10), the solution of Eq. (4.5.12) exists for arbitrary values of β, and the value of this parameter is determined by Eqs. (4.5.13) and agrees with Eq. (4.5.11).

In an analysis of the particular solutions corresponding to a Langmuir soliton and an optical soliton, the frequency and the amplitude are adjustable parameters, while the wave number and the group velocity of the wave are determined in terms of these adjustable parameters. The auxiliary condition imposed on the pulse width reduces the number of independent parameters and makes possible to put the amplitude in a more specific form for a given frequency.

The condition $\Lambda_1 = \Lambda_2$,

$$8\sqrt{2}\pi\beta = q^2\epsilon_T \tag{4.5.15}$$

along with the equality

$$\epsilon_T^2 = A\left(1 - \frac{\beta^2}{\beta_s^2}\right), \quad A = \frac{32\pi n_p T}{N^2 d^2}, \quad \beta_s^2 = \frac{c_s^2}{v_T^2} = \frac{2m}{M} \tag{4.5.16}$$

constitutes an algebraic system of equations. Solving it, we find

$$\beta^2 = B\left(1+\frac{B}{\beta_s^2}\right)^{-1}, \quad \epsilon_T^2 = A\left(1+\frac{B}{\beta_s^2}\right)^{-1},$$
$$2B = \left(\frac{q^2}{8\pi}\right)^2 A. \tag{4.5.17}$$

We can determine the soliton amplitude under the assumption that the wave frequency is close to the frequency of resonant transitions of the atoms ($\delta \to 0$). Assuming $\mu \gg 1$ in (4.5.11), and finding an expression for β_g from (4.5.14),

$$\beta_g^2 \simeq \Delta + \epsilon_m^2/2\epsilon_T^2, \quad \Delta = 1 - \omega_p^2/\Omega^2 \ll 1,$$

we write

$$\beta^2 = \left(\Delta + \frac{\epsilon_m^2}{2\epsilon_T^2}\right)\left[1+\left(\frac{8\pi}{q\epsilon_m}\right)^2\right]^{-2}. \tag{4.5.18}$$

From (4.5.17) and (4.5.18) we find an equation for the soliton amplitude:

$$y^3 + 2(\Delta - \beta^2)y^2 - 2q^2y - \alpha q^3 = 0, \tag{4.5.19}$$

where $y = \epsilon_m^2/\epsilon_T^2$ and $\alpha = q/2\beta^2$. According to (4.5.17), we have $\beta^2 \leq m/M$, so that this quantity should be omitted in comparison with Δ [since the contribution of ions to the plasma waves was ignored in Eq. (4.5.3)].

Since the solutions of Eq. (4.5.19) in the general case are quite complicated, we will write asymptotic expressions for the most interesting limiting cases. When the transition of the atoms is close to the plasma frequency, the solution becomes

$$y = q\begin{cases}\sqrt{2}, & \alpha \ll 1, \quad |\Delta| \ll q, \\ \alpha^{1/3}, & \alpha \gg 1, \quad |\Delta| \ll q\alpha^{1/3}.\end{cases} \tag{4.5.20}$$

Far from resonance

$$|\Delta| \gg q, \quad |\Delta| \gg q\alpha^{1/3},$$

we have the asymptotic behavior

$$y = \Delta\begin{cases}-2 + \dfrac{q^2}{\Delta^2}\left(\dfrac{\alpha q}{4\Delta} - 1\right), & \Delta < 0, \\ \dfrac{q^2}{2\Delta^2}\left[1+\left(\dfrac{2\alpha\Delta}{q}+1\right)^{1/2}\right], & \Delta > 0.\end{cases} \tag{4.5.21}$$

The polarization amplitude for arbitrary values of the parameter μ can be found from (4.5.13); it is

$$a = 2\eta\sqrt{1+\eta^{-2}\tanh^2(\zeta/\Lambda)}\cosh^{-1}(\zeta/\Lambda)\exp i\varphi$$
$$\eta = \mu/(1+\mu^2), \quad \varphi = -\arctan\left[\eta^{-1}\tanh(\zeta/\Lambda)\right]. \tag{4.5.22}$$

The difference between the populations of the atomic energy levels is [15,16]

$$W = -\sqrt{1-|a|^2} = -\frac{1-\mu^2+2\mu^2\tanh^2(\zeta/\Lambda)}{1+\mu^2}, \tag{4.5.23}$$

so that under the condition $\mu > 1$ at the head of the pulse, $\zeta = 0$, the atoms are in an inverted state, $W > 0$. Part of the energy is in a form of the energy of resonant transitions of the atoms; this is the reason for the significant slowing of the wave in the group velocity.

The appearance of a population inversion among the levels may be accompanied by a coherent emission of an electromagnetic wave, since a two-level system is unstable with respect to small electromagnetic perturbations. Clearly, this effect may be accompanied by damping of the Langmuir soliton and by conversion of its energy into electromagnetic radiation.

We conclude by writing yet another nonlinear solution of system (4.5.9), which exists under the conditions

$$\beta = \beta_g, \quad \delta = 0, \quad Q = 3(2\pi/\epsilon_T)^{2/3}$$

and which is bell-shaped [14]:

$$\epsilon = \pm\frac{6\pi}{Q}\left(-1 + \frac{4}{1 + 2Q\zeta^2/3}\right), \quad a = \mp 1. \tag{4.5.24}$$

Expression (4.5.24) corresponds to atomic-level populations $W = 0$ and has nonvanishing asymptotic forms as $\zeta \to \pm\infty$.

§4.6. Landau damping of short pulses.

It is known [7,19] that the energy of the monochromatic Langmuir wave in a plasma gets transformed into the energy of resonance electrons in a time on the order of γ_L^{-1} (γ_L is Landau damping) [7]. The problem is spatially homogeneous and the field amplitude decreases exponentially with time uniformly over the entire length of the plasma. However, the representation of Langmuir perturbations in the form of a plane wave with time-dependent amplitude is possible only for quite large wave packets, when the transit time of the resonance electron through the pulse $\Delta t = \Lambda/|v_{ph} - v_g|$ (Λ is the width of the pulse, v_{ph} and v_g are respectively the phase and group velocities) is sufficiently large: $\Delta t \gg \gamma_L^{-1}$ and the field amplitude along the trajectory of the electron does not change due to convective drift of the oscillations.

When this inequality is not satisfied, the electron gets appreciably displaced along the wave packet and there arises a need to take the form of the pulse into consideration. This becomes especially significant in the case investigated below,

$$\Delta t \ll \gamma_L^{-1}, \tag{4.6.1}$$

when the change in the field energy during collision with the electron is small and the damping of such a short pulse is determined by the successive interaction with a large number of resonance electrons transiting to it [8]. In the reference system of the phase velocity of the wave, in which the resonance electrons are at rest, a situation occurs similar to one of in the propagation of a short optical pulse in the resonance medium [20]; the role of the active atoms is played by the resonance electrons. For $t \gg \Delta t$ the pulse leaves each fixed point of space, but the information about it remains in the spatial distribution of electrons accelerated in the field of the pulse. Therefore, by reconstructing the distribution function along the trajectory of individual electrons it is possible to determine the change in the density of the resonance particles after the passage of the pulse. This approach leads to a differential equation for the integral characteristic of the pulse

$$S(x) = \int_{-\infty}^{\infty} E(t, x)\, dt$$

(area of the pulse [20]). The inverse of the relaxation length of the function $S(x)$ is the spatial scale of damping of the pulse in the reference system of the phase velocity of the wave.

According to [21,22], Landau damping becomes nonlinear for $t > \tau_{\rm ph} = \sqrt{m/eEk}$ due to the development of nonlinear oscillations the resonance electrons in the field of a monochromatic wave. But in our case of a spatially bounded pulse, an additional time parameter Δt appears and when the inequality

$$\Delta t \ll \tau_{\rm ph} \tag{4.6.2}$$

holds, the effects of phase oscillations do not appear during the transit time of the electron through the pulse.

However, the linearity of the trajectory of the electrons is no longer a criterion for the linearity of the decrement. Actually, the normally used series expansion

$$f(v_{\rm ph} - V_{NL}) \simeq f(v_{\rm ph}) - V_{NL} f'(v_{\rm ph}) \tag{4.6.3}$$

takes into consideration the nonlinearity of the distribution function caused by the nonlinear nature of motion of the electron in the wave field [21,22]. For a maxwellian function the condition of applicability of (4.6.3) is

$$v_{\rm ph} V_{NL}/V_T^2 \ll 1. \tag{4.6.4}$$

It is clear that in the region of small thermal velocities $V_T = \sqrt{2T_e/m}$ this condition is violated and distribution function $f(v_{\rm ph} - V_L)$ is nonlinear even for linear correction V_L to the initial velocity of resonance electrons [8]. Precisely this case, when in view of inequality (4.6.2) nonlinear phase oscillations do not appear and the decrement is nonlinear because of appreciable distortion of the distribution function in the field of a short Langmuir pulse, is discussed in this article.

We assume that the phase velocity of the wave is considerably larger than the thermal velocity of the plasma electrons $v_{\rm ph} \gg V_T$ and only a small group of resonance electrons with density $n_R \ll n_p$ require kinetic description. The main part of the electrons can be described by the system of hydrodynamic equations and Poisson's equation:

$$\begin{gathered}\frac{\partial}{\partial x}\left(\hat{F} + \omega_p^2\right)\mathcal{E}(t,x) = 4\pi e \hat{F}\, n_R(t,x),\\ \hat{F} = \left(\frac{\partial}{\partial t} - v_{\rm ph}\frac{\partial}{\partial x}\right)^2 - \frac{3}{2}V_T^2\frac{\partial^2}{\partial x^2},\end{gathered} \tag{4.6.5}$$

where $\omega_p^2 = 4\pi n_p e^2/m$, n_p is plasma density. In comparison with Eq. (4.1.1), a factor 3/2 takes into account kinetic nature of the Landau damping.

We shall seek the solution of (4.6.5) in the form of a pulse with envelope

$$\begin{gathered}\mathcal{E}(t,x) = \operatorname{Re} E(t,x)\exp(ikx),\\ E_t' \ll \omega E, \quad E_x' \ll kE,\end{gathered} \tag{4.6.6}$$

where ω and k are, respectively, the carrier frequency and the carrier wave number which satisfy the dispersion equation

$$\omega^2 = k^2 v_{\rm ph}^2 = \omega_p^2 + (3/2)k^2 V_T^2.$$

From (4.6.5) and (4.6.6) we obtain the equation for the field amplitude slowly varying with time and coordinate:

$$\left[\frac{\partial}{\partial t} + (v_g - v_{ph})\frac{\partial}{\partial x}\right] E(t,x) = -2e\omega_p \int_0^{\lambda} \exp(-ikx) n_R(t,x)\, dx, \tag{4.6.7}$$

where

$$v_g = \sqrt{3\epsilon/2}\, V_T,\ \epsilon = 1 - \omega_p^2/\omega^2,\ v_g v_{\rm ph} = (3/2)V_T^2.$$

The density of the resonance particles is determined by the method of integration of the kinetic equation along the trajectories [22]

$$n_R(t,x) = \int f[v_0(t,x,v)]\, dv, \tag{4.6.8}$$

where the integration is carried over the region $|v - v_{ph}| \le (k\Delta t)^{-1}$ and v_0 is the integral of motion of electron in a self-consistent field:

$$\ddot{x}(\tau) = \frac{e}{m} E[\tau, x(\tau)] \cos[kx(\tau)]. \tag{4.6.9}$$

For short pulses satisfying condition (4.6.2) the trajectory remains linear and is of the form

$$\begin{gathered} x(\tau) = x_0 + v_0(t - t_0), \quad v(\tau) = v_0 + V(\tau), \\ V(\tau) = \frac{e}{m} \int_{t_0}^{t} E[\tau', x(\tau')] \cos[kx(\tau')]\, d\tau'. \end{gathered} \tag{4.6.10}$$

From formulas (4.6.7), (4.6.8), and (4.6.10) we get

$$\begin{aligned} &\left[\frac{\partial}{\partial t} + (v_g - v_{ph})\frac{\partial}{\partial x}\right] E(t,x) \\ &= -\frac{2e\omega_p}{k} \int_0^{2\pi} d\varphi \int_{\Delta v} dv \exp[-i(\varphi + kvt)]\, f[v - V(t,x,\varphi,v)], \end{aligned} \tag{4.6.11}$$

where

$$V(t,x,\varphi,v) = \int_{t_0}^{t} E[\tau, x + v(t-\tau)] \cos(\varphi + kv\tau)\, d\tau.$$

Following [22], we eliminate the time dependence by integrating both sides of (4.6.11) in the limit t_0 to $t \to \infty$. Then considering that for $t \gg \Delta t$ the resonance electron has a weak interaction with the pulse passing by it, we put $t \to \infty$ in the formula for $V(t,x,\varphi,v)$ and take $f(t \to \infty)$ out of the sign of integral over time. After this the integral over velocities is computed using delta function:

$$\pi\delta(kv) = \lim_{t\to\infty} \int_{t_0}^{t} \exp[-ikv(t - t_0)]\, dt.$$

Next, considering the arbitrary nature of parameter t_0 we put $t_0 \to -\infty$ and in the left hand-side of the equation we drop the term $E(\pm\infty, x) = 0$. After these operations from (4.6.11), we get the differential equation

$$V'(x) = \frac{\omega_p^3}{2k^2 n_p (v_{ph} - v_g)} \int_0^{2\pi} f[v_{ph} - V(x)\cos\varphi] \cos\varphi\, d\varphi, \tag{4.6.12}$$

describing the spatial relaxation of the area of the pulse

$$V(x) = \frac{e}{m} \int_{-\infty}^{\infty} E(t, x)\, dt.$$

Thus, computing the increment of the velocity of resonance electrons in the reference system of the phase velocity of the wave, we obtain information about the passing pulse recorded in the phase memory of the accelerated particles. We note that formula (4.6.12) is an analog of the area theorem in optics [20].

We make general formula (4.6.12) specific by assuming the distribution function to be maxwellian:

$$\theta'(x) = \frac{\alpha_0}{2\pi q} \int_0^{2\pi} \exp(q\theta \cos\varphi - \theta^2 \cos^2\varphi) \cos\varphi, \tag{4.6.13}$$

$$\theta = V/V_T, \quad \alpha_0 = \gamma_L/(v_{ph} - v_g), \quad q = v_{\rm ph}/V_T,$$

where $\gamma_L = \sqrt{\pi}\omega_p q^3 \exp(-q^2)$ is Landau damping [7].

As already noted above, if parameter $q\theta \simeq 1$ and increment of electron velocity V is comparable to group velocity v_g, the exponent in (4.6.13) cannot be expended in series and the spatial distribution of the field is nonlinear. Writing the exponent in the form of series of modified Bessel functions [18] and carrying out integration we determine the spatial decrement in the following way:

$$\alpha(x) = \frac{\theta'}{\theta}$$
$$= \frac{\alpha_0}{q\theta} \exp\left(-\frac{\theta^2}{2}\right) \sum_{n=-\infty}^{\infty} (-1)^n I_n\left(-\frac{\theta^2}{2}\right) I_{2n-1}(2q\theta). \tag{4.6.14}$$

Equation (4.6.14) has the trivial solution $\theta' = \theta = 0$; hence Langmuir pulse with zero area propagates in the plasma without change of the area, since the energy of the electron transiting through it does not change.

If $\theta^2 \ll 1$, only one term $n = 0$ could be retained in the sum in (4.6.14):

$$\alpha(\theta) = \alpha_0 \frac{I_1(2q\theta)}{q\theta}. \tag{4.6.15}$$

Therefore, the decrement is linear $\alpha(\theta) \simeq \alpha_0$ only for $2q\theta \ll 1$, i.e., for $V \ll v_g$ in dimensional variables.

For large field amplitudes $2q\theta \gg 1$ functions $I_{2n-1}(2q\theta)$ can be replaced by their asymptotic forms and taken out of the summation sign. The summation of the series thus left leads to the following asymptotic formula:

$$\alpha(\theta) = \frac{\alpha_0}{\sqrt{4\pi q^3\theta^3}} \exp(2q\theta - \theta^2). \tag{4.6.16}$$

We note that by definition decrement is negative and the perturbation developing at the point $x = 0$ gets damped in the region $x < 0$, where the pulse propagates with velocity $v_g - v_{ph} < 0$.

The form of the pulse may be made specific by writing the amplitude of the field in the form

$$E(t, x) = G(x) F[(x + (v_{\rm ph} - v_g)t)/\Lambda],$$
$$S(x) = G(x)\Delta t \int_{-\infty}^{\infty} F(\xi) d\xi, \quad \Delta t = \Lambda/(v_{ph} - v_g). \tag{4.6.17}$$

In the absence of resonance particles $n_R = 0$ and of second derivatives of the amplitude, which take account of the dispersion spreading of the pulse and which were dropped in going from (4.6.5) to (4.6.7), form-factor F is arbitrary and function $G(x)$ is given by Eq. (4.6.14). It is obvious that this wave representation has physical meaning only when inequality (4.6.1) is satisfied, since otherwise the pulse damps earlier than it gets formed. In the reference system of the plasma $x' = x + v_{\text{ph}}t$, where the pulse moves with velocity v_g along x' axis, the wave has space-time damping.

Appendix 4.1

After integrating Eq. (4.2.12), we find the solution (4.2.13),

$$\int (\rho'^2/\mu^2 + 4\rho y + 2y^2 - y^4)^{-1/2} dy = \pm(\zeta - \zeta_0)/\sqrt{2}. \tag{A4.1.1}$$

The form of the analytic expression for the elliptic integral on the left side of (A4.1.1) is determined by the roots of the denominator:

$$y^4 - 2y^2 - 4\rho y - \rho'^2/\mu^2 = 0. \tag{A4.1.2}$$

According to Ref. [17], these roots are expressed in terms of the roots of the cubic equation

$$z^3 - z^2 + (1 + 2\rho'^2/\mu^2)z/4 - \rho^{2/4} = 0, \tag{A4.1.3}$$

which are

$$\begin{gathered} z_{1,2,3} = t_{1,2,3} + 1/3, \\ t_1 = A + B, \quad t_{2,3} = \frac{1}{2}(A+B) \pm i\frac{\sqrt{3}}{2}(A - B), \\ A, B = (v - u \pm \sqrt{Q})^{1/3}, \quad Q = (2u)^3 + (v - u)^2, \\ v = (\rho - 4/27)/8, \quad u = (\rho'^2/\mu^2 - 1/6)/12. \end{gathered} \tag{A4.1.4}$$

In the region of fast oscillation the inequalities $v > 0$ and $u \geq -1/72$ hold. These inequalities make the discriminant Q positive. This case corresponds to a situation in which Eq. (A4.1.3) has one real root z_1 and two complex-conjugate roots z_2, z_3. Equation (A4.1.2) in this case has two real roots and two complex-conjugate roots, which are given by [17]

$$\begin{gathered} \alpha = -\sqrt{z_1} + \sqrt{z_2} + \sqrt{z_3}, \quad \beta = -\sqrt{z_1} - \sqrt{z_2} - \sqrt{z_3}, \\ m \pm in = \sqrt{z_1} \pm (\sqrt{z_2} - \sqrt{z_3}). \end{gathered} \tag{A4.1.5}$$

Using (A4.1.5) and Ref. [9], we can put the integral (A4.1.1) in the form of (4.2.13). The relations presented in (4.2.14) can be found from (4.2.13) by simple transformations.

The derivation of Eq. (4.2.16) for the slowly varying function ρ involves averaging the function $y(\zeta)$ over the period of the nonlinear oscillations, T:

$$\langle y \rangle = T^{-1} \int_0^T y \, d\zeta, \quad T = \int_0^T d\zeta, \tag{A4.1.6}$$

where $y(\zeta)$ is given by (4.2.14). Introducing the new integration variable φ by means of relation

$$\frac{d\zeta}{d\varphi} = \kappa^{-1} \left(1 - k^2 \sin\varphi\right)^{-1/2} \tag{A4.1.7}$$

[k^2 and κ are given by (4.2.13) and (4.2.14)], we find

$$T = \kappa^{-1}\int_0^{2\pi} \left(1 - k^2 \sin\varphi\right)^{-1/2} d\varphi,$$

$$\langle y\rangle = -(t+1/3)^{1/2} + \frac{1}{\kappa T}(2/3 - t + 2R)^{1/2} \times \int_0^{2\pi} \frac{a - \cos\varphi}{1 - a\cos\varphi}\frac{d\varphi}{\left(1 - k^2\sin\varphi\right)^{1/2}}. \tag{A4.1.8}$$

After some simple calculations we find

$$T = 4\kappa^{-1}K(k),$$
$$\langle y\rangle = -(t+1/3)^{1/2} + a^{-1}(2/3 - t + 2R)^{1/2}[1 - \Pi(n,k)/K(k)], \tag{A4.1.9}$$

where $\Pi(n,k)$ is the complete elliptic integral of the third kind, and

$$n = -a^2/(1-a^2) < 0. \tag{A4.1.10}$$

A transformation of the negative parameter $n < 0$ to the interval $0 \div 1$ (Ref. [18]) yields relation (4.2.16).

Appendix 4.2

Using the dimensionless variables (4.3.2), we rewrite Eqs. (4.4.6) in the form

$$Y'' + ivY' + (\kappa + |Y|^2)Y = \frac{2i}{Q}\sum_s e^{-i\psi_s} \tag{A4.2.1}$$

$$\psi_s'' = -Q\,\mathrm{Re}\,\left(Ye^{i\psi_s}\right), \tag{A4.2.2}$$

where

$$Q = \frac{eE_T\lambda_d}{mv_0^2}\frac{k_0}{\Delta k} \simeq 2^{3/2}\left(\frac{n_p}{n_b}\right)^{1/4}\left(\frac{v_T}{v_0}\right)^{5/2}, \quad \sum_s \equiv S^{-1}\sum_{s=1}^{S}.$$

In going from (4.4.6) to (A4.2.2), the terms of order $(\Delta k/k_0)\psi_s'$ in the right-hand side of the equation are omitted.

Multiplying (A4.2.1) by $(Y^*)'$ and putting these results together with complex–conjugate expressions, we find

$$\frac{Q}{4}\left[|Y'|^2 + \left(\kappa + \frac{|Y|^2}{2}\right)|Y|^2\right]' = \mathrm{Re}\left(iY'^*\sum_s e^{-i\psi_s}\right). \tag{A4.2.3}$$

Multiplying (A4.2.2) by ψ_s', we rewrite this expression in the form

$$-\frac{1}{2Q}\sum_s (\psi_s'^2)' = \mathrm{Re}\left[Y^*\sum_s \left(ie^{-i\psi_s}\right)'\right]. \tag{A4.2.4}$$

Adding (A4.2.3) and (A4.2.4), and carrying out the integration, we obtain the energy integral

$$\frac{Q}{2}\left[|Y'|^2+\left(\kappa+\frac{|Y|^2}{2}\right)|Y|^2\right]-\sum_s\left[\frac{1}{Q}\psi_s'^2+2\mathrm{Re}\,\left(iY^*e^{-i\psi_s}\right)\right]=C_1. \tag{A4.2.5}$$

Multiplying (A4.2.1) by Y^* and subtracting these results from complex-conjugate expressions, we find

$$Y^*Y''-Y(Y^*)''+v|Y'|^2=\frac{4i}{Q}\mathrm{Re}\left(Y\sum_s e^{i\psi_s}\right) \tag{A4.2.6}$$

Summing (A4.2.2) and (A4.2.6), and carrying out the integration, we find the momentum conservation law

$$-2\mathrm{Re}\,(iY^*Y')+v|Y'|^2+\frac{4}{Q^2}\sum_s\psi_s'=C_2. \tag{A4.2.7}$$

To go in (A4.2.1) and (A4.2.2) to the linear beam, let us put

$$\psi_s=\psi_{so}+\psi_{s1},\quad \psi_{so}=2\pi s/S,\quad |\psi_{s1}|\ll\psi_{so},$$

and fulfill the expansion of the exponentials in the right-hand sides of equations in series in ψ_{s1}. Introducing the perturbation of beam density,

$$R=\frac{2}{Q}\sum_s\psi_{s1}\,e^{-i\psi_s}, \tag{A4.2.8}$$

we get the equations (4.3.4) and (4.3.5):

$$Y''+ivY'+(\kappa+|Y|^2)Y=R,\quad R''=-Y. \tag{A4.2.9}$$

The limit passing from (A4.2.5) and (A4.2.7) to integrals (4.3.6) and (4.3.7) can be done by means of the expressions

$$\frac{2}{Q^2}\sum_s\psi_s'^2=|R'|^2,\quad \frac{2}{Q^2}\sum_s\psi_s'=\mathrm{Re}\,(iR^*R') \tag{A4.2.10}$$

which follow from (A4.2.2) and (A4.2.9).

CHAPTER 5

ELECTROMAGNETIC SOLITONS IN RETARDING DISPERSIVE MEDIA WITH A CHARGE PARTICLE BEAM

The nonlinear parabolic equation [1]

$$\frac{v_g'}{2}\frac{\partial^2 E}{\partial z^2} + i\left(\frac{\partial E}{\partial t} + v_g\frac{\partial E}{\partial z}\right) - \alpha|E|^2 E = 0, \tag{5.1}$$

can often be used to describe electromagnetic wave propagation in nonlinear dispersive materials. Here E is the complex amplitude of the electric field; the coefficient α characterizes the nonlinearity of the medium; and the group velocities v_g and $v_g' = dv_g/dk$ are determined by the linear index of refraction $ck/\omega = n(\omega)$. For example, Eq. (5.1) can be used to calculate the properties of Langmuir solitons in a plasma which satisfy Lighthill's condition [2] : $\alpha v_g' < 0$.

However, the problem becomes more difficult if resonant particles are present; in this case the refractive index $n(\omega)$ has a singularity and the group velocities v_g and v' are consequently not defined at the resonant frequency. The parabolic equation (5.1) becomes meaningless, and the waveform and the nonlinear dispersion law must be found by solving a self-consistent system of equations [3-10]. This situation occurs for electromagnetic soliton waves in two-level system (in which case the poles $n(\omega)$ correspond to the frequency of the transitions between the levels [3,4]) or to the Cherenkov's soliton in a plasma with an electron beam (see Chap. 4). An analogue of the latter solution is the stationary solitary electromagnetic wave with the carrier frequency and carrier wave number satisfying the conditions of the normal or anomalous Doppler effect [5-10]. Our results obtained in this chapter can be used to calculate the group and phase velocities near resonance and eliminate the singularity in the equations in the linear theory.

§5.1. Propagation of solitons in the presence of the anomalous Doppler effect.

The small monochromatic electromagnetic perturbations are unstable in plasma-beam systems and grow exponentially with time under the conditions of resonance; the wave number k_0 fixed and the wave amplitude is uniform in space along the beam. The nonlinear stage of the interaction of mono-energetic beams with a plasma (the retarding medium) is accompanied with the appearance of oscillations in time of the field amplitude which are caused by the trapping of the beam particles by the field of the wave and their periodic shift from decelerating to accelerating phases. According to [5], a trapping of the beam particles by the wave also is a cause for the existence of a solitary-wave for the field amplitude in the case of narrow wave packet, $\Delta k \ll k_0$.

We consider an arbitrary dispersive retarding medium with refractive index $n(\omega)$ through which a charged particle beam propagates along a constant external magnetic field $\mathbf{B}_0$ Since practically the whole information about the conditions for the existence of nonlinear waves of the kind

$$E(t,x) = \mathrm{Re}\,[A(\xi)\exp(i\omega t - ikz)], \quad \xi = z - ut \tag{5.1.1}$$

is contained in the linear dispersion equation and allowance for the nonlinearity only enables one to determine the maximum amplitude and to find the shape of the wave pulse, we consider first the problem in the linear approximation.

The dispersion equation connecting the frequency ω and the wave number k has the form [5]:

$$\left(\frac{ck}{\omega}\right)^2 = n^2(\omega) - \frac{\omega_b^2(\omega - kv_0)}{\omega^2(\omega - kv_0 + \omega_B)}, \tag{5.1.2}$$

where $\omega_b^2 = 4\pi n_b e^2/m$, $\omega_B = |e|B_0/mc$, and n_b, v_0, and m are the density, initial velocity, and mass of the beam.

We look for a solution of (5.1.2) in the form

$$\omega = \omega_0 + \Delta\omega, \quad k = k_0 + \Delta k,$$

assuming that ω_0 and k_0 satisfy the resonance conditions:

$$ck_0 = \omega_0 n_0(\omega_0), \quad \omega_0 - k_0 v_0 = -\omega_B. \tag{5.1.3}$$

The small corrections to the frequency and wave number are then connected by the relation

$$c^2(2k_0\Delta k + \Delta k^2) = \frac{d}{d\omega_0}(\omega_0^2 n_0^2)\,\Delta\omega + \frac{d^2}{d\omega_0^2}(\omega_0^2 n_0^2)\,\frac{\Delta\omega^2}{2} + \frac{\omega_b^2\omega_B}{\Delta\omega - v_0\Delta k} - \omega_b^2. \tag{5.1.4}$$

in the case of monochromatic wave with a time-dependent amplitude which is excited in the dispersionless medium by a low-density beam when the conditions of the anomalous Doppler effect are satisfied we have

$$\omega_0 = \frac{\omega_B}{\beta_0 n_0 - 1}, \quad \Delta\omega^2 = -\frac{\omega_b^2\omega_B}{2\omega_0 Q}, \quad \Delta k = 0, \tag{5.1.5}$$

where $\beta_0 = v_0/c$ and $Q = d(\omega_0^2 n_0^2)/d\omega_0^2$.

Taking dispersion into account enables us to find the condition for the propagation of finite-width wave packets. Putting $\Delta\omega = u\Delta k$ in (5.1.4) we get

$$\Delta k^2 = -\frac{v_g}{2nc}\frac{\omega_B}{\omega_0}\frac{\omega_b^2}{(u - v_g)(u - v_0)}, \tag{5.1.6}$$

$$1 - \frac{u^2}{v_g^2}\left(1 - \frac{k_0 v_g'}{v_g}\right) + \left(\frac{\omega_b}{c\Delta k}\right)^2 = 0, \tag{5.1.7}$$

where

$$v_g = c\Big[\frac{d}{d\omega_0}(\omega_0 n_0)\Big]^{-1}, \quad v_g' = \frac{dv_g}{dk_0}, \quad 1 - \frac{k_0 v_g'}{v_g} = \frac{v_g^2}{2c^2}\frac{d^2}{d\omega_0^2}(\omega_0^2 n_0^2).$$

From Eq. (5.1.6), which generalizes Eq. (5.1.5), it follows that aperiodic perturbations corresponding in the nonlinear approximation to soliton solutions exist when $u > v_g$, $u > v_0$ or when $u < v_g$, $u < v_0$; in both cases the condition $\beta_0 n_0 > 1$ is fulfilled. Relation (5.1.7) is obtained in the next approximation in the beam density and is the condition that there is no dispersive spreading of the wave packet. One must note that when there is no beam, $\omega_b = 0$, such stationary solutions are not realized and wave packets in the medium with $v_g' \neq 0$ spread out due to dispersion, since Eq. (5.1.4) is equivalent to a parabolic equation (5.1).

We find from Eqs. (5.1.6) and (5.1.7)

$$u_\pm = \frac{v_g + v_0 \pm \sqrt{(v_g - v_{\rm ph})^2 + (v_0^2 - v_{\rm ph}^2)(k_0 v_g'/v_g)}}{2 + v_g^{-1}(v_0 - v_{\rm ph})(1 - k_0 v_g'/v_g).} \tag{5.1.8}$$

We shall look for the solution of the nonlinear problem in the form of a circularly polarized plane wave with a slowly varying amplitude (5.1.1), using the presence of the small parameter

$$\delta = |\Delta k(u - v_0)|/\omega_B \ll 1. \tag{5.1.9}$$

The calculations necessary to derive the set of slow equations are then in principle not different from those for the amplitudes and phases which depend only on the time [5]. However, now the equations depend on the parameter u which we can determine by retaining in the equation for the field the next terms in the expansion in the small parameter (5.1.9):

$$\begin{aligned}(v_g - u)\frac{dE}{d\xi} &= -2\pi e n_b \frac{v_g}{c n_0}\left[1 + \frac{v_0 n_0}{c}\frac{(c - n_0 u)(v_0 - v_{||})}{(c - n_0 v_0)(v_0 - u)}\right] v \\ &+ i\frac{2\pi e n_b v_g}{k_0 c^2}\left\{\frac{v_g c}{2n_0(v_g - u)}\left[1 - \frac{u^2}{v_g^2}\left(1 - \frac{k_0 v_g'}{v_g}\right)\right] - \frac{c(v_0 - u)}{c - n_0 v_0}\right\}\frac{dv_\perp}{d\xi}.\end{aligned} \tag{5.1.10}$$

One sees easily that the expression in braces vanishes if the group velocity u satisfies Eqs. (5.1.8). The coefficient of $v_g - u$ in the principle term in (5.1.10) is thereby determined and the correction to unity turns out to be small and can be dropped, $v_0 - v_{||} \simeq \delta v_0$.

A set of equations for the slowly varying amplitudes can be found from the analogous formulas (5.1.7) by means of replacement: $dt \to d\xi/(v_{||} - u)$. In the dimensionless variables

$$v_\perp/c = a\exp(i\vartheta), \quad E/B_0 = \mathcal{E}\exp(i\varphi), \quad b = v_{||}n_0/c - 1,$$
$$d\zeta = \omega_B d\xi/(v_0 - u), \quad \Omega = \omega_0/\omega_B, \quad q^2 = \delta^2\Omega$$

the set of nonlinear equations takes the form

$$\begin{aligned} a' &= -b\mathcal{E}\cos\eta, \quad b' = n_0^2 a\mathcal{E}\cos\eta, \\ \mathcal{E}' &= -q^2 a\cos\eta, \quad \eta' = 1 - \Omega b - \tan\eta\,(\ln\mathcal{E}a)', \end{aligned} \tag{5.1.11}$$

and is the same as Eqs. (5.1.7) and (5.1.9), if we change in the latter to the nonrelativistic limit, $\beta_0 \ll 1$ and $n_0 > 1/\beta_0$. However, this agreement is formal, as we are considering a totally different kind of solution.

We look for a soliton-like solution of equations (5.1.11) satisfy the conditions

$$\mathcal{E} = \eta = a = 0, \quad b_0 = \Omega^{-1}. \tag{5.1.12}$$

Using the integrals of motion of Eqs. (5.1.11)

$$n_0^2 a^2 + b^2 = \Omega^{-2}, \quad \mathcal{E}^2 = 2q^2(\Omega^{-1} - b)/n_0^2,$$
$$\sin\eta = -(1 - \Omega b)^2 / 2a\mathcal{E}\Omega n_0^2, \tag{5.1.13}$$

we get the following equation for the field amplitude:

$$\mathcal{E}' = \delta\mathcal{E}\sqrt{1 - \mathcal{E}^4/\mathcal{E}_m^4}, \quad \mathcal{E}_m^2 = (2\delta)^3/n_0^2. \tag{5.1.14}$$

The solution of Eq. (5.1.14) has the form of soliton:

$$\mathcal{E} = \mathcal{E}_m \cosh^{-1/2}(2\delta\zeta). \tag{5.1.15}$$

Note that the parameter δ which determines the width and amplitude of the envelope pulse is given by Eq. (5.1.9) of the linear theory.

The physical reason for the existence of a stationary wave in a non-equilibrium medium can be understood by considering the process of the interaction of resonance particles with the wave in the soliton frame of reference where the beam moves with the speed $v_{\|} - u$. Since the frequency of the field is in resonance, a particle incident on the trailing edge of the wave will then start to slow down in the longitudinal direction, and its transverse energy increases. This process, accompanied by the transformation of the energy of the longitudinal motion into wave energy, stops when the particle reaches maximum of the wave amplitude, since at that point the difference in phases between the transverse velocity and the electrical field vectors, due to the detuning of the anomalous Doppler resonance as the result of the longitudinal retardation, reaches the value $\eta_{\mathrm{m}} = \pi/2$. After this the particle reaches a retarding phase, loses transverse energy, and accelerates along the magnetic field to reach its initial velocity.

We thus find that, on the one hand, the beam constantly replenishes the energy of the wave, and, on the other hand, there is no exchange of energy of the beam particles with the wave over the time which is larger than the time during which the particle passes through the field pulse. As a result, the energy density of a resonance wave pulse propagating in non-equilibrium medium remains unchanged.

Let us consider a helicon soliton in a plasma when there is an ion beam present. The refractive index of the plasma in this case equal to

$$n^2(\omega) = \omega_e^2/\omega\omega_{Be} \gg 1, \quad \omega_{Bi} \ll \omega \ll \omega_{Be},$$
$$\omega_e^2 = 4\pi e^2 n_p/m, \quad \omega_{Be} = |e|B_0/mc, \quad \omega_{Bi} = (m/M)\omega_{Be}. \tag{5.1.16}$$

(ω_e is the plasma frequency, ω_{Be} and ω_{Bi} are the electron and ion gyrofrequencies of the plasma). Substituting (5.1.16) in (5.1.5), we get the equation for the carrier frequency, ω_0,

$$\frac{\omega_{Bi}}{\omega_0} - \frac{v_0}{c_A}\sqrt{\frac{\omega_{Bi}}{\omega_0}} + 1 = 0, \tag{5.1.17}$$

where $c_A = B_0^2/\sqrt{4\pi e n_p M}$ is the Alfven velocity. The solution of Eq. (5.1.17) is

$$\sqrt{\frac{\omega_{Bi}}{\omega_0}} = \frac{v_0}{2c_A} - \sqrt{\frac{v_0^2}{4c_A^2} - 1} \simeq \frac{c_A}{v_0} \ll 1. \tag{5.1.18}$$

We find from Eqs. (5.1.16), (5.1.18) and (5.1.8), we get [4)]

$$
\begin{gathered}
v_g = 2c/n_0, \quad u_+ = v_0 + c/n_0, \quad u_- = c/n_0, \\
n_0 = \frac{c}{c_A}\sqrt{\frac{\omega_{Bi}}{\omega_0}}.
\end{gathered}
\tag{5.1.19}
$$

The characteristic sizes of the pulses are then of the same magnitude:

$$
|\Delta k_\pm| = \sqrt{\frac{n_b}{n_p}}\frac{\omega_{Bi}}{c_A}, \qquad k_0 = \sqrt{\frac{\omega_0}{\omega_{Bi}}}\frac{\omega_{Bi}}{c_A}. \tag{5.1.20}
$$

The maximum values of the electric field amplitude of each of the waves are equal to

$$
(E_\pm)_m = \left(4\frac{n_b}{n_p}\right)^{3/4}\frac{B_0}{n_0}\begin{cases} \left(\dfrac{\omega_0}{\omega_{Bi}}\right)^{3/4}, & u = u_+, \\ \left(\dfrac{\omega_0}{\omega_{Bi}}\right)^{-3/4}, & u = u_-. \end{cases} \tag{5.1.21}
$$

We note that the ion beam excites in the plasma short-wavelength plasma oscillations, if its velocity v_0 exceeds the thermal velocity v_{Te}. The condition for the applicability of Eqs. (5.1.16) to (5.1.20) is thus the inequality

$$
v_{Te} \geq v_0 \geq 2c_A. \tag{5.1.22}
$$

The propagation of the helicon soliton in a nonlinear plasma along an ion beam is considered in Ref. [6]. According to [6], the refractive index of the plasma in this case equal to $n^2(\mathcal{E}) = n_0^2 + \alpha\mathcal{E}^2$ and increases in proportion to $\mathcal{E}^2$ because of $\alpha > 0$. As a result, the phase velocity of the wave, $v_{\text{ph}}(\mathcal{E}) = c/n(\mathcal{E})$, in a center of a pulse decreases so that at this point the additional difference in phases between the transverse velocity and the electric field vector occurs.

The results obtained are of interest in connection with the possibility to generate electromagnetic pulses in the circumterrestrial plasma by solar corpuscular currents (plasma of solar wind), and also for the transmission of weakly damped regular electromagnetic signals in a plasma along charged particle beams. It is necessary in the latter case that the time growth rate $\Delta\omega = u|\Delta k|$ exceeds the damping of the wave in a plasma.

§5.2. Helicon soliton with trapped electrons in an ion-beam plasma.

In this section we represent a possible modification of found in §5.1 solution,

$$
\begin{gathered}
E_x + E_y = E_0 \cosh^{-1/2}(2|\Delta k|\xi)\exp(i\omega_0 t - ik_0 z), \\
n_0 E_0 = \left(\frac{2q}{\sqrt{\Omega}}\right)^{3/2} B_0, \quad \Omega = \frac{\omega_0}{\omega_{Bi}} = \left(\frac{n_0 v_0}{c} - 1\right)^{-1}, \\
q^2 = \frac{\omega_b^2}{2\omega_{Bi}^2}\frac{v_g}{c n_0}\frac{v_0 - u}{v_g - u}, \quad \Delta k^2 = -\frac{q^2}{\Omega}\left(\frac{\omega_{Bi}}{v_0 - u}\right)^2, \\
\omega_0 - k v_0 = -\omega_{Bi}, \quad n_0 = c k_0/\omega_0,
\end{gathered}
\tag{5.2.1}
$$

[4)] Note that in our case the electric field vector rotates in synchronism with the transverse velocity of ion beam. Therefore, in contrast to Ref. [11], we consider a resonance ion helicon when the plasma electrons ensure a slowing of the phase velocity of the wave.

in connection with the trapping of plasma electrons by the uniform RF-field of the soliton pulses [7].

Since the electron gyrofrequency is "negative", the gradient force

$$F_z = -\frac{d\Phi}{dz'}, \quad \Phi = \frac{e^2 E^2(z')}{2m\omega'(\omega' - \omega_{Be})} \simeq -\frac{e^2 E^2(z')}{2m\omega'\omega_{Be}} \tag{5.2.2}$$

draws electrons toward the field maximum if the wave frequency in the frame of reference of the group velocity is positive, $\omega' = \omega_0 - k_0 u > 0$ and $\omega \simeq \omega_{Bi} \ll \omega_{Be}$ (**see Appendix 5.1**).

In tern, the presence of the trapped electrons is accompanied by an increase in the refractive index of the plasma in the pulse propagation region. Since the field amplitude reaches saturation because the beam and wave go out of phase as a result of the longitudinal slowing of the ions in the wave field, it is clear that the simultaneous decrease in the phase velocity near the field maximum should "streach out" the synchronization of the beam with the wave and should increase the maximum soliton amplitude. In this paper we show that there is a certain optimum number density of the trapped electrons at which the decrease in the wave phase velocity is synchronized with the slowing of the beam, and the soliton amplitude is significantly higher than in Eq. (5.2.1).

We follow Bohm and Gross (Ref. [12]) in choosing the longitudinal part of the electron distribution function in a circularly polarized plane wave with a coordinate-dependent amplitude. It is an arbitrary function of integral of motion

$$W_{||} = \Phi - e\varphi + \frac{mv_z^2}{2}, \tag{5.2.3}$$

where φ takes into account the Coulomb field of a trapped ion bunch. Since the trapped particles correspond to $W_{||} < 0$, we put

$$f(z', \mathbf{v}_z) = \frac{2\rho_m}{\pi v^2(\Phi_m, \varphi_m)}\sqrt{v^2(\Phi, \varphi) - v_z^2}, \tag{5.2.4}$$

where $v^2(\Phi, \varphi) = 2|\Phi - e\varphi|/m$, ρ_m and φ_m are the electron density and potential at the bottom of the RF-potential well, Φ_m.

Integrating (5.2.4) over the velocity, $v_z \leq v(\Phi, \varphi)$, we find the trapped-particle density to be

$$\rho_e = \rho_m \frac{\Phi - e\varphi}{\Phi_m - e\varphi_m}. \tag{5.2.5}$$

The Coulomb potential φ describes the electric field which is produced by the trapped electrons and which is screened by the plasma flow in the coordinate system of the wave. The heavy ions are moving uniformly at a velocity $-u$, and the field dependence of the plasma electron density n_e (untrapped particles) is determined by the system of integrals of motion,

$$\frac{mV_z^2}{2} + \frac{\omega'}{\omega}\Phi - e\varphi = \frac{mu^2}{2}, \quad n_e V_z = -\rho_p u, \tag{5.2.6}$$

where V_z is the electron velocity in the coordinate system of the wave, n_p is the unperturbed plasma density, and $(\omega'/\omega)\Phi$ is the RF-potential which acts on an untrapped electron.

Using the charge neutrality of the plasma, $n_e - n_p + \rho_e = 0$, we can express φ in terms of Φ. Assuming $|(\omega'/\omega)\Phi - e\varphi| \ll mu^2$, we find

$$\rho_e = \rho_m \frac{\Phi}{\Phi_m}, \quad e\varphi = \left(\frac{\omega'}{\omega}\Phi_m + \frac{\rho_m}{n_p}mu^2\right)\frac{\Phi}{\Phi_m}. \tag{5.2.7}$$

Note that equations (5.2.7) are rigorous, not based on the assumption that $e\varphi$ is small in comparison with Φ.

The distribution function (5.2.4) can be used to find the current density produced by the trapped particles as a function of the wave amplitude. When substituded into Maxwell's equation, this current adds up with the plasma current and gives rise to an nonlinear change in the refracting index of the plasma:

$$n^2(E) = n_0^2\left(1 + \frac{\omega'}{\omega}\frac{\rho_m}{n_p}\frac{E^2}{E_m^2}\right), \tag{5.2.8}$$

where E_m is the maximum electric field amplitude.

Since the presence of the trapped particles can be taken into account by means of a nonlinear increment in the refracting index, the problem is mathematically equivalent to that of the propagation of free solitons (5.1.14) in a nonlinear plasma (see Ref. [3] in [7]) and the corresponding equation for the electric field amplitude of the wave can be presented in the form:

$$\begin{gathered}\mathcal{E}' = \mathcal{E}\sqrt{1 - \frac{\mathcal{E}^4}{C^2}(B + A\mathcal{E}^2)^2}, \\ A = \frac{5\hat{\alpha}}{3Q}n_0^2, \quad B = n_0^2 - \frac{\alpha}{Q}q^2, \quad C = 8\delta^3, \\ \zeta = |\Delta k|\xi, \quad \delta = q/\sqrt{\Omega},\end{gathered} \tag{5.2.9}$$

where the prime means the derivative with respect to the variable ζ, and the remaining variables are equal to

$$\alpha = \frac{\rho_m}{n_p}\frac{n_0^2}{\mathcal{E}_m^2}\left(1 - \frac{u}{v_{ph}}\right), \quad \hat{\alpha} = \frac{\rho_m}{2n_p}\frac{n_0^2}{\mathcal{E}_m^2}\left(1 - \frac{u}{v_g}\right),$$

$$v_g = 2v_{ph}, \quad Q = \frac{n_0^2}{2}, \quad n_0^2 \gg 1.$$

To proceed, we rewrite Eq. (5.2.9) in the form

$$\begin{gathered}y' = y\sqrt{1 - y^4\left(1 - \frac{a}{y_m^2} + b\,\frac{y^2}{y_m^2}\right)^2}, \\ y = \frac{\mathcal{E}}{\mathcal{E}_0}, \quad a = \frac{\sqrt{\Omega^3}}{4q}\frac{\rho_m}{n_p}\left(1 - \frac{u}{v_{ph}}\right), \quad b = \frac{5\rho_m}{3n_p}\left(1 - \frac{u}{v_g}\right),\end{gathered} \tag{5.2.10}$$

where $\mathcal{E}_0^2 = 8\delta^3/n_0^2$ corresponds to a free soliton.

Taking into account that the maximum field amplitude satisfy the condition $y'(y_m) = 0$, we represent the right-hand side of Eq. (5.2.10) in the form [5)]

$$y' = y\sqrt{1 - \frac{y^4}{y_m^4}\left(\frac{1-ab}{1+b} + by^2\right)^2}, \quad y_m^2 = \frac{1+a}{1+b}. \tag{5.2.11}$$

---

5) In the case of the wave with a trapped electrons $\rho_m > 0$ $(b>0)$ this equation

$$X' = X[1 - (X/X_m)^4(1 - X_m^2 + X^2)^2]^{1/2}, \quad X = \sqrt{b}y$$

depends on one and only parameter, $X_m = \sqrt{b}y_m$.

It follows from (5.2.11) that the nature of the slution depends on one parameter ab. If $ab \ll 1$, the last term on the right-hand side of Eq. (5.2.11) can be omitted, and the soliton has the form in (5.2.1). At $|ab-1| \simeq 1$ this effect is important, and the shape of the soliton is different from (5.2.1).

The optimum case $ab \to 1$, in which the nonlinear decrease in the phase velocity to a large extent cancels the phase excursion of the wave due to the longitudinal slowing of the beam, and the ions which have passed through the soliton pulse remain in phase with the wave longer than in case (5.2.1). Setting $ab = 1$ in (5.2.11), we find

$$y(\zeta) = y_m \cosh^{-1/4}(4\zeta),$$
$$y_m = \sqrt{a}, \quad |ab - 1 \ll 1.$$

Substituting the parameters a and b from Eqs. (5.2.10), we obtain

$$\mathcal{E}_m^2 = \frac{2q^2}{n_0^2}\frac{\rho_m}{n_p}\left(1 - \frac{u}{v_{\text{ph}}}\right), \quad v_g = 2v_{\text{ph}},$$
$$\frac{5\sqrt{\Omega^3}}{12q}\frac{\rho_m^2}{n_p^2}\left(1 - \frac{u}{v_{\text{ph}}}\right)\left(1 - \frac{u}{v_g}\right) = 1. \tag{5.2.12}$$

Note that these formulas differ from the analogous expressions of Ref. [7] which found in the absence of dispersia by the term "2" and $v_g \neq v_{\text{ph}}$.

It follows from (5.2.12) that group velocity of the wave is

$$1 - \frac{u}{v_{\text{ph}}} = \frac{1}{2}(\sqrt{1+8d} - 1), \quad d = \frac{12}{5}\frac{q}{\sqrt{\Omega^3}}\frac{n_p^2}{\rho_m^2}. \tag{5.2.13}$$

An obvious inequality $u > 0$ corresponds $d < 1$,

$$\rho_m > \rho^* = \sqrt{\frac{12q}{5\Omega^{3/2}}}\, n_p. \tag{5.2.14}$$

Comparing (5.2.12) and (5.2.13), we find

$$n_0^2\mathcal{E}_m^2 = \sqrt{\frac{12q^5}{5\Omega^{3/2}}}\frac{\sqrt{1+8d} - 1}{\sqrt{d}}. \tag{5.2.15}$$

The condition for the gradient approximation, $\dot{E} \ll \omega' E$, places a lower limit on the difference between the group, u, and phase $v_{\text{ph}} = \omega_0/k_0$ velocities (**Appendix 5.1**):

$$\left(1 - \frac{u}{v_{\text{ph}}}\right)^{3/2} \gg \frac{eE_m k_0}{m\sqrt{\omega_B \omega_0^3}}. \tag{5.2.16}$$

The effects discussed above are of interest in connection with astrophysical applications since they represent a possible mechanism for the acceleration of electrons in the interplanetary medium and the plasma environment of the earth, in which trehe are streams of solar protons. Since the velocity of the electrons trapped at the bottom of the potential well varies over the range

$$u_{\mp} = u \mp \sqrt{\frac{2|\Phi_m|}{m}},$$

we find the maximum increase in the energy of the accelerated particles to be

$$\Delta W = \frac{m}{2}(u_+^2 - u_-^2) \simeq \frac{2}{\omega_p}\frac{ecE_m}{\sqrt{1 - u/v_{\text{ph}}}}, \tag{5.2.17}$$

where u and E_m are given by (5.2.13) and (5.2.15), respectively.

§5.3. Solitary pulses in retarding plasma-filled wave guide with an electron-ion beam.

The model of an unbounded beam is physically correct if the waveguide is filled with a sufficiently dense plasma, which neutralizes the Coulomb charge of the beam. An alternative method for achieving charge and current neutrality in the system, which is realized for a small density of the background plasma (or its absence), is the accompanying ion beam. The longitudinal polarization electric field, which appears upon electron shift with respect to ions, impedes the electron-beam retardation and diminishes the role of the high-frequency nonlinearity. The time of synchronous interaction between electrons and the wave increases, and, therefore, the radiation-energy density also increases.

According to Ref. [8], the polarization field partially compensates for the action of the longitudinal component of the Lorentz force (dephasing of electrons with transverse electric field of the wave) also in the case of resonant electromagnetic soliton which propagates along the electron-ion beam in a low-density plasma $n_p \leq n_b$ (n_p and n_b are the plasma density and beam density). By analogy with §5.1 this solution is a stationary envelope wave which the carrier frequency and the carrier wave number satisfy the conditions of the anomalous Doppler effect. The difference between the group velocity of the wave and beam velocity and a trapping of the beam particles by the wave is a cause for the existence of a solitary-wave for the field amplitude. As a result, there is no exchange of energy of the beam particles with the wave over the time which is larger than the time during which the particle passes through the field pulse.

We assume that an electron-beam that propagates along a magnetic field B_0 in a retarding plasma-filled system with superlight velocity $v > v_{\text{ph}} = c/N$, where N is the effective refractive index and c is the light velocity. Since the presence of the accompanying ion beam does not changes a high-frequency characteristic of plasma-beam system we can use the dispersion equation (connecting the frequency ω and the wave number k) in the form [8]:

$$\left(\frac{ck}{\omega}\right)^2 = N^2 - \frac{\omega_b^2(\omega - kv_0)}{\omega^2(\omega - kv - \omega_B)},$$
$$N^2 = n_0^2 - \frac{\omega_p^2}{\omega(\omega + \omega_B)}. \tag{5.3.1}$$

The asymptotics of nonlinear solutions describing the peripheral soliton regions are determined by the dispersion equation (5.3.1), in which we must assume

$$k = k_0 + \Delta k, \quad \omega = \omega_0 + u\Delta k, \quad \omega_0 - k_0 v_0 + \omega_B = \delta\omega_0, \tag{5.3.2}$$

where u is the group velocity of the wave and $|\delta| \ll 1$. Expanding in powers of $|\Delta k|/k_0 \ll 1$, we obtain

$$\begin{gathered} 2c\omega_0(N - N_0 u/v_g)(u - v_0)\Delta k^2 + \omega_0^2[(N^2 - N_0^2)(u - v_0) \\ + 2c(N - N_0 u/v_g)\delta]\Delta k + \omega_0^3(N^2 - N_0^2)\delta + \omega_b^2\omega_B = 0, \\ N = ck_0/\omega_0, \quad N_0^2 = n_0^2 - \omega_p^2/\omega_0(\omega_0 + \omega_B), \\ v_g = c[d/d\omega_0(\omega_0 N_0)]^{-1}. \end{gathered} \tag{5.3.3}$$

In a more general case under study, in which $\delta \neq 0$ the space growth rate

$$\Delta k^2 = -\frac{\omega_b^2\omega_B v_g}{2\omega_0 c(u - v_0)(N_0 u - N v_g)} + \frac{\omega_0^2\delta^2}{(u - v_0)^2} \tag{5.3.4}$$

is defined by the condition

$$2c(N - N_0u/v_g)\delta + (N^2 - N_0^2)(u - v_0) = 0, \tag{5.3.5}$$

which corresponds to the phase resonance of the beam with the wave and couples the group and phase velocities in both the linear and the nonlinear instability stages [9]. In the limit case $\delta = 0$ and $N = N_0$ formula (5.3.4) coincides with the maximum space growth rate (5.1.6).

1. Nonlinear system of "slow" equations.

A self-consistent system of equations involves the Maxwell equations for transverse electric $E_\perp = E_x + iE_y$ and magnetic $B = B_x + iB_y$ fields in a plasma

$$\begin{gathered}\left(\frac{\partial}{\partial t} + i\omega_B\right)\left[\left(\frac{n_0^2}{c^2}\frac{\partial^2}{\partial t^2} - \frac{\partial^2}{\partial z^2}\right)E_\perp + \frac{4\pi e}{c^2}\frac{\partial}{\partial t}(n_e v_\perp)\right] \\ + \frac{\omega_p^2}{c^2}\frac{\partial E_\perp}{\partial t} = 0, \qquad \frac{1}{c}\frac{\partial B}{\partial t} = -\frac{\partial E_\perp}{\partial z},\end{gathered} \tag{5.3.6}$$

($\omega_B = eB_0/(mc)$ and $\omega_\mathrm{p} = (4\pi e^2 n_\mathrm{p}/m)^{1/2}$ are the cyclotron and Langmuir electron frequencies and n_p is the plasma density) and a system of nonlinear hydrodynamic equations for the velocity $v_\perp = v_x + iv_y$ and the density n_e of beam electrons

$$\begin{gathered}\left(\frac{\partial}{\partial t} + v_z\frac{\partial}{\partial z}\right)v_\perp = \frac{e}{m}\left(E_\perp - \frac{v_z}{c}B\right) - i\omega_B v_\perp, \\ \left(\frac{\partial}{\partial t} + v_z\frac{\partial}{\partial z}\right)v_z = \frac{e}{m}\left[\frac{1}{c}\,\mathrm{Re}(v_\perp B^*) + E_z\right], \\ \frac{\partial n_e}{\partial t} + \frac{\partial}{\partial z}(n_e v_z) = 0.\end{gathered} \tag{5.3.7}$$

The longitudinal polarization (low-frequency) electric field E_z satisfies the Poisson equation, which, with allowance for the equations of longitudinal motion of plasma electrons and ion beam, is presented in the form

$$\begin{gathered}\frac{\partial^2 E}{\partial t^2} + \omega_p^2 E_z = 4\pi e\frac{\partial}{\partial t}(n_e v_z - n_i v_i), \\ \left(\frac{\partial}{\partial t} + v_i\frac{\partial}{\partial z}\right)v_i = -\frac{e}{M}E_z, \qquad \frac{\partial n_i}{\partial t} + \frac{\partial}{\partial z}(n_i v_i) = 0.\end{gathered} \tag{5.3.8}$$

We shall seek the solution of Eqs. (5.3.6)-(5.3.8) in the form of a plane wave with amplitude depending on variable $\xi = z - ut$,

$$\begin{gathered}E_\perp = E(\xi)\exp(i\Phi), \quad v_\perp = V(\xi)\exp(i\Phi), \\ E_z(\xi) = -d\varphi/d\xi, \quad n_e = n_e(\xi), \quad n_i = n_i(\xi), \quad v_i = v_i(\xi),\end{gathered} \tag{5.3.9}$$

where $\Phi = \omega_0 t - k_0 z$, and ω_0 and k_0 are determined by Eq. (5.3.5). For slowly varying

$((d/d\xi \ll k_0))$ complex amplitudes we obtain the following system of nonlinear equations:

$$(v_z - u)\Big[2(N - N_0 u/v_g)E' - i\frac{\omega}{c}(N^2 - N_0^2)E\Big] = -\frac{4\pi e n_b}{c}(v_0 - u)V,$$

$$(v_z - u)V' + i\Big[\omega(1 - N\frac{v_z}{c}) + \omega_B\Big]V = \frac{e}{m}\Big(1 - N\frac{v_z}{c}\Big)E,$$

$$(v_z - u)V' = \frac{e}{m}\Big[\frac{N}{c}\mathrm{Re}\,(VE^*) - \varphi'\Big], \tag{5.3.10}$$

$$\varphi'' + \frac{\omega_p^2}{u^2}\varphi = -4\pi e n_b(v_0 - u)\left(\frac{1}{v_z - u} - \frac{1}{v_i - u}\right),$$

$$(v_i - u)v_i' = \frac{e}{M}\varphi',$$

where

$$n_e(v_z - u) = n_b(v_0 - u),$$

$$N = ck_0/\omega_0, \quad N_0^2 = n_0^2 - \omega_p^2/\omega_0(\omega_0 + \omega_B),$$

$$v_g = cN_0\left[N_0^2 - \omega_p^2\omega_B/2(\omega_B + \omega_0)^2\omega\right]^{-1},$$

n_b is the unperturbed density of the beam.

The energy and momentum integrals of Eq. (5.3.10) are

$$\frac{1}{2}mn_b\left(|V|^2 + v_z^2\right) + \frac{1}{2}Mn_b v_i^2 + \frac{(N_0 u/v_g - N)c}{4\pi(u - v_0)}|E|^2$$

$$+\frac{u}{(u - v)}\left(\frac{e^2 n_p}{2mu^2}\varphi^2 + \frac{\varphi'^2}{8\pi}\right) = W_0, \tag{5.3.11}$$

$$n_b(mv_z + Mv_i) + \frac{N(N_0 u/v_g - N)}{4\pi(u - v_0)}|E|^2 + \frac{1}{(u - v_0)}\left(\frac{e^2 n_p}{2mu^2}\varphi^2 + \frac{\varphi'^2}{8\pi}\right) = P_0,$$

where $W_0 = n_b(M + m)v_0^2/2$, and $P_0 = n_b(M + m)v_0$ are the initial values of the energy and momentum density of the electron-ion beam.

Taking into account that the nonlinear variation of longitudinal velocity of a low-density electron beam is not large $v_z = v_0 + \delta v$, $|\delta v| \ll v_0$, we consider the inequality

$$|\delta v_{e,i}| \ll |v_0 - u|,$$

$$\delta v_e = -\frac{1}{\mu}\delta v_i - \frac{N_0(1 - u/v_g)}{4\pi n_b m(v_0 - u)}|E^2|, \quad \delta v_i = \frac{\mu e\varphi}{m(v_0 - u)}$$

and expand Eqs. (5.3.10) in powers of these small parameters.

Moving to dimensionless variables

$$\zeta = \omega\xi/(v_0 - u), \quad \mathcal{E} = eE/mc\omega, \quad \beta = V/c,$$

$$\Phi = e\varphi/mc^2, \quad \mu = m/M, \quad \beta_\alpha = v_\alpha/c,$$

we obtain

$$\mathcal{E}' - i\delta\mathcal{E} = -\frac{q^2\delta}{N^2 - N_0^2}\beta,$$

$$\beta' + i\left[\delta + N\Phi + \frac{N^2(N^2 - N_0^2)}{2q^2\delta}|\mathcal{E}|^2\right]\beta = (1 - N\beta_0)\mathcal{E}, \tag{5.3.12}$$

$$\Phi'' + \Omega^2 \Phi = -\frac{N^2 - N_0^2}{2\delta} N |\mathcal{E}|^2,$$

$$\Omega^2 = q^2 \left[1 + \mu + \frac{n_p}{n_b} \left(1 - \frac{\beta_0}{\beta_u} \right)^2 \right]. \tag{5.3.13}$$

In the derivation of Eqs. (5.3.12) and (5.3.13) we use condition (5.3.5) and omit terms of the subsequent order of smallness in parameter $\delta = (\omega_0 + \omega_B - kv_0)/\omega_0$, which are proportional to Φ^2 and Φ'^2. At the same time we allow for the small ion addition μ to the dimensionless Langmuir frequency of the beam $q = \omega_b/\omega_0$, which turns out to be substantial in the absence of plasma.

Eliminating β from the equation for the field, we write Eq. (5.3.12) in the form

$$\mathcal{E}'' + \left(N\Phi + \frac{N^2(N^2 - N_0^2)}{2q^2\delta} |\mathcal{E}|^2 \right) (i\mathcal{E}' + \delta) - \Gamma^2 \mathcal{E} = 0, \tag{5.3.14}$$

where the growth rate

$$\Gamma^2 = \left[q^2 \delta (N\beta_0 - 1)/(N^2 - N_0^2) \right] - \delta^2$$

matches the formula of the linear theory (5.3.4) (if Eq. (5.3.5) is fulfilled).

Let us represent the complex amplitude of the field in the form $\varepsilon = A\exp(i\vartheta)$ and separate the real and imaginary parts of Eq. (5.3.14)

$$A'' - A\vartheta'^2 - (\vartheta' - \delta) \left[\frac{N^2(N^2 - N_0^2)}{2q^2\delta} A^2 + N\Phi \right] A - \Gamma^2 A = 0,$$

$$A\vartheta'' + \left[2\vartheta' + N\Phi + \frac{N^2(N^2 - N_0^2)}{2q^2\delta} A^2 \right] A' = 0. \tag{5.3.15}$$

Integrating Eq. (5.3.15), we obtain the following equations for the field amplitude:

$$A'^2 = (\Gamma^2 + 2\delta\vartheta' - \vartheta'^2) A^2, \tag{5.3.16}$$

$$2\vartheta' = -\frac{N^2(N^2 - N_0^2)}{4q^2\delta} A^2 - N\Phi - \frac{\delta}{N^2 - N_0^2} \frac{1}{A^2} (\Phi'^2 + \Omega^2 \Phi^2), \tag{5.3.17}$$

which together with the equation for potential (5.3.13) form the closed system of slow equations describing the propagation of nonlinear resonant electromagnetic pulses in a plasma retarding system with an electron-ion beam.

2. Electromagnetic soliton.

Let us consider the asymptotic solution in the limiting case of small growth rates, in which $\Phi'' \ll \Omega^2$ $(\Gamma^2 \ll q^2)$ and the second derivative of the potential in Eq. (5.3.13) can be omitted

$$\Phi = -\frac{N^2 - N_0^2}{2\delta\Omega^2} NA^2. \tag{5.3.18}$$

Eliminating variable Φ from Eq. (5.3.17) and moving to the variable

$$a^2 = A^2(N^2 - N_0^2)N^2/2q^2\delta,$$

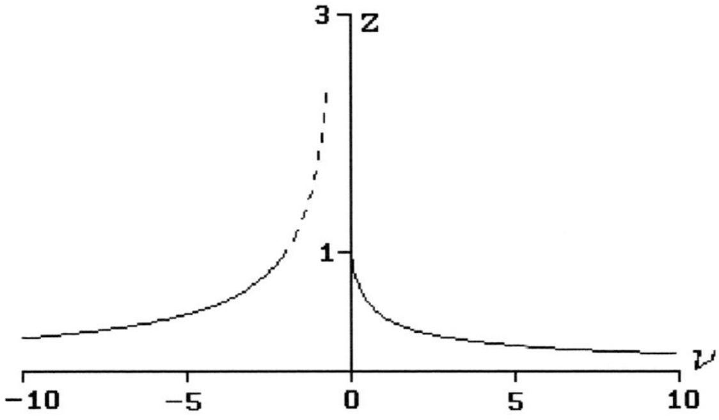

Fig. 5.3.1. The solution of Eq. (5.3.23). Two branches of a curve correspond to $\nu_{\pm}=Z^{-3/2}(-\sqrt{Z}\pm 1)$.

we find

$$\vartheta' = -\frac{\kappa a^2}{4}, \qquad \kappa = \frac{\mu + \dfrac{n_p}{n_b}(1-\dfrac{\beta_0}{\beta_u})^2}{1+\mu+\dfrac{n_p}{n_b}(1-\dfrac{\beta_0}{\beta_u})^2}. \tag{5.3.19}$$

Substituting Eq. (5.3.19) into Eq. (5.3.16), we obtain

$$a' = a\sqrt{\Gamma^2 - \frac{\delta\kappa}{2}a^2 - \frac{\kappa^2}{16}a^4}, \tag{5.3.20}$$

which is integrated in the elementary functions

$$\kappa a^2 = \frac{4\Gamma^2}{\delta + \sqrt{\Gamma^2+\delta^2}\cosh(2\Gamma\zeta)} \tag{5.3.21}$$

The maximum amplitude of the soliton at point $\zeta = 0$ equals

$$a_m^2 = (4\Gamma/\kappa)(\sqrt{1+\delta^2/\Gamma^2} - \delta/\Gamma) \tag{5.3.22}$$

and defines the nonlinear dispersion law.

Returning to the field amplitude A_m, we obtain

$$\nu^2 Z^3 + 2\nu Z^2 + Z - 1 = 0, \tag{5.3.23}$$

where $Z = \delta(N^2 - N_0^2)/q^2(N\beta_0 - 1)$, and $\nu = \kappa N^2(N\beta_0 - 1)A_m^2/8\delta^3$.

The plot of the function $Z(\nu)$ is shown in Fig. 5.3.1. The soliton-existence region is bounded by the inequality $Z < 1$, which results from condition $\Gamma^2 > 0$.

The asymptotic solutions of Eq. (5.3.23)

$$Z(\nu) = \begin{cases} 1, & \nu \ll 1, \\ \pm\nu^{-2/3}, & |\nu| \gg 1 \end{cases} \tag{5.3.24}$$

correspond to the linear and strongly nonlinear approximation. In the latter case the nonlinear dispersion equation has the form

$$N^2 = N_0^2 + \frac{4q^2\delta(N\beta_0 - 1)^{1/3}}{(N_0^2 A_m^2 \kappa)^{2/3}}. \tag{5.3.25}$$

The expression for the group velocity of the wave packet from Eqs. (5.3.5) and (5.3.25) is written as

$$\beta_u = \beta_g \frac{\beta_0 \sigma - \kappa^{3/2}}{\beta_g \sigma - \kappa^{3/2}}, \quad \sigma = \frac{2q^2(N\beta_0 - 1)^{1/3}}{N_0(N_0 A_m)^{4/3}}. \tag{5.3.26}$$

and for the arbitrary parameters of the retarding system and the beam it represents an algebraic equation for β_u, where function $\kappa(\beta_u)$ determined by Eq. (5.3.19).

In the dense plasma ($n_p \gg n_b$, $\kappa \simeq 1$) when the influence of the polarization field on the nonlinear retardation of beam particles by the wave is insignificant and the resulting dephasing between the beam and wave leads to a fast transition of electrons from the retarding phase to the accelerating one, Eq. (5.3.21) coincides with that in Eq. (5.3.15). However, as the plasma density decreases, its screening effect on the beam attenuates, and the longitudinal electric field that arises partially compensates the action of the longitudinal component of the Lorentz force and slows down the electron retardation. As a result, the coefficient κ in Eq. (5.3.19) decreases and the phase synchronism of the resonance particles with the wave is delayed. At the same time the ion beam is retarded and its energy and momentum transform into the wave energy and momentum. Note that our consideration is valid if the characteristic time of variation of the field amplitude is small compared with the growth rate of beam instability: $u\Delta k \gg (\omega_b^2 \omega_p)^{1/3}$.

In a rarefied plasma ($n_p \ll \mu n_b$, $\kappa \simeq \mu$), where the neutralization of the electron current is due mainly to the ion beam, the polarization field is so large that electron retardation due to radiation occurs in synchronism with ions, since in the considered case of small growth rates ($\Gamma^2 \ll q^2$) the electron beam is "frozen" into the ion one. In other words, a peculiar mass anisotropy emerges, since electrons act as particles with mass m as they travel across the magnetic field, while their effective mass equals M as they travel along the field. As a result, due to the large ion mass, the soliton amplitude becomes greater by a factor of $(M/m)^{1/2}$ than that in a dense plasma.

It is noteworthy that in the limiting cases of dense and rarefied plasmas, the parameter κ is independent of β_u, and Eq. (5.3.26) explicitly determines the group velocity. In addition, the solution of the algebraic equation (5.3.26) is $\beta_u = \beta_0$ for $\beta_0 = \beta_g$ and $\kappa = \mu/(1+\mu)$.

3. Numerical integration.

The above "inertialess" approximation $\Phi'' \ll \Omega^2$ is violated for $\Gamma^2 \simeq q^2$, when we must allow for longitudinal oscillations of an electron beam with respect to an ion beam. Since in this case the dynamics of an electron beam in the electromagnetic pulse field becomes much more complicated, Eqs. (5.3.13) and (5.3.16) can be studied only by numerical methods.

For numerical integration it is convenient to use new s

$$\Psi = \frac{N}{\delta}\Phi, \quad b^2 = \frac{N(N^2 - N_0^2)}{2q^2\delta^2}A^2, \quad Q^2 = \frac{\Gamma^2}{\delta^2}, \quad \zeta_1 = \delta\zeta$$

and present the system of equations in the form

$$\Psi'' + \frac{\Omega^2}{\delta^2}\Psi = -\frac{q^2}{\delta^2}b^2, \tag{5.3.27}$$

$$b'^2 = (Q^2 + 2\vartheta' - \vartheta'^2)b^2, \quad 2\vartheta' = -\frac{b^2}{2} - \frac{W}{b^2} - \Psi,$$
$$W = \frac{\delta^2}{2q^2}\left(\Psi'^2 + \frac{\Omega^2}{\delta^2}\Psi^2\right). \tag{5.3.28}$$

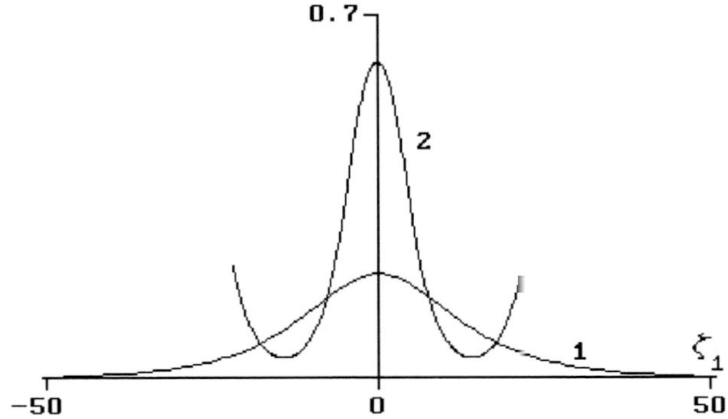

Fig. 5.3.2. Deviation of numerical solution from soliton as the parameter Q increases: (1) Q^2=0.01 and (2) Q^2=0.1.

The system of nonlinear equations (5.3.27) and (5.3.28) describes the coupled transverse and longitudinal oscillations of electrons and ions. Numerical study of Eqs. (5.3.27) and (5.3.28) shows that for $Q^2 \ll q^2/\delta^2$ the solutions of Eq. (5.3.21) are realized. However, as the parameter $(Q\delta/q)^2$ grows, the role of longitudinal fields increases and their influence results in the failure of soliton type solutions. Figure 2 shows the distortion of the soliton pulse ($q^2/\delta^2 = 0.5$, $\Omega^2/\delta^2 = 1$) as the parameter Q^2 increases: $Q^2 = 0.01$ (curve 1) and $Q^2 = 0.1$ (curve 2). Generally, a regular stationary wave is excited in the system, which has no zero asymptotics for $|\zeta_1| \to \infty$. Under these conditions, soliton solutions can be obtained only for certain period ratios for Langmuir and electromagnetic oscillations.

The results of numerical integration of Eqs. (5.3.27) and (5.3.28) are shown in Fig.5.3.3 in the form of solitary pulses for the field b (curve 1) and potential Ψ (curve 2) for $\Gamma^2 = 4$, $\Omega^2/\delta^2 = 1$ with variation of parameter q^2/δ^2: 0.089507 (a), 0.157287 (b), and 0.508552 (c). At the periphery of the electromagnetic pulse ($|\zeta_1| \to \infty$), where the field amplitude has the form $b \sim \exp(-Q|\zeta_1|)$, the role of the Coulomb field is insignificant. The nonlinear retardation of an electron beam at the pulse head is accompanied by the separation of charges and the appearance of longitudinal Langmuir oscillations of the electron beam with respect to the ion beam. Under these conditions, the soliton solutions exist if the extrema of functions b and Φ coincide. The amplitude of nonlinear electromagnetic oscillations increases with Q^2. The passage to the limit $Q^2 \to 0$ corresponds to the above obtained solution of Eq. (5.3.21).

Table 1

q^2/δ^2	0.089507	0.118728	0.157287	0.209972	0.242566
N_Ψ/N_f	2/5	3/8	1/3	3/10	2/7
q^2/δ^2	0.306216	0.373077	0.395710	0.422106	0.451650
N_Ψ/N_f	1/4	2/9	3/14	1/5	3/16
q^2/δ^2	0.471476	0.508552	0.545492	0.558581	0.573992
N_Ψ/N_f	2/11	1/6	2/13	3/20	1/7
q^2/δ^2	0.602349	0.624752	0.647073		
N_Ψ/N_f	2/15	1/8	2/17		

In the Table 1 we show the numerical values of the parameters of the problem N_Ψ/N_b (N_Ψ and N_b are the integers of potential and field amplitude oscillations) under which the reciprocal solutions with $b(\pm\infty) = \Phi(\pm\infty) = 0$ are realized.

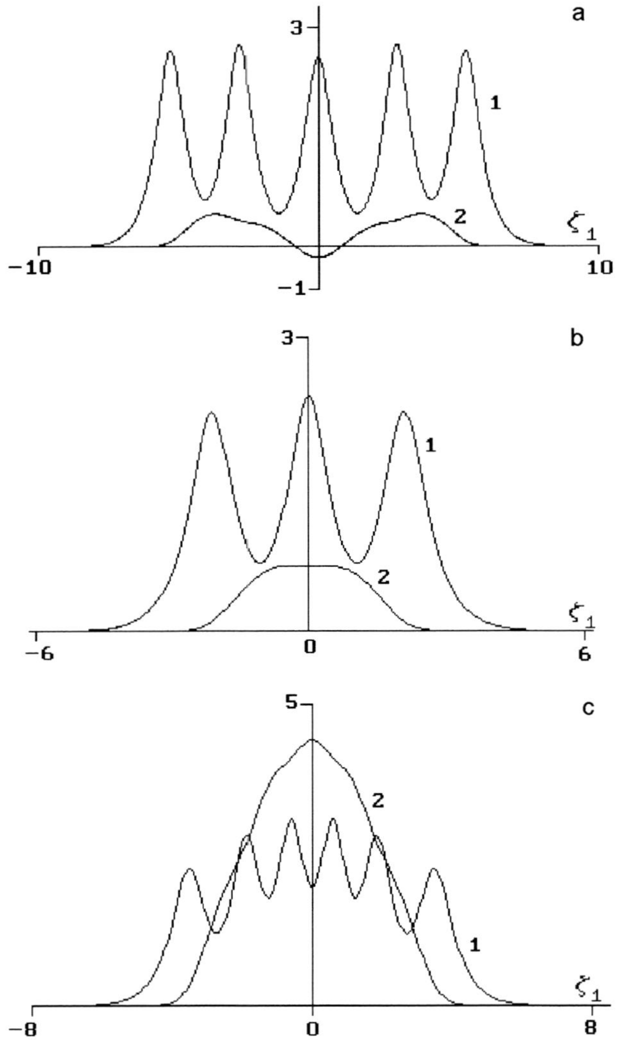

Fig. 5.3.3. Profiles of the field amplitude b (curve 1) and potential Ψ (curve 2) for different values of the parameter q^2/δ^2: 0.089507 (a); 0157287 (b) and 0.508552 (c).

The appearance of nonlinear oscillations of the field amplitude $b(\zeta_1)$ shown in Fig. 5.3.4 is qualitatively explained by analogy with the problems of mechanics, by writing Eq. (5.3.28) in the form

$$\begin{gathered} b'^2 + U(b) = E, \\ U(b) = (\Psi^2/4 + \Psi + W/4 - Q^2)b^2 \\ +(1 + \Psi/2)b^4/2 + b^6/16 + W^2/4b^2, \\ E = -W(\Psi/2 + 1), \end{gathered} \tag{5.3.29}$$

where $U(b)$ is the effective potential pit and E is the particle energy.

The evolution of the function $U(b)$ with shift from the periphery to the pulse center (see Fig. 5.3.3 c) is shown in Fig. 5.3.4. The infinite motion of the particle (Fig. 5.3.4a), which corresponds to an exponential growth of the amplitude, is replaced by a finite one when the second turning point appears (Fig. 5.3.4b). The steepening of potential-pit walls

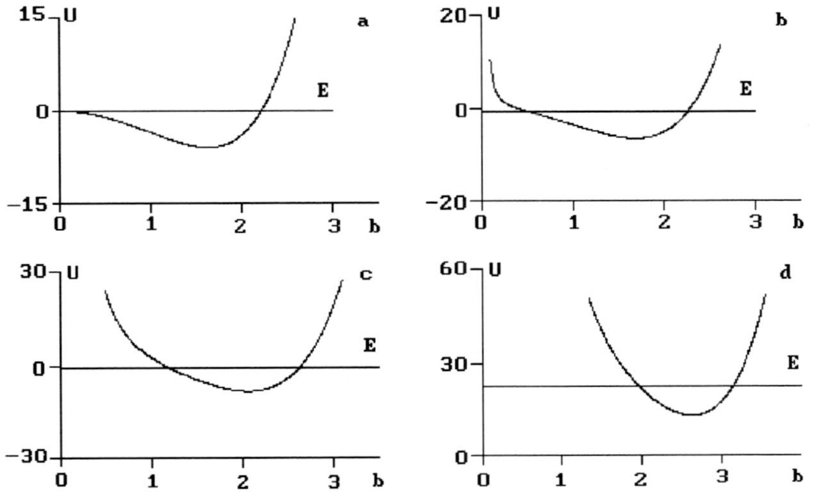

Fig. 5.3.4. Evolution of the potential well $U(b)$ with shift from the periphery to center of soliton for $\zeta_1=-8$ (a); -4 (b); -2 (c) and 0 (d).

is accompanied by particle fall on the pit bottom and decrease in the nonlinear-oscillation period (Fig.5.3.4c,d).

4. Adiabatic approximation.

As is evident from numerical integration (see Fig. 5.3.3c), the transition to $Q^2 \gg 1$ ($\Gamma^2 \gg q^2$) region is accompanied by an increase in the period of Langmuir oscillations as compared with the inverse nonlinear electron frequency, which allows us to study Eqs. (5.3.27) and (5.3.28) in the adiabatic approximation.

Since the potential is slightly changed during the period of nonlinear oscillations, Eq. (5.3.28) can be integrated under constant Ψ and Ψ'. After the standard operations we find

$$b^2(\zeta_1,\Psi) = \frac{x_1+x_2}{2} - \frac{x_1-x_2}{2}\,\frac{\lambda - \mathrm{cn}(\sqrt{gh/4}\,\zeta_1,k)}{1-\lambda\mathrm{cn}(\sqrt{gh/4}\,\zeta_1/2,k)}, \tag{5.3.30}$$

where $\mathrm{cn}(x,k)$ is the elliptic cosine (see **Appendix 5.2**).

The period of nonlinear oscillations is equal to

$$T = \frac{4}{\sqrt{fg}}K(k). \tag{5.3.31}$$

Averaging Eq. (5.3.27) over the nonlinear-oscillation period (fast variable ζ_1), we find (**Appendix 5.2**)

$$\begin{gathered}\Psi'' + \frac{\Omega^2}{\delta^2}\Psi = -\frac{q^2}{\delta^2}\langle b^2\rangle,\\ \langle b^2\rangle = \frac{x_1+x_2}{2} + \frac{x_1-x_2}{2}\,\frac{f+g}{f-g}\left[1-\frac{\Pi(-\gamma,k)}{K(k)}\right].\end{gathered} \tag{5.3.32}$$

Equation (5.3.32) describes the dynamics of the Coulomb potential at the nonlinear stage of beam–wave interaction. The asymptotic formulas of the adiabatic approximation are compared with numerical-integration results in Fig. 5.3.5, where $\Gamma^2 = 4$, $\Omega^2/\delta^2 =$

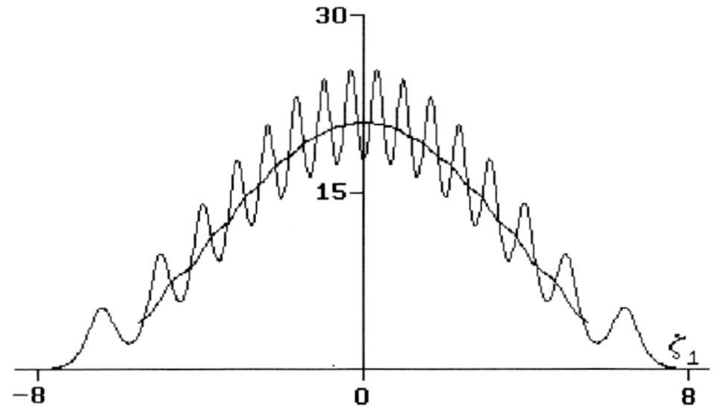

Fig. 5.3.5. Numerical-integration results (oscillation curve) and the asymptotic formulas of adiabatic approximation (5.3.32).

1, and $q^2/\delta^2 = 0.810326$; the oscillating curve $b(\zeta_1)$ is found by numerical integration of Eqs. (5.3.27) and (5.3.28), and the monotonic $\langle b^2(\zeta_1)\rangle^{1/2}$ corresponds to Eq. (5.3.32).

In conclusion it must be noted that transition to the rarefied plasma is accompanied, as in Sec. 2, by a significant increase in the amplitude of the solitary pulse with "built-in" Langmuir oscillations.

§5.4. Slow soliton in a plasma in the presence of the normal Doppler effect.

As an example of an actual model of non-equilibrium medium in which a resonant electromagnetic soliton exist, we have already considered a charged particle beam which is propagating through an arbitrary dispersive retarding medium (plasma) along an external magnetic field $\mathbf{B}_0$. The goal of this section is to study the condition of propagation of the solitary electromagnetic wave in a plasma without beam under cyclotron resonance conditions. In comparison with the above approximation assuming that the density of resonant beam particles is small and the retarding medium remains linear (see §5.1 - §5.3), the dispersion law and the group velocity of envelope wave which is determined by the resonant plasma electrons become nonlinear [9].

We begin with a discussion of the physical mechanism with the appearance of a resonant electromagnetic soliton in a plasma. The leading edge of the pulse causes an electron initially at rest to "uncoil" across $\mathbf{B}_0$ and to absorb energy from the field. At the same time, it is accelerated along the magnetic field by the Lorentz force $\mathbf{F} = (e\mathbf{B}_0/cB_0)[\mathbf{vB}]$ ($\mathbf{v}$ is the velocity of the electron). As a result, a nonlinear Doppler frequency shift $\omega^* = \omega(1 - v_{||}/v_{\text{ph}})$ (v_{ph}) is the phase velocity of the wave) arises and a relativistic decrease in the gyrofrequency $\omega_B^* = \omega_B(1\text{-}\mathbf{v}_\perp^2/2c^2)$ occur.

According to first formula (5.1.13), an increase in the longitudinal velocity is $v_{||} = v_\perp^2/2v_{\text{ph}}$. Therefore, in the case of the slow wave, $v_{\text{ph}} = c/n_0 < c$ (n_0 is the refractive index of retarding medium), the electric field exceeds the magnetic one, $|\mathbf{B}| = n_0|\mathbf{E}|$, and the nonlinear frequency shift, $\omega_B^* - \omega^* = (n_0^2 - 1)\omega v_\perp^2/2c^2$, disrupts the phase resonance of the electron with the wave. Consequently, after an electron reaches the maximum of the soliton field it enters a region of retarding phases, and it gives up all its energy to the trailing edge of the pulse. The wave is a stationary wave, since the charge in the energy of an electron which passes through the field pulse vanishes.

A self-consistent system of equations involves the Maxwell equations for the transverse electric, $E_\perp = E_x + iE_y$, and the magnetic, $B = B_x + iB_y$, fields in a plasma

$$\begin{gathered}\left(\frac{\partial^2}{\partial z^2} - \frac{n_0^2}{c^2}\frac{\partial^2}{\partial t^2}\right) E_\perp = \frac{4\pi e}{c^2}\frac{\partial}{\partial t}(n_e v_\perp), \\ \frac{\partial E_\perp}{\partial z} = -\frac{1}{c}\frac{\partial B}{\partial t},\end{gathered} \tag{5.4.1}$$

where the transverse velocity, $v_\perp = v_x + iv_y$, the longitudinal velocity, $v_{||}$, and the plasma density n_e satisfy the set of relativistic hydrodynamic equations

$$\begin{gathered}\left(\frac{\partial}{\partial t} + v_{||}\frac{\partial}{\partial z}\right)\gamma v_\perp = \frac{e}{m}\left(E_\perp - \frac{v_{||}}{c}B\right) - i\omega_B v_\perp, \\ \left(\frac{\partial}{\partial t} + v_{||}\frac{\partial}{\partial z}\right)\gamma v_{||} = \frac{e}{m}\left[\frac{1}{c}\operatorname{Re}(v_\perp B^*) + E_{||}\right], \\ \frac{\partial n_e}{\partial t} + \frac{\partial}{\partial z}(n_e v_{||}) = 0.\end{gathered} \tag{5.4.2}$$

The longitudinal electrostatic field $E_{||}$ arising due to a charge separation under the action of electromagnetic radiation pressure, obeys the Poisson equation

$$\frac{\partial}{\partial z}\left(n_0^2 E_{||}\right) = 4\pi e(n_e - n_i), \tag{5.4.3}$$

which should be complemented by the equations for the density n_i and the velocity v_i of plasma ions

$$\frac{\partial n_i}{\partial t} + \frac{\partial}{\partial z}(n_i v_i) = 0, \quad \left(\frac{\partial}{\partial t} + v_i\frac{\partial}{\partial z}\right) v_i = -\frac{e}{M}E_{||}. \tag{5.4.4}$$

We seek a solution of Eqs. (5.4.1) - (5.4.6) in the form

$$\begin{gathered}E_\perp(\xi) = E(\xi)\exp(i\Phi), \quad v_\perp(\xi) = v(\xi)\exp(i\Phi), \\ v_{||} = v_{||}(\xi), \quad \Phi = kz - \omega t, \quad \xi = z - v_g t,\end{gathered} \tag{5.4.5}$$

where the coordinate z runs along the direction of the magnetic field $\mathbf{B_0}$, and $v_g = \beta_g c$ is the wave group velocity of the pulse. Following Refs. [8,9], we find a system of nonlinear total differential equations for the slowly varying complex amplitudes of the field and the electron velocity:

$$\begin{gathered}2(N - \beta_g n_0^2)E' + i(\omega/c)(N^2 - n_0^2)E = -4\pi e n_e v/c, \\ (v_{||} - v_g)(\gamma v)' + i[\omega_B - \gamma(\omega - kv_{||})]v = (e/m)(1 - Nv_{||}/c)E, \\ (v_{||} - v_g)(\gamma v_{||})' = (e/m)[(N/c)\operatorname{Re}(vE^*) + E_{||}], \\ n_0^2 E_{||}' = 4\pi e(n_e - n_i), \quad (v_i - v_g)v_i' = -(e/M)E_{||}, \\ n_e(v_g - v_{||}) = n_p v_g, \quad n_i(v_g - v_i) = n_p v_g,\end{gathered} \tag{5.4.6}$$

where the prime means the derivative with respect to the ξ, n_p, and $N = ck/\omega$ are the unperturbed density and the refractive index of the plasma, and $\gamma = (1-|v|^2/c^2-v_{||}^2/c^2)^{-1/2}$.

The energy and momentum integrals of Eq. (5.4.6) which satisfy homogeneous conditions on the functions and its derivatives are

$$\begin{gathered}mc^2\gamma + \frac{Mv_i^2}{2} + \frac{1}{4\pi n_p}\left[\left(n_0^2 - \frac{N}{\beta_g}\right)|E|^2 + \frac{n_0^2 E_{||}^2}{2}\right] = mc^2, \\ m\gamma v_{||} + Mv_i + \frac{1}{4\pi n_p c}\left[N\left(n_0^2 - \frac{N}{\beta_g}\right)|E|^2 + \frac{n_0^2 E_{||}^2}{2\beta_g}\right] = 0.\end{gathered} \tag{5.4.7}$$

Using the dimensionless variables

$$\mathcal{E} = eE/mc\omega, \quad \mathcal{E}_{||} = eE_{||}/mc\omega, \quad \beta = v/c,$$
$$\beta_{||} = v_{||}/c, \quad \beta_i = v_i/c,$$
$$q^2 = 4\pi e^2 n_p/m\omega^2, \quad \mu = m/M, \quad \delta = \omega_B/\omega - 1$$

we represent the equations (5.4.6) in the form

$$2(N/\beta_g - n_0^2)\mathcal{E}' + i(N^2 - n_0^2)\gamma(\beta_{||}/\beta_g - 1)\mathcal{E} = q^2\beta\gamma, \tag{5.4.8}$$

$$(\beta\gamma)' + i(\delta + 1 - \gamma + N\gamma\beta_{||})\beta\gamma = \gamma(1 - N\beta_{||})\mathcal{E}, \tag{5.4.9}$$

$$(\beta_{||}\gamma)' = \gamma[N\mathrm{Re}(\beta\mathcal{E}^*) + \mathcal{E}_{||}], \tag{5.4.10}$$

$$n_0^2\mathcal{E}_{||}' = -q^2\beta_g\gamma(\beta_{||} - \beta_i)(\beta_g - \beta_i)^{-1}, \tag{5.4.11}$$

$$\beta_i' = -\mu\gamma(\beta_g - \beta_{||})(\beta_g - \beta_i)^{-1}\mathcal{E}_{||}, \tag{5.4.12}$$

where the prime means the derivative $d\tilde{\xi} = [\omega/\gamma(v_{||} - v_g)]d\xi$.

We rewrite the energy and momentum integrals (5.4.7) as

$$\begin{aligned} 1 - \gamma(1 - N\beta_{||}) &= \frac{N^2-1}{q^2}\left(\frac{N}{\beta_g} - n_0^2\right)|\mathcal{E}|^2 \\ &- \left(\frac{N}{\beta_g} - 1\right)\frac{n_0^2\mathcal{E}_{||}^2}{2q^2} - \frac{\beta_i}{\mu}\left(N - \frac{\beta_i}{2}\right), \\ \gamma(1 - \beta_{||}/\beta_g) &= 1 - \frac{1}{q^2}\left(\frac{N}{\beta_g} - 1\right)\left(\frac{N}{\beta_g} - n_0^2\right)|\mathcal{E}|^2 \\ &- \left(1 - \frac{1}{\beta_g^2}\right)\frac{n_0^2\mathcal{E}_{||}^2}{2q^2} + \frac{\beta_i}{\mu}\left(\frac{1}{\beta_g} - \frac{\beta_i}{2}\right), \end{aligned} \tag{5.4.13}$$

and we shall use the relations (5.4.13) below to reduce the equations (5.4.8) and (5.4.9).

In the derivation of Eqs. (5.4.6) we have made use of the approximation which holds asymptotically in the small parameters

$$\begin{aligned} &|(N^2 - n_0^2)/(N/\beta_g - n_0^2)| \ll 1, \\ &q^2/|N/\beta_g - n_0^2| \ll 1, \quad |\delta| \ll 1. \end{aligned} \tag{5.4.14}$$

In the linear approximation ($\beta_{||} \ll 1$, $\gamma = 1$), it follows from (5.4.8) and (5.4.9) that [6)]

$$\begin{aligned} &\mathcal{E}'' - \delta^2(Z^{-1} - 1)\mathcal{E} = 0, \quad Z = \delta(N^2 - n_0^2)/q^2, \\ &N^2 - n_0^2 = 2\delta(N/\beta_g - n_0^2). \end{aligned} \tag{5.4.15}$$

Here, the latter equality corresponds to the correct choice of the linear group velocity [8,9]. The necessary monotonic solutions of equation (5.4.15) that define the solitary pulse shape at $|\tilde{\xi}| \to \pm\infty$, do exists when $Z < 1$.

[6)] Note that in the special case of the plane wave $\mathcal{E}''=0$ and $Z=1$ ($N^2=n_0^2+q^2/\delta$) the dispersion relation in the dimension variables is equal to: $N^2=n_0^2-\omega_p^2/\omega(\omega-\omega_B)$.

To proceed, we assume that the propagation of the solitary wave in a plasma is not accompanied by the essential charge separation [9]. According to (5.4.10), this approximation realizes if the longitudinal component of the Lorentz force greatly exceeds the electrostatic force,

$$N\,\mathrm{Re}(\beta\mathcal{E}^*) \gg \mathcal{E}_{||}. \tag{5.4.16}$$

By setting $\mathrm{E}_{||} = 0$, with allowance for the latter relation (5.4.15), we can rewrite equations (5.4.8), (5.4.10), and (5.4.13) as

$$\begin{aligned}\mathcal{E}' - i\delta[1-(N/\beta_g - 1)\alpha|\mathcal{E}|^2]\mathcal{E} &= \frac{q^2}{2(N/\beta_g - n_0^2)}\beta\gamma,\\ (\beta\gamma)' + i[\delta + (N^2-1)\alpha|\mathcal{E}|^2]\beta\gamma &= [1-(N^2-1)\alpha|\mathcal{E}|^2]\mathcal{E},\end{aligned} \tag{5.4.17}$$

where

$$\alpha = (N/\beta_g - n_0^2)/q^2 = (N^2 - n_0^2)/2q^2\delta. \tag{5.4.18}$$

Eliminating $\beta\gamma$ from the first equation, we represent (5.4.17) in the form

$$\begin{aligned}&\mathcal{E}'' - \Gamma^2\mathcal{E} + i\alpha[(N^2-1+g\delta)|\mathcal{E}|^2\mathcal{E}' + g\mathcal{E}(|\mathcal{E}|^2)']\\ &- \alpha[\delta(1-N^2) + g\delta^2 + (1-N^2)(\Gamma^2+\delta^2)]\mathcal{E}|\mathcal{E}|^2]\\ &+ g\delta(1-N^2)\alpha^2|\mathcal{E}|^4\mathcal{E} = 0,\end{aligned} \tag{5.4.19}$$

where

$$\Gamma^2 = (q^2/2g) - \delta^2, \quad g = N/\beta_g - 1.$$

We set $\mathcal{E} = a\exp(i\vartheta)$ and separate real and imaginary parts in (5.4.19):

$$\begin{aligned}a'' - \Gamma^2 a - a\vartheta'^2 - [\vartheta'(N^2-1+g\delta) + \delta(1-N^2)\\ +g\delta^2 + (\Gamma^2+\delta^2)(1-N^2)]\alpha a^3 + g\delta(1-N^2)\alpha^2a^5 = 0,\\ a\vartheta'' + 2a'\vartheta' + (N^2-1+3g\delta)\alpha a^2a' = 0.\end{aligned} \tag{5.4.20}$$

Integrating the second equation (5.4.20),

$$\vartheta' = -(N^2-1+3g\delta)\alpha a^2/4, \tag{5.4.21}$$

and eliminating ϑ' from first one, we find the field amplitude equation

$$a'' - \Gamma^2 a + B\alpha a^3 + A\alpha^2 a^5 = 0, \tag{5.4.22}$$

$$\begin{aligned}A &= (1/16)(N^2-1+3g\delta)[3(N^2-1)+g\delta] - g\delta(N^2-1),\\ B &= (N^2-1)(\delta+\Gamma^2+\delta^2) - g\delta^2.\end{aligned}$$

We rewrite the relations (5.4.15) as

$$N^2 - 1 = n_0^2 - 1 + \frac{q^2Z}{\delta}, \quad g = n_0^2 - 1 + \frac{q^2Z}{2\delta^2},$$

and we reduce the coefficients in Eq. (5.4.22) by means of the latter relations. Omitting $\sim\delta$ and $\sim(n_0^2-1)$ in comparison with unity, we find

$$A = \frac{3F^2}{16}, \quad B = \delta F, \quad F = n_0^2 - 1 + \frac{q^2Z}{2\delta}. \tag{5.4.23}$$

Substituting (5.4.23) in (5.4.22) and integrating, we obtain the equation

$$a' = \frac{a}{4}\sqrt{16\Gamma^2 - 8\delta\alpha F a^2 - (\alpha F a^2)^2}, \tag{5.4.24}$$

which is integrated in the elementary functions

$$\alpha F a^2 = \frac{4\Gamma^2}{\delta + \sqrt{\Gamma^2 + \delta^2}\cosh(2\Gamma\tilde{\xi}\,)}. \tag{5.4.25}$$

The maximum of soliton field a_m at the point $\tilde{\xi} = 0$ is

$$\alpha F a_m^2 = 4\left(\sqrt{\Gamma^2 + \delta^2} - \delta\right). \tag{5.4.26}$$

Using $\alpha = Z/2\delta^2$ and $\Gamma^2 + \delta^2 = \delta^2/Z$, and substituting $F(Z)$ from Eq. (5.4.23), we get a nonlinear dispersion law:

$$Z\left[1 + Z\left(n_0^2 - 1 + \frac{q^2 Z}{2\delta}\right)\frac{a_m^2}{8\delta^3}\right]^2 = 1, \tag{5.4.27}$$

which determines the dependence of plasma refractive index N on the field amplitude a_m.
In the non-relativistic approximation of Ref. [9]:

$$n_0^2 - 1 \gg \frac{q^2}{2\delta}Z, \tag{5.4.28}$$

the equation (5.4.27) reduces to the form

$$v^2 Z^3 + 2vZ^2 + Z - 1 = 0, \tag{5.4.29}$$

$$Z = \delta(N^2 - n_0^2)/q^2, \quad v = (n_0^2 - 1)a_m^2/8\delta^3.$$

The function $Z(v)$ is plotted in Fig. 5.3.1. The asymptotes of Eq. (5.4.29) are

$$Z = \begin{cases} 1, & v \ll 1, \\ v^{-2/3}, & v \gg 1. \end{cases}$$

The first of these relations is the same as (5.4.15); the second becomes applicable near the resonance ($\delta = 0$) and is very nonlinear:

$$N^2 = n_0^2 + \frac{4q^2\delta}{(n_0^2 - 1)^{2/3} a_m^{4/3}}. \tag{5.4.30}$$

In this spatial case, when the wave phase velocity $v_{\rm ph}$ and the group velocity v_g are

$$v_{\rm ph} = \frac{c}{n_0}, \quad v_g = v_{\rm ph}\left[1 + \frac{2q^2/n_0^2}{(n_0^2 - 1)^{2/3} a_m^{4/3}}\right]^{-1}, \tag{5.4.31}$$

the solitary wave (5.4.25) transforms into a soliton

$$y = y_m \operatorname{sech}\left(\frac{y_m}{2}\tilde{\xi}\right), \tag{5.4.32}$$

$$y = (n_0^2 - 1)\alpha a^2, \quad y_m = 4\Gamma = 2(n_0^2 - 1)^{1/3} a_m^{2/3}.$$

The transverse $\gamma\beta$ and longitudinal $\gamma\beta_{||}$ momentums of plasma electrons can be found from Eqs. (5.4.8) - (5.4.10) in the limiting case $\mathcal{E}_{||} \to 0$. Substituting (5.4.32) into the relations

$$\begin{aligned} y' &= \frac{y}{2}\sqrt{y_m^2 - y^2}, \\ \exp(2i\vartheta) &= \sqrt{1 - \frac{y^2}{y_m^2}} - i\frac{y}{y_m}, \end{aligned} \tag{5.4.33}$$

we define all components of the slow ($n_0 > 1$) solitary electromagnetic wave

$$\begin{aligned} \mathcal{E} = y_m\sqrt{\frac{y}{8(n_0^2 - 1)}}\exp(i\vartheta), &\quad \beta\gamma = \sqrt{\frac{2y}{n_0^2 - 1}}\exp(3i\vartheta), \\ \gamma\beta_{||} = \frac{n_0 y}{n_0^2 - 1}, &\quad \vartheta = -\frac{1}{2}\arcsin\frac{y}{y_m}. \end{aligned} \tag{5.4.34}$$

The acceleration of electrons with respect to the ions is accompanied by the appearance in the plasma of a polarized electric field $\mathcal{E}_{||}$ which satisfies the equation (5.4.11). Integrating Eqs. (5.4.11) with the help of (5.4.33) in the case of $\delta = 0$, we find

$$\mathcal{E}_{||} = -\frac{2q^2}{n_0(n_0^2 - 1)}\arcsin\frac{y}{y_m}. \tag{5.4.35}$$

Returning to the physical variable ξ by means of the equation $d\xi/d\tilde{\xi} = \gamma(v_{||} - v_g)/\omega$, we rewrite the Eq. (5.4.33) in the form

$$\begin{aligned} \frac{dy}{d\xi} &= \frac{\omega}{2c\gamma(\beta_g - \beta_{||})}y\sqrt{y_m^2 - y^2}, \\ \gamma(\beta_g - \beta_{||}) &= \beta_g(1 - y/y_g), \quad y_g^{-1} = 1 + \frac{8q^2}{y_m^2(n_0^2 - 1)}, \end{aligned} \tag{5.4.36}$$

at which the singularity in the derivative at the point $\beta_| = \beta_g$ arises. Integrating (5.4.36), we get

$$\frac{\omega y_m}{2v_g}\xi = \operatorname{arccosh}\frac{y_m}{y} - \frac{y_m}{y_g}\arcsin\frac{y}{y_m}. \tag{5.4.37}$$

The wave takes the form of a soliton ($y_m \ll y_g$) which transforms into a cusp in the limit $y_m/y_g \to 1$ (see Ref. [9]).

It follows from the equations (5.4.10), (5.4.32), and (5.4.33) that the inequality (5.4.16) can be written as

$$\begin{aligned} \sqrt{y_m^2 - y^2} &\gg \frac{4q^2\gamma}{n_0^2}\frac{1}{y}\arcsin\frac{y}{y_m}, \\ \gamma &= 1 + \frac{y}{n_0^2 - 1}. \end{aligned} \tag{5.4.38}$$

Since the term $\mathcal{E}_z$ has been omitted from the right side of Eq. (5.4.10), a condition for the applicability of the solution found above is violated in a narrow field amplitude interval near y_m (at the center of pulse where the longitudinal component of the Lorentz force changes its sign).

It is shown above that the nonlinear saturation of the filed amplitude arises as a result of the detuning of the phase resonance between the plasma electrons and the wave due to the longitudinal slowing down of the particles under the action of the Lorentz force. Since the magnetic field is proportional to the refractive index of the plasma, $n_0 = c/v_{ph}$, this nonlinear mechanism becomes effective for the slow electromagnetic wave $n_0 \gg 1$ when the magnetic field greatly exceeds the electric one.

In the absence of the external retarding medium, $n_0 \to 1$, a relativistic decrease in the gyrofrequency compensates a Doppler frequency shift, and the efficiency of the defocusing nonlinear mechanism is $n_0 - 1$ times weaker than in a case the slow wave. In the region of parameters $n_0^2 - 1 \ll q^2 Z/\delta$, the relation (5.4.27) determines the "relativistic" dispersion law:

$$v_R^2 Z^5 + 2v_R Z^3 + Z - 1 = 0, \tag{5.4.39}$$

$$Z = \delta(N^2 - n_0^2)/q^2, \quad v_R = q^2 a_m^2/16\delta^4.$$

A dispersion becomes nonlinear in a case of $v_R \simeq 1$ which corresponds to $a_m^2 \simeq 16\delta^4/q^2$. However, as the field amplitude grows, $a_m^2 \simeq \delta^3$, the longitudinal electrostatic field increases and their influence results in the failure of soliton type solutions must be taken into account [10].

Note that the solution derived here is qualitatively analogous to a resonant light pulse (see §5.6) and is characterized by an intense exchange of energy between electron and the wave, with the result that some of the energy of the pulse goes into electron kinetic energy, and the wave group velocity is found to be below the velocity of light. In contrast with a Langmuir soliton [1], which arises under conditions such that dispersion compensates for the self-contraction of a slightly nonlinear wave packet, the solution derived here for the resonant case $|\omega - \omega_B| \ll \omega_B$ is highly nonlinear and exists in a cold plasma in the absence of dispersion.

§5.5. Autoresonance soliton in a plasma.

In the case if the electromagnetic wave propagates along an external magnetic field $\mathbf{B_0}$ in a plasma with the velocity of light the acceleration of electrons does not accompanied by the detuning of the cyclotron resonance condition (autoresonance). Therefore, a nonlinear saturation of the field amplitude arises as a result of charge separation in a plasma due to the electromagnetic radiation pressure when a longitudinal electric field occurs.

We consider the set of equation (5.4.8) - (5.4.12) and the energy and momentum integrals of set (5.4.13) in the absence of a retarding medium, $n_0 = 1$, when the terms taking into account an "electromagnetic nonlinearity" (proportional to $N^2 - 1 \ll 1$) can be omitted. Assuming that the inequalities $\beta_g \gg \beta_{||} \gg \beta_i$ hold and introducing the Coulomb potential by means of expressions

$$\mathcal{E}_{||} = -\tilde{\varphi}', \quad \tilde{\varphi} = \beta_i/\mu,$$

we can rewrite this equations in the form:

$$\begin{aligned} &\mathcal{E}'' - i\tilde{\varphi}\mathcal{E}' - \delta^2(Z^{-1} - 1 + \tilde{\varphi}/\delta)\mathcal{E} = 0, \\ &\tilde{\varphi}'' + q^2\tilde{\varphi} + \frac{1}{2\beta_g}(\tilde{\varphi}')^2 = \frac{N^2 - 1}{2\delta}|\mathcal{E}|^2. \end{aligned} \tag{5.5.1}$$

By changing the scale

$$\varphi = \tilde{\varphi}/\delta, \quad d\zeta = \delta d\tilde{\xi} \simeq -(\delta\omega/v_g)d\xi$$

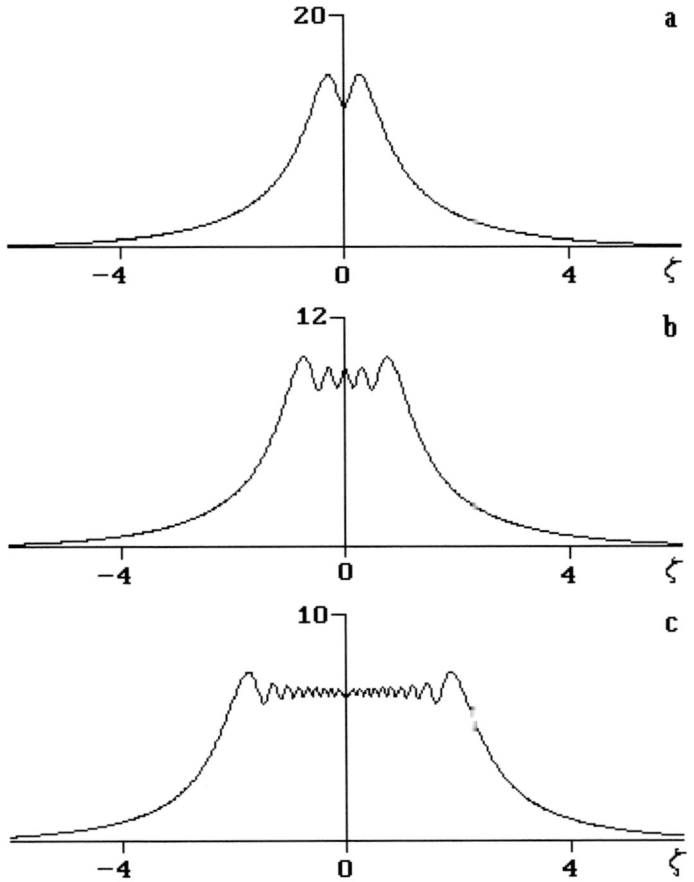

Fig. 5.5.1. The shape of a solitary pulse $a(\zeta)$ at $\Gamma^2=0.5$ and $q^2/\delta^2 = 19.533178$ (a), 5.868868 (b), 1.623496 (c).

one can eliminate the small parameter $\delta \ll 1$ from equations (5.5.1) because the term $(\tilde{\varphi}')^2/2\beta_g \sim \delta$ may be omitted. Thus, set (5.5.1) is simplified:

$$\begin{gathered} \mathcal{E}'' - i\varphi\mathcal{E}' - (\Gamma^2 + \varphi)\mathcal{E} = 0, \\ \varphi'' + \left(\frac{q}{\delta}\right)^2 \varphi = \frac{N^2 - 1}{2\delta^4}\,|\mathcal{E}|^2, \end{gathered} \tag{5.5.2}$$

where $\Gamma^2 = Z^{-1} - 1 > 0$.

Let us represent the field complex amplitude in the form:

$$\sqrt{\frac{N^2 - 1}{2\delta^4}}\,\mathcal{E} = a\exp(i\vartheta). \tag{5.5.3}$$

Substituting (5.5.3) into equations (5.5.2) and splitting the result into real and imaginary parts, we get

$$\begin{gathered} a'' - (\Gamma^2 - \varphi\vartheta' + \vartheta'^2 + \varphi)a = 0, \\ a\vartheta'' + (2\vartheta' - \varphi)a = 0, \quad \varphi'' + (q/\delta)^2\varphi = a^2. \end{gathered} \tag{5.5.4}$$

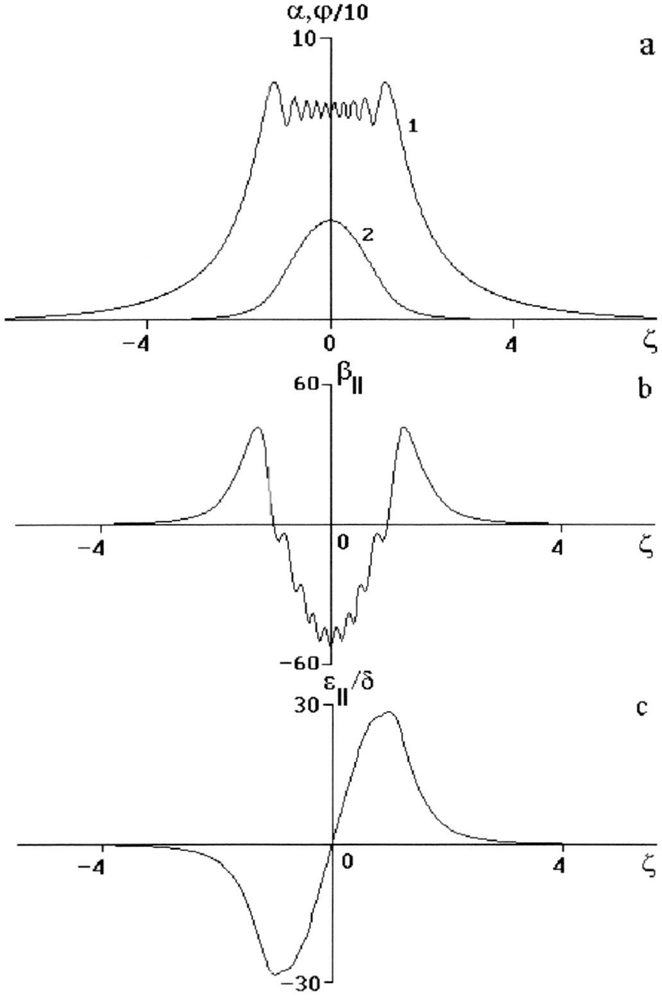

Fig. 5.5.2. The shape of a soliton pulse $a(\zeta)$ at Γ^2=0.5 and q^2/δ^2 = 2.959586: (a) field amplitude (1) and potential $\varphi/10$ (2); (b) longitudinal velocity $\beta_{\|}\equiv q^2\gamma\beta_{\|}/\delta^3$ and (c) longitudinal field $\mathcal{E}/\delta=-\varphi'$.

The integrals of (5.5.4) are

$$2\vartheta' = \varphi - [\varphi'^2 + (q/\delta)^2\varphi]/2a^2, \tag{5.5.5}$$

$$a'^2 = (\Gamma^2 + 2\vartheta' - \vartheta'^2)a^2. \tag{5.5.6}$$

Set (5.5.4) describes the interaction between two oscillators and has soliton solutions at certain relations between their periods. found numerically, the symmetrical solutions satisfying the conditions $a(\zeta \to \pm\infty) = \varphi(\zeta \to \pm\infty) = 0$ are presented in Fig. 5.5.1.

At the pulse periphery, $\zeta \to -\infty$, the field amplitude grows exponentially with a linear growth rate $a \sim \exp(\Gamma\zeta)$, but the Coulomb potential remains small $|\varphi| \ll 1$. However, the longitudinal acceleration of plasma electrons with the Lorentz's force (see Fig. 5.5.2) is accompanied by charge separation $|\varphi| \simeq 1$. According to (5.5.5), this causes the oscillations of the electron transverse velocity and field to fall out of synchronism. Reflexive (soliton)

Table 2

q^2/δ^2	19.533178	11.723112	7.529148	5.868868	4.846759
N_a	2	3	4	5	6
q^2/δ^2	4.151652	3.646700	3.262488	2.959586	2.714262
N_a	7	8	9	10	11
q^2/δ^2	2.511220	2.340174	2.193946	2.067372	1.956640
N_a	12	13	14	15	16
q^2/δ^2	1.858872	1.771854	1.693854	1.623496	
N_a	17	18	19	20	

solutions exist if the $a(\zeta)$ and $\varphi(\zeta)$ functions have coinciding extremums at the center of the pulse $\zeta = 0$. Figure 5.5.1. illustrates the field pulse shape. The Table 2 presents the number of field amplitude oscillations, N_a, as a function of parameter q^2/δ^2 at $\Gamma^2 = 0.5$.

The changes in the character of the solutions and the appearance of nonlinear field amplitude oscillations have a simple qualitative explanation. Eliminating ϑ' from (5.5.6) by using (5.5.5), one can represent this equation in the form

$$\begin{gathered} a'^2 + U(a) = E, \\ U(a) = (\varphi^2/4 - \varphi - \Gamma^2)a^2 + (W/2a)^2, \\ E = (\varphi/2 - 1)W, \quad W = (1/2)[\varphi'^2 + (q/\delta)^2\varphi]. \end{gathered} \tag{5.5.7}$$

This equation has analogs in mechanics [1]. Figure 5.5.3 illustrates the evolution of the potential well $U(a)$ shape upon crossing the point $(\varphi_0/2 - 1)^2 = \Gamma^2 + 1$. The necessary condition $U(a_0) < E$ for turning points occurrence has the form

$$\sqrt{(\varphi_0/2 - 1)^2 - \Gamma^2 - 1} < (\varphi_0/2 - 1)$$

and is satisfied when $\varphi_0/2 - 1 > \sqrt{\Gamma^2 + 1}$.

The field amplitude turning points a_n are determined by the condition $a_n'(a_n) = 0$, and from (5.5.6) one gets

$$\vartheta_n' = 1 \pm \sqrt{\Gamma^2 + 1}. \tag{5.5.8}$$

If one introduces, by analogy with (5.5.3), a transverse electron velocity phase $\chi = \vartheta - \eta$ between vectors $\mathbf{E}_\perp$ and $\mathbf{v}_\perp$, it follows from equation (5.4.9) that

$$\vartheta_n' = 1 + \sqrt{\Gamma^2 + 1}\sin\chi_n, \tag{5.5.9}$$

and the values $\chi_n = n\pi/2$ correspond to turning points.

A nonlinear dispersion equation determining the dependence of both the plasma refractive index N and wave phase velocity $\beta_{\text{ph}} = N^{-1}$ on the oscillation amplitude can be derived as follows. One expresses ϑ_n' in (5.5.8) in terms of the field amplitude a_m and φ_m ($a_m' = \varphi_m' = 0$) external values at the pulse center $\zeta = 0$:

$$(N^2 - 1)\left(1 - \frac{\varphi_m}{2} + \frac{q^2\varphi_m^2}{4\delta^2 a_m^2}\right)^2 = \frac{q^2}{\delta}. \tag{5.5.10}$$

The wave group velocity is determined by (5.4.15) and, in view of (5.5.10), is equal to

$$\beta_g^{-1} = 1 + \frac{q^2}{2\delta^2}\left(1 - \frac{\varphi_m}{2} + \frac{q^2\varphi_m^2}{4\delta^2 a_m^2}\right)^{-2}. \tag{5.5.11}$$

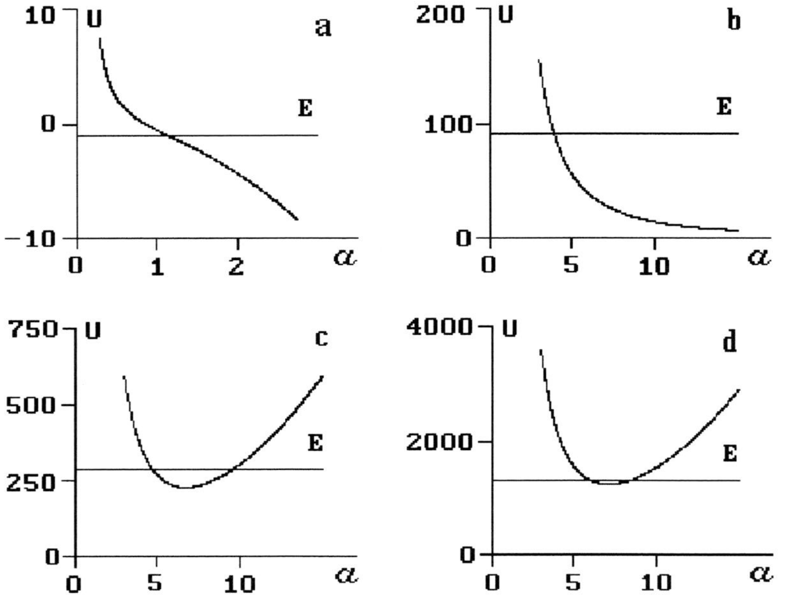

Fig. 5.5.3. Potential well $U(a)$ evolution at ζ = -2.2 (a), -1.6 (b), -1.5 (c), -1.2 (d) for Γ=0.5 and $q^2/\delta^2 = 2.959586$.

It follows from (5.5.10) that nonlinearity parameter is

$$G = \frac{\tilde{\varphi}_m}{\delta} = \frac{\beta_{\rm im}}{\delta\mu}, \tag{5.5.12}$$

where $\tilde{\varphi}_m = \delta\varphi_m$ and $\beta_{\rm im}$ the Coulomb potential and the ion velocity at the solitary pulse center, respectively. In a week electromagnetic field $G \ll 1$, (5.5.10) and (5.5.11) coincide with the formulas of the linear theory, and for $G \geq 1$, the phase and group velocities of the wave depend upon the oscillation amplitude.

At $\varphi_m \simeq 1$, it follows from the third equation (5.5.4) that $a_m \simeq q/\delta$; and one can use formula (5.5.3) and equations (5.4.8) - (5.4.10) to obtain the following estimates at the order of magnitude accuracy level

$$\mathcal{E}_m \simeq \delta^{3/2}, \quad |\beta_m| \simeq \delta^{1/2}, \quad \beta_{\|m} \simeq \delta. \tag{5.5.13}$$

The numerical study of set (5.5.4) showed that the number of field amplitude nonlinear oscillation increases (Fig. 5.5.1c) with the decrease in plasma density (parameter q) if the inequality

$$\frac{q}{\delta} \ll \Gamma \tag{5.5.14}$$

is satisfied, i.e., the Langmuir frequency is small as compared with the nonlinear frequency of field amplitude oscillations. An analytical solution of set (5.5.4) may be obtained by using the WKB approximation [14].

Integrating (5.5.7), we get, in this approximation,

$$a^2 = \frac{W}{2\Omega^2}\left[\frac{\varphi}{2} - 1 + \sqrt{\Gamma^2+1}\,\sin\left(2\int_0^\zeta \Omega d\zeta\right)\right],$$
$$\Omega = \sqrt{\varphi/2 - 1)^2 - \Gamma^2 - 1}, \tag{5.5.15}$$

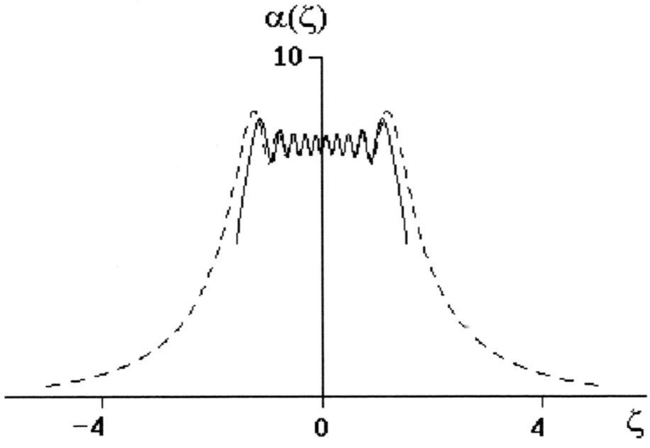

Fig. 5.5.4. Function a of $a(\zeta)$: the dashed curve vepresents the numerical calculations, the solid curve shows the results of the adiabatic theory.

where the point $\zeta = 0$ corresponds to the pulse center [7])

The equation for the potential (5.5.4) is simplified by the substitution

$$a^2 \to \langle a^2 \rangle = \frac{W(\varphi - 2)}{(\varphi - 2)^2 - 4\Gamma^2 - 4} \tag{5.5.16}$$

The averaging is performed over an oscillation period of (5.5.15), and after integrating (5.5.7), one gets the relation

$$\varphi'^2 + \left(\frac{q}{\delta}\right)^2 \varphi^2 = C_m \sqrt{\left(\frac{\varphi}{2} - 1\right)^2 - \Gamma^2 - 1}\,. \tag{5.5.17}$$

The integration constant

$$C_m = \left(\frac{q}{\delta}\right)^2 \varphi_m^2 \left[\left(\frac{\varphi_m}{2} - 1\right)^2 - \Gamma^2 - 1\right]^{-1/2}, \tag{5.5.18}$$

corresponds to the pulse center $\zeta = 0$ where the potential φ_m reaches its maximum value. Figure 5.5.4 shows that the results calculated according to formulas (5.5.15) and (5.5.17) are in satisfactory agreement with the curves obtained by integrating numerically the initial equations (5.5.4).

It should be noted that the potential φ_m at the pulse center is an independent parameter of the analytical solution obtained. It determines the frequency of nonlinear oscillations Ω in (5.5.15) at a fixed q/δ value. On the other hand, when integrating numerically, it is convenient to specify, from the soliton existence condition , the q/δ and Γ parameters determining the value of φ_m. In the parameter region bounded by inequality (5.5.14) and corresponding to the adiabatic approximation, these points lie on the surface shown in Fig. 5.5.5.

---

[7]) Note that the first equation (5.5.4) is reduced to the standard form of the **WKB** approximation by changing the $a = \sqrt{Y}$,

$$Y'' + 4\Omega^2 Y = 2W(\varphi/2 - 1), \quad \Omega' \ll \Omega^2,$$

and its solution satisfying the integral (5.5.8) coincides with (5.5.15).

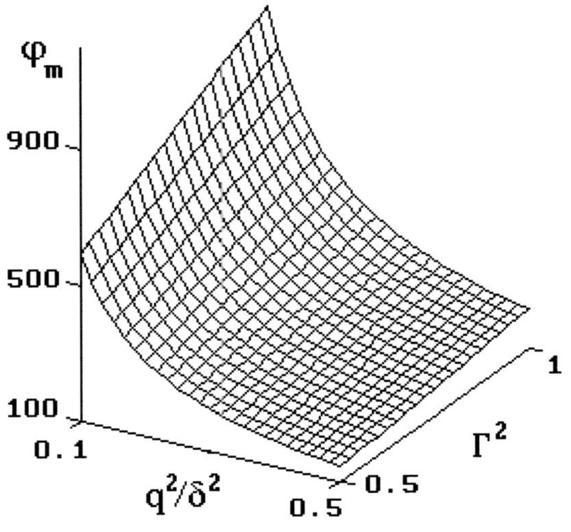

Fig. 5.5.5. Potential at the pulse center φ_m in the region of parameters $q/\delta\Gamma<1$.

Thus, powerful electromagnetic pulses propagating in a plasma along a strong magnetic field under cyclotron resonance conditions are shown to take the form of solitons with "built-in" Langmuir oscillations. Nonlinear saturation of the of field amplitude arises as a result of charge separation in a plasma due to the electromagnetic radiation pressure. A discrete set of carrier frequencies for soliton envelope solutions is determined by the ratio of the frequency of nonlinear longitudinal electron oscillations in a field of an electromagnetic wave to the plasma Langmuir frequency. With plasma density decrease, the discrete frequency spectrum is shown, by numerical integration, to become a continuous one. An asymptotic analytical solution is obtained, in this case, describing the physics of electromagnetic radiation interaction with electrons and ions of a plasma under strong nonlinearity condition.

§5.6. Resonant solitons in a nonlinear two-level system.

The electromagnetic solitons found above in a plasma with a admixture of the resonance particles are qualitatively similar to that discussed in a nonlinear optical medium [3,4]. The technique of seeking analytically of the nonlinear solutions of ordinary differential equations, developed in §5.4, is used in this section to get the conditions of the existence of solitons in a two-level system [4]. A shape of envelope solitons and a dispersion law for these solitons are derived below in the case of a pronounced nonlinearity of the atoms.

Following to Ref. [4], we start with the following system of nonlinear equations:

$$\begin{aligned} \frac{\partial^2 E}{\partial z^2} - \frac{1}{c^2}\frac{\partial^2 E}{\partial t^2} &= \frac{4\pi}{c^2}\frac{\partial^2 P}{\partial t^2}, \\ \frac{\partial^2 E}{\partial t^2} + \Omega^2 P &= -\frac{2Nd_0^2\Omega}{\hbar} WE, \end{aligned} \tag{5.6.1}$$

where E and P are the electric field and polarization vectors; Ω, N, and d_0 are the resonant frequency, the density, and the dipole moment of the atoms; the population difference for the two levels is given by [15]

$$W = -\sqrt{1 - \left(\frac{P}{Nd_0}\right)^2 - \left(\frac{1}{Nd_0\Omega}\frac{\partial P}{\partial t}\right)^2} \tag{5.6.2}$$

In the linear approximation, Eq. (5.6.1) implies the dispersion equation

$$\left(\frac{ck}{\omega}\right)^2 = 1 + \frac{q^2}{\delta} \equiv n_0^2(\omega), \quad q^2 = \frac{8\pi N d_0^2}{\hbar\Omega}, \quad \delta = \frac{\Omega^2}{\omega^2} - 1 \tag{5.6.3}$$

for the harmonic $\exp(i\omega t - ikz)$. This equation has the singularity at $\delta = 0$, i.e., when the wave frequency coincides with the natural frequency Ω of the medium. The coefficients

$$v_g = cn_0\left(1 + \frac{q^2}{\delta^2}\right)^{-1}, \quad v_g' = -\frac{c^2}{\omega^2}\left(\frac{v_g}{cn_0}\right)^3 \frac{q^2}{\delta^2}\left(1 + 3n_0^2\right) \tag{5.6.4}$$

in equation (5.1) are therefore undefined for $\delta = 0$.

We seek a solution of Eqs. (5.6.1) in the nonlinear approximation as a wave envelope

$$\begin{gathered} E = Nd_0 \operatorname{Re} \mathcal{E}(\xi) \exp i\Phi, \quad P = Nd_0 \operatorname{Re} a(\xi) \exp i\Phi, \\ \Phi = \omega t - kz, \quad \xi = \Omega(z - \beta ct) \end{gathered} \tag{5.6.5}$$

with slowly varying amplitudes. This assumption is valid if

$$q^2 \ll 1, \quad |\delta| \ll 1, \quad |n^2 - 1| \ll 1, \tag{5.6.6}$$

where $n = ck/\omega$, ω and k are the frequency and wave number, and $u = \beta c$ is the wave velocity.

If we substitute (5.6.5) into (5.6.1) and (5.6.2), in equalities (5.6.6) permit us to omit the second derivatives of the amplitudes and decrease the order of the system of nonlinear equation, which we express in terms of total derivatives with respect to the variable ξ:

$$2i(n-\beta)\mathcal{E}' + (n^2 - 1)\mathcal{E} = 4\pi a, \quad -2i\beta a' + \delta a = (q^2/4\pi)\sqrt{1-a^2}\,\mathcal{E}. \tag{5.6.7}$$

Excluding a from the second equation in (5.6.7), we get the following nonlinear equation for the dimensionless complex field amplitude:

$$\begin{gathered} 4\beta(n-\beta)\mathcal{E}'' + 2i[\delta(n-\beta) - \beta(n^2-1)]\mathcal{E}' + \delta(n^2-1)\mathcal{E} \\ = (q/4\pi)^2[16\pi^2 - 4(n-\beta)^2|\mathcal{E}'|^2 - (n^2-1)^2|\mathcal{E}|^2 - 2i(n-\beta)(\mathcal{E}^*\mathcal{E}' - \mathcal{E}^*\mathcal{E}^{*\prime})]^{1/2}\mathcal{E}. \end{gathered} \tag{5.6.8}$$

This equation can be simplified considerably if the coefficient of $\mathcal{E}$ vanishes and $\mathcal{E}$ is real-valued; this corresponds to a wave with a resonant group velocity, in which case no phase shift accompanies changes in amplitude. We see easily that such a solution exists for $n = 1$ and $\delta = 0$; however, the velocity u remains undetermined. We will therefore assume that the expression in square brackets in Eq. (5.6.8) vanishes, so that

$$\beta = n[1 + (n^2-1)/\delta]^{-1}. \tag{5.6.9}$$

We see by comparing (5.6.9) and (5.6.4), that Eq. (5.6.9) is equivalent to $u = v_g$. However, the group velocity $v_g = c\beta_g$ now depends on the field amplitude and thus cannot coincide with the linear result (5.6.4).

If we recall that (5.6.9) gives the coefficient of the highest derivative in (5.6.8), we can introduce the new variables

$$y = \frac{\mathcal{E}}{4\pi}|n^2 - 1|, \quad \tau = \frac{|\delta|}{2\beta}\xi, \quad Q = \frac{q^2}{\delta(n^2-1)} \tag{5.6.10}$$

and recast Eqs. (5.6.8) in the form

$$y'' + y = Q\sqrt{1 - y^2 - y'^2}\, y, \tag{5.6.11}$$

where the primes denote derivatives with respect to τ.

We seek a soliton solution of Eq. (5.6.11), i.e., a solution satisfying $y = y' = 0$. We find after integrating that

$$y' = y\sqrt{Q - 1 - Q^2y^2/4}. \tag{5.6.12}$$

A soliton solution will exist if $Q > 1$; in this case the right-hand side of (5.6.12) has a "turning point" and we have the solution

$$y = \frac{2}{Q}\sqrt{Q-1}\,\text{sech}\left(\sqrt{Q-1}\,\tau\right). \tag{5.6.13}$$

It is convenient to express the intermediate parameter Q in terms of the maximum field amplitude $\mathcal{E}_m$. Recalling (5.6.10), we find that

$$\mathcal{E}(\xi) = \mathcal{E}_m \text{sech}\left(\frac{q^2\mathcal{E}_m}{16\pi\beta}\right), \quad \sqrt{Q-1} = \frac{q^2\mathcal{E}_m}{8\pi\delta}. \tag{5.6.14}$$

Equation (5.6.14) embodies the a nonlinear dispersion law for the soliton in a two-level system. Indeed, if we substitute (5.6.10) for Q we get the equation

$$n^2 = 1 + \frac{q^2}{\delta}\left[1 + \left(\frac{q^2\mathcal{E}_m}{8\pi\delta}\right)^2\right]^{-1}, \tag{5.6.15}$$

which generalizes the linear dispersion equation (5.6.3).

We find the group and phase velocities of the wave at resonance by substituting (5.6.15) into (5.6.9) and taking the limit $\delta \to 0$:

$$u = c\left(1 + \frac{8N\hbar\Omega}{E_m^2}\right)^{-1}, \quad v_{\text{ph}} = c, \tag{5.6.16}$$

where $E_m = Nd_0\mathcal{E}_m$. In terms of dimensional variables, Eq. (5.6.14) for the soliton pulse has the form

$$\mathcal{E} = \mathcal{E}_m \text{sech}\left[\frac{dE_m}{2\hbar}\left(t - \frac{z}{u}\right)\right]. \tag{5.6.17}$$

We observe that the wave field depends on the frequency mismatch δ from resonance only through the group velocity u. On other hand, the polarization amplitude of the medium and the population difference W in (5.6.2) depend strongly on δ. At peak pulse field $\xi = 0$ we have the relations

$$P_m = 2Nd_0\frac{\mu}{1+\mu}, \quad W_m = -\frac{1-\mu}{1+\mu},$$

where $\mu = d_0E_m/\hbar\omega|\delta|$. This implies that if $\mu > 1$ at the center of the soliton, the population of the two-level system will be inverted, W_m¿0. In particular, population inversion is always possible for a resonant soliton, $\delta \to 0$.

We consider the process of the interaction of resonance particles with the wave in the frame of reference of the group velocity, $z' = z - ut$. The non-homogenious HF-electric field of a circularly polarized plane wave propagating along the magnetic field $\mathbf{B}_0$ has the form

$$E_x + iE_y = E(z')\exp(i\omega' t - ik_0 z'), \tag{A5.1.1}$$

where $\omega' = \omega_0$-$k_0 u$. Using Maxwell's equation, $\mathbf{B} = i(c/\omega)\mathrm{curl}\,\mathbf{E}$, we can represent the magnetic field of the wave as

$$B_x + iB_y = -\frac{c}{\omega'}\left[-ik_0 E(z') + \frac{dE}{dz'}\right]\exp(i\omega' t - ik_0 z'), \tag{A5.1.2}$$

We assume that the motion of the particle is nonrelativistic so that a transverse part of the Lorentz force can be neglect,

$$\frac{d}{dt}(v_x + iv_y) + i\Omega_B(v_x + iv_y) = \frac{q}{m}(E_x + iE_y) \tag{A5.1.3}$$

(here $\Omega_B = eB_0/mc$ is the hyrofrequency and $z' = z(t)$). A solution of Eq. (A5.1.3),

$$v_x + iv_y = -\frac{iq}{m\omega'(\omega' + \Omega_B)}E(z)\exp\left(i\omega' t - ik_0 z\right), \tag{A5.1.4}$$

founded in an adiabatic approximation,

$$\dot{E} \ll \omega' E, \quad k_0\dot{z} \ll \omega', \tag{A5.1.5}$$

is used below to analyze the trapping of charge particle by the uniform RF-field.

A change in the longitudinal velocity, $v_z = \dot{z}$, determines by the Lorentz force

$$\frac{dv_z}{dt} = \frac{q}{mc}(v_x B_y - v_y B_x). \tag{A5.1.6}$$

By means of expression

$$v_x B_y - v_y B_x) = \mathrm{Re}\,(v_x + iv_y)(iB_x + B_y),$$

we rewrite the above equation as

$$m\frac{dv_z}{dt} = -\frac{d\Phi}{dz}, \quad \Phi = \frac{q^2|E(z)|^2}{2m\omega'(\omega' + \Omega_B)}. \tag{A5.1.7}$$

Integrating (A5.1.7), we get the integral of motion:

$$W_{||} = \frac{mv_z^2}{2} + \Phi(z). \tag{A5.1.8}$$

Asuming that the width of the pulse much exceeds a length of the wave, $dE/dz \ll k_0 E$, and comparing Eqs. (A5.1.4) and (A5.1.8), we obtain the inequality

$$\left(\frac{\omega'}{\omega_0}\right)^{3/2} \gg \frac{qE_m k_0}{m\sqrt{\Omega_B\omega_0^3}} \tag{A5.1.9}$$

(E_m is the maximum field amplitude) which places a lower limit the difference between the group and phase velocity.

Appendix 5.2

To obtain Eq. (5.3.30), we integrate Eq. (5.3.29) with a upper limit

$$\frac{\zeta_1}{2} = \int_{x_2}^{x_1} \frac{dx}{\sqrt{E-U}}, \tag{A5.2.1}$$

where $x = b^2$. The roots of the expression in denominator (A5.2.1) are

$$x_{1,2} = \sqrt{y_1} \pm 2\mathrm{Re}\sqrt{y_2}, \quad x_{3,4} = d + if,$$
$$d = -\sqrt{y_1}, \quad f = 2\mathrm{Im}\sqrt{y_2},$$

such that y_1 and y_2 are determined by the expressions

$$y_1 = -\frac{A_1}{6} + B + C, \quad R = \frac{1}{27}\left(B_2 - \frac{A_1^2}{6}\right)^3 + \frac{1}{4}\left(C_2 - \frac{1}{6}A_1B_2 + \frac{2}{216}A_2^3\right)^2,$$

$$y_2 = -\frac{A_1}{6} - \frac{B+C}{2} + i\frac{\sqrt{3}}{2}(B-C), \quad B, C = \left[-C_2 + \frac{1}{6}A_1B_2 - \left(\frac{A_1}{6}\right)^3 \pm \sqrt{R}\right]^{1/3},$$

$$B_2 = (A_1^2 - 4C_1)/16, \quad C_2 = -B_1/64,$$
$$A_1 = B - 3A^2/8, \quad B_1 = A^3/8 - AB/2 + C,$$
$$C_1 = -3A^4/256 + A^2B/16 - AC/4 + D,$$
$$A = 8(1 + \Psi/2), \quad B = 16(\Psi^2/4 + \Psi + W - Q^2),$$
$$C = 16W(1 + \Psi/2), \quad D = 4W^2.$$

Since equation $E - U = 0$ has two real and two complex roots, to reduce integral (A5.2.1) to the normal Legendre form, we must use the substitution [13]

$$x = \frac{x_1 + x_2}{2} - \frac{x_1 - x_2}{2}\frac{\lambda - \cos\alpha}{1 - \lambda\cos\alpha},$$
$$\lambda = (g-h)/(g+h), \quad g, h = \sqrt{(x_{1,2} - d)^2 + f^2}. \tag{A5.2.2}$$

With allowance for the latter relation, the integral on the right side in (A5.2.1) is transformed into

$$\int_{x_1}^{x} \frac{dx}{\sqrt{E-U}} = \frac{1}{\sqrt{gh}}F(\alpha, k),$$

$$k^2 = \frac{1}{2}\left(1 - \frac{(x_1 - d)(x_2 - d) + f^2}{gh}\right), \quad \cos\alpha = 1 - \frac{2(1-\lambda)(x_1 - x)}{x_1(1-\lambda) - x_2(1+\lambda) + 2x\lambda}, \tag{A5.2.3}$$

where $F(\alpha, k)$ is the incomplete elliptic integral of the first kind. Substituting Eq. (A5.2.3) into (A5.2.1) and allowing for (A5.2.2), we obtain Eq. (5.3.30).

The average value of the square of the field amplitude is defined by the formula

$$\langle x \rangle = \frac{1}{T}\int_{x_1}^{x_2} \frac{x\,dx}{\sqrt{E-U}}. \tag{A5.2.4}$$

Calculating integral (A5.2.4)

$$\langle x \rangle T = \frac{1}{\sqrt{gh}}\left[\frac{gx_1 - hx_2}{g-h}K(k) - \frac{x_1 - x_2}{2}\frac{g+h}{g-h}\Pi(-\gamma, k)\right],$$

where $K(k)$ and $\Pi(-\gamma, k)$ are the complete elliptic integral of the first and third kind, $\gamma = (f-g)^2/(4fg)$, we obtain Eq. (5.3.32).

CHAPTER 6

COLLECTIVE ACCELERATION BY IONS BY A RELATIVISTIC ELECTRON BEAM

In connection with the creation of powerful relativistic electron beam, it has become possible to realize in practice the ideas of Veksler [1], who proposed to accelerate heavy ions by using the inverse Cherenkov effect, which arises in a moving medium. The gist of this phenomenon is that an electron beam moving relative to an ion bunch produces, in the reference frame connected with the ions, an electric field that is stationary with respect to the coordinates and has a geometry that depends on the shape of the bunch. This field is proportional to the density of the electron beam and to the number of particles in the bunch. Therefore, by using strong-current beams for the acceleration and increasing the number of particles in the bunch, it is possible to accelerate ions to appreciable energies.

The nonlinear theory of collective acceleration by ions by a relativistic electron beam (REB) presented in the paper [2] yields the characteristic parameters of the process, viz., the energy acquired by the ions, the acceleration time, and the maximum density of the accelerated ion bunch. The nonlinear saturation of the electron-ion instability caused by the phase bunching of the accelerated ion and the trapping of the resonant bunches by the wave. Sinchronization of the accelerated ions with the charge density wave in the electron beam can be provided with the help of the magnetic field diminishing along the acceleration direction [3-5]. The mechanism of the ion acceleration by the relativistic beam in the converging wave guard was suggested in [6], and acceleration by the modulated electron beam was discussed in [7,8]. More references on collective ion acceleration by an electron beam can be found in [5,8].

§6.1. Nonlinear phase oscillations of trapped ions in a REB.

Let a relativistic electron beam move with velocity v_0 relative to an infinite chain of ion bunches located at a distance ℓ from each other. If the linear dimensions of the bunches $\delta\ell$ are small compared with the distance between them, $\delta\ell \ll \ell$, then the nonlinear effects due to interaction with the electric field and connected with the motion of the bunches as a unit can be taken into account by replacing each bunch by a charged layer with a surface charge density σ [2]

$$n_i(t,x) = q\sigma \sum_{s=-\infty}^{\infty} \delta\left[x - s\ell - x_s(t)\right], \quad \frac{d}{dt}\frac{M\dot{x}_s}{\sqrt{1-\dot{x}_s^2/c^2}} = qE(t,x_s), \tag{6.1.1}$$

where $x_s(t)$ isis the coordinate of each bunch, and q and M are the charge and mass of the ion. .

We consider that a relativistic electron beam remains linear over all time of the ion acceleration,

$$|\tilde{v}| = |v_e - v_0| \ll v_0, \quad |\tilde{n}| = |n_e - n_0| \ll n_0, \tag{6.1.2}$$

so that the linearized hydrodynamic equations can be used.

The electric field satisfy the Poisson equation

$$\frac{\partial E}{\partial x} = 4\pi\left(-en_e + qn_i + qN\right), \tag{6.1.3}$$

where N is the density of the immovable ions which support a charge neutrality of the electron beam. Eliminating the electron density n_e, we represent the electric field equation in the form:

$$\frac{\partial}{\partial x}\left[\left(\frac{\partial}{\partial t} + v_0\frac{\partial}{\partial x}\right)^2 E + \omega_0^2 E\right] = 4\pi q\left(\frac{\partial}{\partial t} + v_0\frac{\partial}{\partial x}\right)^2 n_i, \tag{6.1.4}$$

where

$$\omega_0^2 = \frac{4\pi n_0 e^2}{m\gamma_0^3}, \quad \gamma_0 = \frac{1}{\sqrt{1-\beta_0^2}}, \quad \beta_0 = \frac{v_0}{c}.$$

Taking into account the spatial periodicity of the system, we seek the solution of Eq (6.1.4) in the form:

$$\begin{gathered} E(t,x) = \operatorname{Re}\mathcal{E}(t)\exp\left[i\vartheta(t) - ikx\right], \quad k = 2\pi/\ell, \\ \omega_0 = kv_0, \quad \dot{E} \ll \omega_0 E, \quad \dot{\vartheta} \ll \omega_0\vartheta \end{gathered} \tag{6.1.5}$$

where $\mathcal{E}(t)$ and $\vartheta(t)$ are functions that very slowly varying during the period of plasma oscillations. Substituting (6.1.5) in (6.1.4) and averaging over the spatial period of the system, we obtain a system of ordinary nonlinear differential equations with respect to time

$$\begin{gathered} \dot{\mathcal{E}} = 2q\sigma\omega_0\cos\left(\vartheta - kx\right), \quad \dot{\vartheta} = -\left(2q\sigma\omega_0/\mathcal{E}\right)\sin\left(\vartheta - kx\right), \\ \frac{d}{dt}\frac{M\dot{x}}{\sqrt{1-\dot{x}^2/c^2}} = q\mathcal{E}\cos\left(\vartheta - kx\right). \end{gathered} \tag{6.1.6}$$

In deriving (6.1.6) we imply that the ions bunches move identically over all spatial periods of the system so that a subscript s can be dropped. We suppose that a charge neutrality condition

$$en_e = q\left(\frac{\sigma}{\ell} + N\right) \tag{6.1.7}$$

is fulfilled and the electrostatic field is not taken into account when an ion acceleration by the high-frequency field occurs.

Expressing $\dot{x}$ in terms of $\mathcal{E}$ from the first and third equations (6.1.6)

$$\dot{x} = \frac{c\mathcal{E}^2}{\sqrt{\left(4c\sigma\omega_0 M\right)^2 + \mathcal{E}^4}}. \tag{6.1.8}$$

and changing over to the $\Phi = kx - \vartheta$, we represent the system of equations (6.1.6) in the form:

$$\begin{gathered} \dot{\mathcal{E}} = 2q\sigma\omega_0\cos\Phi, \\ \dot{\Phi} = -2q\sigma\omega_0\frac{1}{\mathcal{E}}\sin\Phi + \frac{ck\mathcal{E}^2}{\sqrt{\left(4c\sigma\omega_0 M\right)^2 + \mathcal{E}^4}}. \end{gathered} \tag{6.1.9}$$

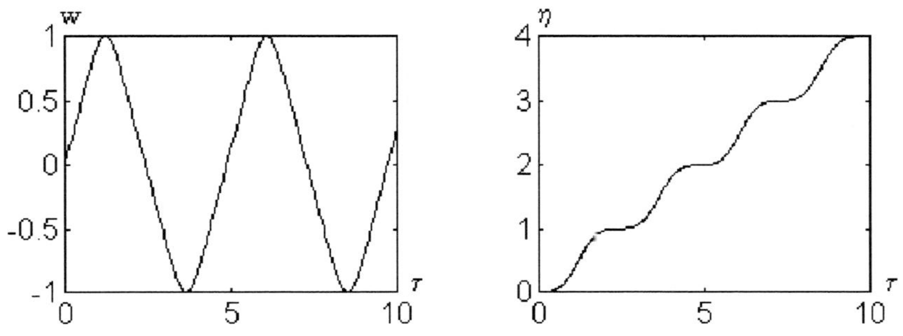

Fig. 6.1.1. Field amplitude $w=\mathcal{E}/\mathcal{E}_m$ and relative phase $\eta=\Phi/\pi$ versus time $\tau=t/T$.

Integrating (6.1.9), we can find the dependence of Φ on $\mathcal{E}$:

$$\sin\Phi = \frac{1}{4\pi\sigma\beta_0}\frac{1}{\mathcal{E}}\left[\sqrt{(4c\sigma\omega_0 M)^2+\mathcal{E}^4}-4c\sigma\omega_0 M\right], \tag{6.1.10}$$

where $\Phi(0)=\mathcal{E}(0)=0$.

Since, according to the first equation of the system (6.1.9), the maximum of the field amplitude $\mathcal{E}_m$ corresponds to $\Phi_m=\pi/2$. It follows that equation for $\mathcal{E}_m$ can be obtained by putting in (6.1.10) $\sin\Phi_m=1$:

$$\left(\frac{\mathcal{E}_m}{4\pi\sigma}\right)^3+\beta_0^2\frac{\mathcal{E}_m}{4\pi\sigma}-\frac{\omega_0^2}{\omega_1^2}=0,\quad \omega_1^2=\frac{4\pi e^2\sigma}{M\ell}. \tag{6.1.11}$$

The solution of this equation at $\omega_0>\omega_1$ can be represented with a high degree of accuracy in the form:

$$\mathcal{E}_m=4\pi\sigma\left(\frac{\omega_0}{\omega_1}\right)^{2/3}\left[1+\frac{\beta_0^2}{3}\left(\frac{\omega_1}{\omega_0}\right)^{4/3}\right],\quad \frac{\omega_1^2}{\omega_0^2}\ll 1 \tag{6.1.12}$$

(when $\omega_0=\omega_1$ and $\beta_0=1$, the next higher-order addition to unity is equal to 1/81).

The maximum ion beam velocity $\dot{x}_m$ is obtained by substituting $\mathcal{E}_m$ from (6.1.12) in (6.1.8):

$$\frac{\dot{x}_m}{c}=\frac{2\beta_0(\omega_1/\omega_0)^{2/3}}{\sqrt{1+4\beta_0^2(\omega_1/\omega_0)^{4/3}}}. \tag{6.1.13}$$

The characteristic acceleration time is determined by formulas (6.1.9) and (6.1.12), and turn out to be

$$T\simeq\frac{2}{\omega_1^{2/3}\omega_0^{1/3}}. \tag{6.1.14}$$

When $t>T$, the ions shift into the region of decelerating phases of the field, and a process opposite to acceleration begins, with the ions transferring energy to the electron beam. Numerical solution of set (6.1.9) for non-relativistic ions is represented ($4c\sigma\omega_0 M\gg\mathcal{E}^2$) is represented in Fig. 6.1.1.

Since we have assumed above that the characteristic time of variation of the amplitude and of the phase of the field greatly exceeds the time of plasma oscillation, it follows that the formulas obtained above are valid only when $(\omega_1/\omega_0)^{2/3}\ll 1$ and, according with (6.1.12), the energy acquired by the ion beam is small compared with the energy of the electron beam. Extrapolating the formulas obtained above into the region of higher ion densities $\omega_1\simeq\omega_0$, we see that the efficiency of acceleration increases.

A change in the velocity of beam electrons which determines by the equation of motion is $\tilde{v} \simeq e\mathcal{E}/m\gamma_0^3\omega_0$. Substituting $\mathcal{E} \simeq \mathcal{E}_m$ from formula (6.1.12), we obtain

$$\tilde{v}_m \simeq 2v_0 \left[\frac{q}{e} \frac{M}{m\gamma_0^3} \left(\frac{\sigma}{n_0 \ell} \right)^2 \right]^{1/3} . \tag{6.1.15}$$

which satisfy the inequality (6.1.2) under approximation considered by us. The relativistic factor remains linear ($\gamma - \gamma_0 \ll \gamma_0$) under the condition

$$2\beta_0 \frac{\tilde{v}_m}{c} \ll \frac{1}{\gamma_0^2}. \tag{6.1.16}$$

which restricts an applicability of the above model in the region of high energies.

The acceleration process investigated above admits of a simple physical explanations. In the reference frame connected with the electron beam, the ions produce a modulated beam moving through the electron-ion plasma with a velocity $-v_0$. Such a beam excites plasma oscillations and is decelerated during the time of (6.1.15). In the laboratory frame, this effect perceived as acceleration of the ions in the direction of motion of the electron beam.

§6.2. Heavy positive ion bunch.

Let us consider now the interaction of a "heavy" bunch with the relativistic electron beam, when main nonlinear effect limiting the maximum energy of the ions is the deceleration of the electron beam in the field produced by the ions. We use for the calculation the model of "trapped particles" in which it is assumed that the bunch is in the potential well produced by the field of the wave and moves together with the wave [2,9].

1. Equilibrium ion bunch.

The initial system of equations consists of the hydrodynamic equation of motion and the continuity equation for the relativistic beam electrons

$$\left(\frac{\partial}{\partial t} + v \frac{\partial}{\partial x} \right) \gamma v = \frac{e}{m} \frac{\partial \varphi}{\partial x}, \quad \frac{\partial n}{\partial t} + \frac{\partial n v}{\partial x} = 0, \tag{6.2.1}$$

the kinetic equation for the ions trapped by the Langmuir wave

$$\frac{\partial f}{\partial t} + u \frac{\partial f}{\partial x} - \frac{q}{M} \frac{\partial \varphi}{\partial x} \frac{\partial f_0}{\partial u} = 0, \tag{6.2.2}$$

and Poisson's equation for the self-consistent potential $\varphi(t, x)$:

$$\frac{\partial^2 \varphi}{\partial x^2} = 4\pi e(n - n_0) - 4\pi q n_i, \quad n_i = \int f \, du, \tag{6.2.3}$$

where v and n are the velocity and the density of the electron beam is accompanied by the ion beam with velocity v_0 and density n_0, $\gamma = \left(1 - v^2/c^2\right)^{-1/2}$; $f(t, z, u)$ is the distribution function, and M and q are the mass and the charge of the trapped ions.

Integrating Eq. (6.2.1), in the reference frame connected with the ions ($\partial/\partial t = 0$), we find the energy and current density integrals

$$mc^2\gamma - e\varphi = mc^2\gamma_0, \quad n_e v = n_0 v_0. \tag{6.2.4}$$

In the system of coordinates in which the wave is at rest, the distribution function of the trapped ions does not depend on time and is an arbitrary function of the parameter $E = -2q\varphi/M - u^2$ $(q > 0, \varphi < 0)$. Following the method used by Bohm and Gross (see Refs. [10,11]), we put

$$f(x,u) = -\frac{NM}{\pi q \Phi}\sqrt{-\frac{2q\varphi}{M} - u^2}, \tag{6.2.5}$$

where N is the ion density at the bottom of the well, i.e., at $\Phi = \max\varphi$.

The density of the trapped particles can be found by integrating the distribution function with respect to the velocities between the limits $-\sqrt{-2q\varphi/M}$ and $\sqrt{-2q\varphi/M}$:

$$n_i(\varphi) = N\frac{\varphi}{\Phi}. \tag{6.2.6}$$

Substituting $n(\varphi)$ and $n_i(\varphi)$ in Poisson's equation and introducing dimensionless variable, we obtain the following equation for the potential

$$w'' = \frac{\beta_0(1+w)}{\sqrt{(1+w)^2 - \gamma_0^{-2}}} - 1 - \nu\frac{w}{W}, \tag{6.2.7}$$

where

$$w = \frac{e\varphi}{mc^2\gamma_0}, \quad W = \frac{e\Phi}{mc^2\gamma_0}, \quad \nu = \frac{qN}{en_0},$$

$$\omega_e^2 = \frac{4\pi n_0 e^2}{m\gamma_0}, \quad \beta_0 = \frac{v_0}{c},$$

and the prime means the derivative with respect to the variable $\xi = \omega_e x/c$.

The first integral of this equation, satisfying the conditions $w = 0$, $w' = 0$, has the form

$$w'^2 = 2\beta_0\left[\sqrt{(1+w)^2 - \gamma_0^{-2}} - \beta_0\right] - 2w - \nu\frac{w^2}{W}. \tag{6.2.8}$$

The maximum amplitude W of the potential can be found by putting $w'(W) = 0$:

$$W = -\frac{\beta_0^2\nu}{(1+\nu/2)^2 - \beta_0^2}. \tag{6.2.9}$$

Substituting W from (6.2.9) into (6.2.8), it can be proved that in the present case a solitary wave solution can exist only for positively charged trapped ions $(q > 0)$.

Using (6.2.9), it is possible to estimate the optimum density of ionic bunch $\nu_{\max}$ and the depth of potential well $W_{\max}$ in this case. Putting $W'(\nu) = 0$, we find

$$W_{\max} = -\frac{\gamma_0\beta_0^2}{1+\gamma_0}, \quad \nu_{\max} = \frac{2}{\gamma_0}. \tag{6.2.10}$$

Note that according to (6.2.10), an ion density in the bunch exceeds an electron density in a frame of reference of the beam by factor of two and optimum depth of the potential well is less than $mv_0^2\gamma_0/2$.

If the density of the trapped ions is small ($\nu \ll \gamma_0^{-2}$), Eq. (6.2.8) can be integrated and the analytical dependence of the potential on the coordinates can be found:

$$w = -\beta_0^2\gamma_0^2\nu\cosh^{-2}\left(\frac{\sqrt{\nu}}{2\beta_0}\xi\right) \tag{6.2.11}$$

(see **Appendix 6.1**).

The electric field ($E = -mc^2\gamma_0 dw/dx$) is equal to

$$E = -\gamma_0^3 \nu^{3/2} \frac{mv_0\omega_e}{e} \sinh\left(\frac{x}{L}\right) \cosh^{-3}\left(\frac{x}{L}\right), \tag{6.2.12}$$

where the size of ion bunch is determined by the parameter $L = 2v_0/\omega_e\sqrt{\nu}$. For large densities of the trapped particles ($\nu \simeq \gamma_0^{-2}$), the typical dimension of the ion bunch is of the same order of magnitude as

$$E_{NL} \simeq \frac{mv_0\gamma_0\omega_e}{e}, \tag{6.2.13}$$

and it is proportional to $\sqrt{n_0}$ (as distinct from the analogous formula, $E_L \sim n_0$, which is predicted by the linear theory [1]).

The electrostatic equilibrium investigated above admits of a simple physical explanation. In the coordinate region $x < 0$ the velocity of electron beam decreases under the action of retarding electric field $-eE < 0$. Therefore according to (6.2.4), its density increases and becomes bigger than a density of the background moving ions at the bunch center, $\xi = 0$, so that arising surplus negative charge can be compensated only by the positive charge of ionic bunch. When $x > 0$, the electrostatic force changes sign, and a process opposite to deceleration begins, with the ions transferring energy to the electron beam.

The collective acceleration of ions by means of moving potential well was investigated numerically in Ref. [30].

2. Electron-ion instability.

We now investigate the stability of these waves against shortwave perturbations:

$$k \gg L^{-1} = \omega_e\sqrt{\nu}/2v_0. \tag{6.2.14}$$

After linearizing the system (6.2.1) - (6.2.3) with respect to small perturbations of the density and of the velocity of the plasma electrons, of the potential, and of the distribution function of the trapped ions, and assuming the perturbations to depend on time and on the coordinate as $\exp(ikx - i\omega t)$, we obtain the following dispersion equation [9] which defines the dependence of the frequency ω on the wave number k and on the parameters of the solitary wave:

$$1 - \frac{\omega_0^2}{(\omega - kv_0)^2} - \frac{4\pi q N}{k^2\Phi}\left(1 - \frac{1}{\sqrt{1 - k^2U^2/\omega^2}}\right) = 0,$$
$$\omega_0^2 = \frac{4\pi e^2 n_0}{m\gamma_0^3}, \quad U^2 = -\frac{2q\varphi}{M}, \tag{6.2.15}$$

If we assume $kU/\omega \ll 1$ (the condition required for this approximation to be valid will be introduced later), Eq. (6.2.15) can be simplified:

$$1 - \frac{\omega_0^2}{(\omega - kv_0)^2} - \frac{4\pi q^2 N}{M\omega^2}\frac{\varphi}{\Phi} = 0. \tag{6.2.16}$$

If in Eq. (6.2.16) we put $\varphi/\Phi \simeq 1$, it becomes fully equivalent to the dispersion relation describing the interaction of an unbounded, compensated electron beam with an ion plasma [12]. Indeed, in the frame of reference of the wave, the electrons and ions of the plasma move with constant velocity through an ion background. In this case, under the present shortwave approximation, the dimensions of the pulse do not affect the dispersion properties of the system.

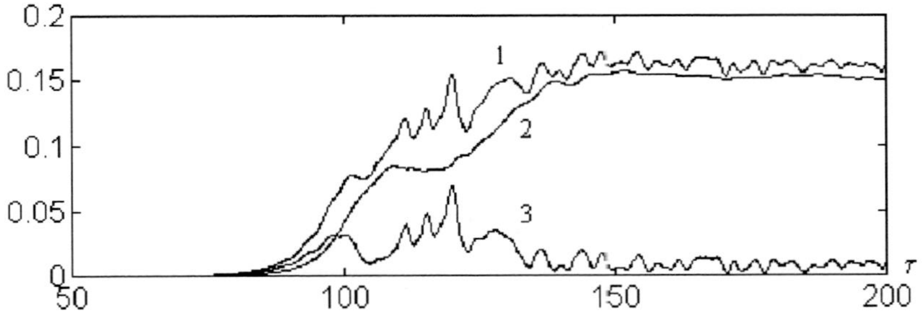

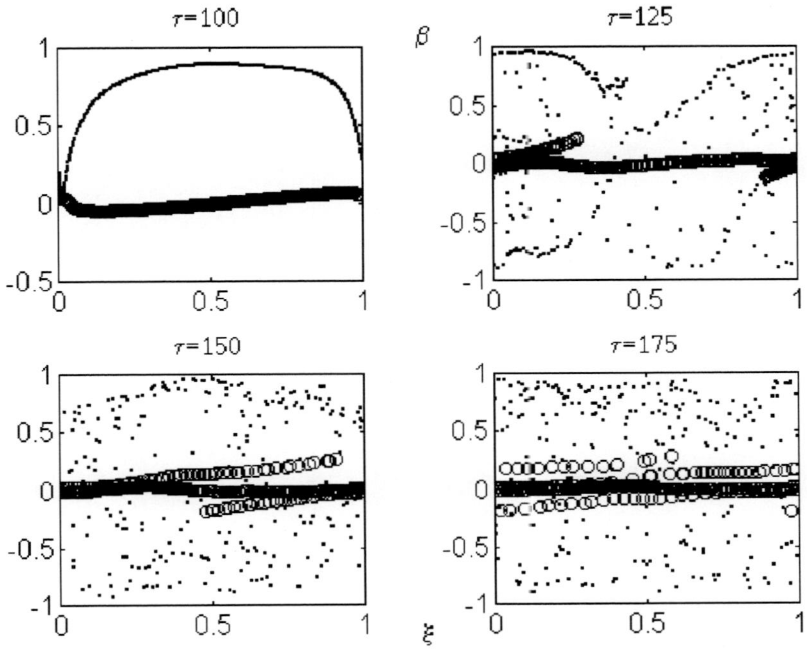

Fig. 6.2.1. Energy density lost by the beam (1), the energy density of ions (2), and the field energy (3) as a function of time $\tau=\omega_0 t$ for a non-relativistic beam $\gamma_0=1.75$ (in units of $8\pi n_0 mc^2\gamma_0$).

Fig. 6.2.2. Phase plane ($\xi=kx$, $\beta=v/c$) of electrons (points) and ions (circles) at the different moments for $\gamma_0=1.75$.

We shall now investigate Eq. (6.2.16), and note that the last term of this equation is small compared with the first ($\sim m/M$). Putting $\omega_0 = kv_0 \gg \omega$, we obtain the following expression for the frequency ω:

$$\left(\frac{\omega}{\omega_e}\right)^3 = -\frac{\omega_i^2}{\omega_e^2}, \quad \omega_i^2 = \frac{4\pi q^2 N}{M\omega^2}\frac{\varphi}{\Phi}, \tag{6.2.17}$$

which has complex roots

$$\omega = \frac{1 \pm i\sqrt{3}}{2^{4/3}}\,\omega_i^{2/3}\omega_0^{1/3}. \tag{6.2.18}$$

It is easy to see that the frequency at which the instability sets in is $\omega \sim \omega_e\,(m/M)^{1/3}$.

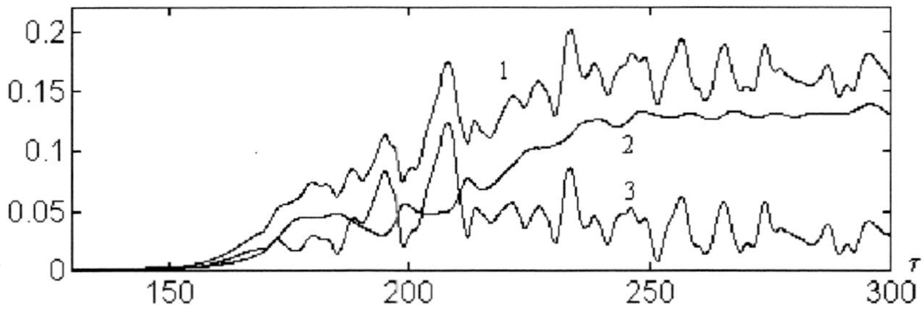

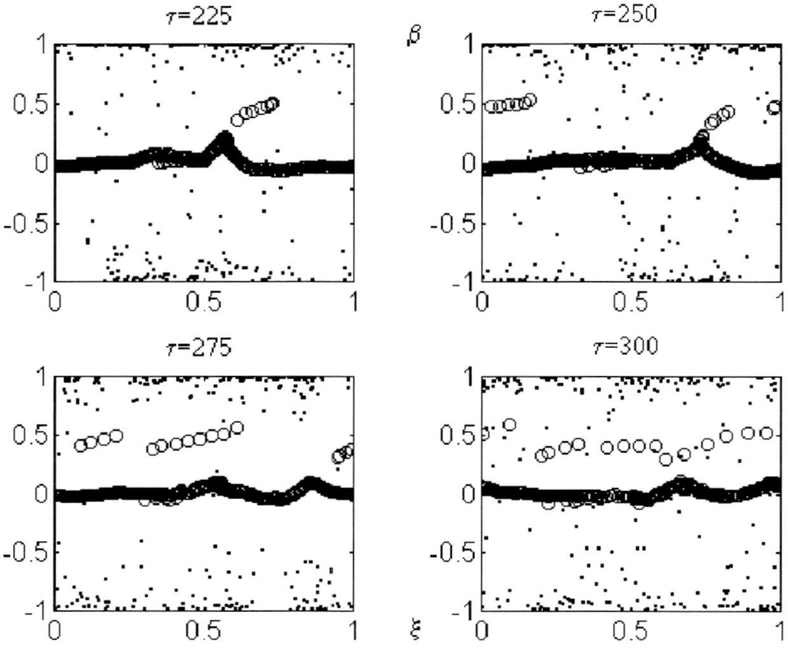

Fig. 6.2.3. Energy density lost by the beam (1), the energy density of ions (2), the field energy (3) as a function of time $\tau=\omega_0 t$ for a relativistic beam $\gamma_0=5$ (in units of $8\pi n_0 mc^2\gamma_0$).

Fig. 6.2.4. Phase plane ($\xi=kx$, $\beta=v/c$) of electrons (points) and ions (circles) at the different moments for $\gamma_0=5$.

Therefore the quantity kU/ω ($U \simeq q\Phi/M$) remains small if the following inequality holds

$$\left(\frac{em}{qM}\right)^{1/3} \ll \gamma_0 \left(\frac{N}{n_0}\right)^{7/6}. \tag{6.2.19}$$

An exponential growth of the small hydrodynamic perturbations (6.2.17) results in a slowing of the electron beam and a passing of the electron-ion system in the kinetic stage of instability. Particle-in-cell simulations of the ion plasma and electron beam, founded on the Vlasov-Poisson set of equations, was used by Y.A.Volkov to analyze the nonlinear dynamics of electron-ion instability. The results of these numerical calculations are represented in

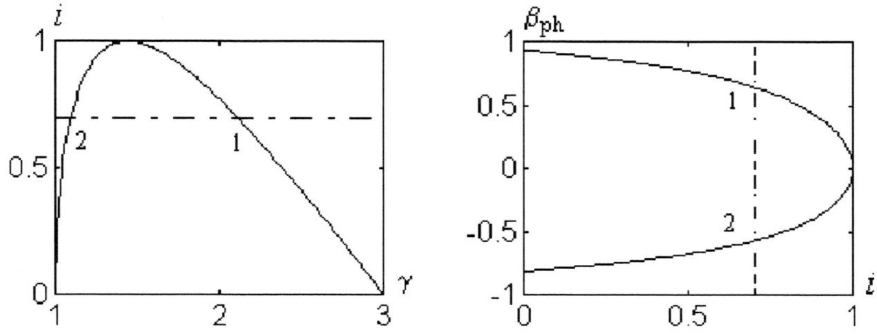

Fig. 6.3.1. Beam current $i=I/I_L$ as a function of electron relativistic factor, γ. The numerals 1 and 2 denote the equilibrium states of the beam existing for $i < 1$.
Fig. 6.3.2. Wave phase velocity $\beta_{ph}=v_{ph}/c$ as a function of beam current, $\gamma_0=3$.

Figs. 6.2.1 - 6.2.4. It is assumed that at the initial time $t = 0$, the electrons of monoenergetic beam and the plasma ions ($q = e$, $M/m = 10^2$, $N_e = N_i = 256$) are distributed uniformly along the coordinate.

§6.3. Nonlinear evolution of a slow wave in a tubular REB.

The wave guide properties of a tubular relativistic electron beam (REB) are analyzed for a wave guide with conducting walls, with consideration of the positive ions captured by the electrostatic field of the beam to form a background plasma. The partial neutralization of the electron beam for a specified injection energy. The system is unstable, and the development of the electron-ion mode leads to a significant nonlinear reduction in the phase velocity of the slow wave as a result of the relationship between the wave guide properties of the system (the limit current) and the amplitude of the oscillations.

It is well known that there exists a limit current for a tubular beam REB moving in a vacuum wave guide with conducting walls along a powerful magnetic field [13-16]:

$$\begin{gathered} I_L = I_0 \left(\gamma_0^{2/3} - 1\right)^{3/2}, \quad I_0 = -mc^3 \left(2e\Lambda\right)^{-1}, \\ \varphi_L = -\frac{mc^2}{e}\left(\gamma_0 - \gamma_L\right), \quad \gamma_L = \gamma_0^{1/3}, \quad \Lambda = \ln\frac{R}{r_b}, \end{gathered} \tag{6.3.1}$$

where $mc^2\gamma_0$ is the beam injection energy, R and r_b are the radii of the tube and the beam; $-e$ is the electron charge, φ_L and $mc^2\gamma_L$ are the potential and the energy of the beam, corresponding to the limit current I_L.

The curve showing the beam current I as a function of the electron energy is shown in Fig. 6.3.1. The self-consistent system of electrostatic equations has no solutions for $-I > -I_L$, while for smaller current $-I < -I_L$ there are two solutions, with point 1 on the curve corresponding to a greater electron energy and a smaller potential than point 2.

As was demonstrated in [14], in this wave guide system there exists a slow electromagnetic wave, $v_{\text{ph}} < c$, which is of interest from the standpoint of ion acceleration. However, the relationship between the phase velocity of the wave and the current of the beam is such (see Fig. 6.3.2) that in the region of greatest interest for acceleration $1 - I/I_L \ll 1$ the derivative

$$v'_{\text{ph}}(I) \sim (1 - I/I_L)^{-1/2}$$

exhibits a singularity and the random oscillations of the current under the conditions of the experiment may lead to disruption of the stable acceleration regime [17].

However, this result, obtained in linear approximation, is valid for perturbations which do not alter the wave guide properties of the REB. As will be demonstrated below, the situation changes in strong fields, when the phase velocity of the wave is a function of the oscillation amplitude and diminishes significantly in comparison with linear theory [17]. We have examined an REB whose charge is partially neutralized by the charge of the ions captured in an electrostatic potential well in the readout system of the wave guide. The existence of a directed electron flow relative to the background ion plasma is accompanied by electron-ion pumping oscillations. The nonlinear stabilization of the instability arises as a result of the relationship between the limit beam current and the oscillation amplitude and, as follows therefrom, it also depends on the nonlinear diminution of the growth rate [18]. The solution has the form of a soliton, and in the process of the nonlinear evolution of the wave the phase velocity initially diminishes to zero, and then changes sign, when the beam loses velocity and changes from the stable state 1 to a state of unstable equilibrium 2 (see Fig. 6.3.2). [8)].

1. Limit current of electron-ion beam in the presence of high-frequency field.

We will assume that the ions are compensated within the tubular REB, so that the charge density of the electron-ion system is equal to

$$\rho(t,r) = \frac{1}{2\pi r}\left[-eN_e(t,z) + qN_{tr}(t,z)\right]\delta(r-r_b), \tag{6.3.2}$$

where N_e and N_{tr} denote the beam electron density and the density of the captured ions per unit length (the running density). To describe the REB we will use a system of one-dimensional equations

$$\frac{\partial p_e}{\partial t} + v_e\frac{\partial p_e}{\partial z} = e\frac{\partial \varphi}{\partial z} - eE, \quad \frac{\partial N_e}{\partial t} + \frac{\partial}{\partial z}(v_e N_e) = 0, \tag{6.3.3}$$

where φ is the electrostatic potential, while $E = \mathcal{E}\cos\Phi$, $\Phi = \omega t - kz$ is the external high-frequency field; $p_e = mv_e\gamma$ is the electron momentum, and $\gamma = \sqrt{1+p^2/m^2c^2}$.

The ions captured in the potential well are "kinetic", and for the determination of N_{tr} we have to specify their distribution function. Following §6.2, we will assume

$$f(u,\varphi) = -\frac{N_i M}{\pi q \varphi}\sqrt{u^2(\varphi) - u^2}, \tag{6.3.4}$$

where N_i is the ion density at the bottom of the well φ, $u^2(\varphi) = -2q\varphi/M$, q and M are the charge and mass of the ion. Having integrated (6.3.4) over the velocities, we obtain

$$N_{tr} = N_i\frac{\varphi}{\varphi}. \tag{6.3.5}$$

Let us substitute (6.3.2) and (6.3.5) into the Poisson equation and for determination of the potential at the great distance from the injection plane let us the usual asymptote

$$\langle\varphi\rangle = -2e\Lambda\langle N_e\rangle\left(1 - \frac{2qN_i\Lambda}{\varphi}\right)^{-1}, \tag{6.3.6}$$

[8)] Note that the electrostatic instability arises at the beam entry in the wave guide and it formally absent for an unbounded tube, since the dispersion equation has no complex roots [14].

where $\langle N_e \rangle$ is the beam density averaged over the space period $\lambda = 2\pi/k$. In comparison with Refs. [13-16], formula (6.3.6) takes into consideration the partial neutralization of the electron charge of the beam by the ions.

Let us represent the density and velocity in Eqs. (6.3.3) in the form

$$N_e = \langle N \rangle + \tilde{N}, \quad v_e = \langle v \rangle + \tilde{v}$$

and let us express the small oscillating additions in terms of the electric field E_z and the background portion of the velocity $v = \langle v \rangle$:

$$\tilde{v} = \frac{\tilde{p}}{m\gamma^3}, \quad \tilde{N} = \frac{k\langle N \rangle}{\omega - kv}\tilde{v}, \quad \tilde{p} = \frac{e\mathcal{E}}{\omega - kv}\sin\Phi, \tag{6.3.7}$$

where $\gamma = \left(1 - v^2/c^2\right)^{-1/2}$. Using (6.3.7), after averaging (6.3.3) over the space oscillations, we find

$$\gamma - \frac{e\varphi}{mc^2} + \frac{\langle \tilde{p}^2 \rangle}{2m^2c^2\gamma^3} = \gamma_0, \quad -e\langle N \rangle \left(v - \frac{\langle \tilde{v}^2 \rangle}{v} \right) = I,$$
$$v = \frac{p}{m\gamma}\left(1 - \frac{3}{2}\frac{\langle \tilde{p}^2 \rangle}{m^2c^2\gamma^4}\right) \tag{6.3.8}$$

(it is assumed that $\omega \ll kv$).

From formulas (6.3.6) and (6.3.8) we obtain an expression for the beam current which takes into consideration the neutralizing ion background and high-frequency modulation

$$I = \left(I_0 + \frac{qN_i mc^3}{e\varphi} \right) \frac{p}{mc\gamma} \left(\gamma_0 - \gamma - \frac{\langle \tilde{p}^2 \rangle}{2m^2c^2\gamma^3} \right)$$
$$\times \left[1 - \frac{\langle \tilde{p}^2 \rangle}{m^2c^2\gamma^4} \left(\frac{3}{2} + \frac{m^2c^2}{p^2} \right) \right]. \tag{6.3.9}$$

In the absence of oscillations $\tilde{p} = 0$, maximizing (6.3.9) with respect to the momentum p, we obtain

$$I = I_L - qN_i v_L, \quad v_L = c\left(1 - \gamma_0^{-2/3}\right)^{1/2},$$
$$\varphi = \varphi_L, \quad \gamma = \gamma_L = \gamma_0^{1/3}. \tag{6.3.10}$$

It is easy to see that consideration of the ions partially neutralizing the charge of the electrons leads to an increase in the limit current in comparison with (6.3.1). At the same time, the beam potential and electron energy corresponding to the maximum current I coincide with (6.3.1). The maximum ion energy, corresponding to the potential φ, is equal to

$$\frac{1}{2} Mu^2(\varphi) = \frac{q}{e} mc^2 \left(\gamma_0 - \gamma_0^{1/3} \right). \tag{6.3.11}$$

If we regard the high-frequency addition to (6.3.9) to be a small perturbation, we find the limit current as a function of the oscillation amplitude:

$$\tilde{I} = I\left(1 - \frac{3}{2}\frac{\langle \tilde{p}^2 \rangle}{p_L^2\gamma_L^2}\right), \quad p_L = mv_L\gamma_L, \tag{6.3.12}$$

and the expansion of the right-hand side of (6.3.9) in powers of $\delta p = p - p_L$ ($|\delta p| \ll p_L$) leads to the formula

$$\delta p/p_L = \pm\gamma_L \left[(2/3)\left(\tilde{I} - I\right)/I\right]^{1/2}. \tag{6.3.13}$$

Let us note that plus and minus signs the right-hand side correspond to the high-frequency and low-frequency currents (points 1 and 2 in Fig. 6.3.1).

2. Nonlinear evolution of the wave phase velocity under the condition of electron-ion instability.

According to (6.2.12) the limit current of the beam diminishes with an increase in the oscillation amplitude. However, the presence of directed electron motion relative to the background ion plasma is accompanied by the development of electron-ion instability [12], so that the wave is unsteady.

To derive the equations determining the relationship between the field amplitude and time let us use the dispersion equation for a tubular REB in a conducting vacuum wave guide [14], considering the contribution of the captured ions in accordance with Eq. (6.2.15):

$$1 = 2\Lambda\left(1 - \frac{\omega^2}{c^2k^2}\right)\left[\frac{e^2\langle N\rangle}{mv^2\gamma^3}\left(1 + \frac{2\omega}{kv}\right) + \frac{qN_i}{\varphi}\left(1 - \frac{i\alpha}{\sqrt{1-\alpha^2}}\right)\right], \tag{6.3.14}$$

where $kR \ll 1$ $\alpha = \omega/ku(\varphi) < 1$ (**Appendix 2 and 3**).

Following [17], we will expand the right-hand side of (6.3.14) in powers of δp and we will retain the first terms of the expansion

$$\frac{\omega}{kv_L}(1 \pm i\epsilon) = \frac{3}{2}\frac{\delta p}{p_L},$$
$$\epsilon = \frac{qN_iv_L}{|I|}\frac{1}{\sqrt{2\mu}}, \qquad \alpha = \frac{\omega}{kv_L}\frac{1}{\sqrt{2\mu}}, \qquad \mu = \frac{q}{e}\frac{m\gamma_0}{M}. \tag{6.3.15}$$

If the ion density is sufficiently small, $\epsilon \ll 1$, and if the field amplitude varies "slowly" within the oscillation period, formulas (6.3.12) and (6.3.15) can be represented in the form of the nonlinear differential equation for the growing mode:

$$\frac{dy}{d\tau} = y\sqrt{1-y^2},$$
$$y = \frac{\mathcal{E}}{\mathcal{E}_{NL}}, \quad \mathcal{E}_{NL} = \left(\frac{4}{3}\frac{\Delta I}{I}\right)^{1/2}\frac{kp_L^2}{me}, \quad \delta = kv_L\epsilon\left(\frac{3}{2}\frac{\Delta I}{I}\right)^{1/2}, \tag{6.3.16}$$

where δ is the linear theory growth rate, $\tau = \delta t$, while $\mathcal{E}_{NL}$ corresponds to the maximum wave amplitude, $\Delta I = I - I$.

Equation (6.3.16) has a soliton solution:

$$y(\tau) = \cosh^{-1}\left(\text{arccosh} y_0^{-1} - \tau\right), \tag{6.3.17}$$

where y_0 is the initial perturbation [9]. Here the nonlinear evolution of the wave phase velocity $v_{\text{ph}} = \omega/k$ with time is described by the formula

$$v_{\text{ph}} = v_L\gamma_L\left(\frac{3}{2}\frac{\Delta I}{I}\right)^{1/2}\tanh^{-1}\left(\text{arccosh} y_0^{-1} - \tau\right), \tag{6.3.18}$$

---

9) Strictly speaking, a soliton arises only for a wave number k satisfying the condition $k^2h^2 + (2\Delta I/I_L) \times (7/6\gamma_L^2 - 1) = 0$, when correction factors of the following order with respect to the small parameter $\Delta I/I \ll 1$ in Eq. (6.3.16) vanish when $\delta p = 0$.

where $v_L \gamma_L = c \left(\gamma_0^{2/3} - 1 \right)^{1/2}$.

At the initial moment $\tau = 0$ this expression coincides with the formula for linear theory [17] when $y_0 \ll 1$, while when $\tau_m = \operatorname{arccosh} y_0^{-1}$ the phase velocity of the wave vanishes. Subsequently, $v_{\text{ph}}(\tau)$ changes sign and increases in absolute magnitude to the initial value for the case in which $\tau \gg 1$. However, there is a simultaneous change in sign for δp [see formula (6.3.12)], while the electron flux assumes low velocity. The possibility of observing this state of the flux under experimental conditions depends on the relationship between the values of δ and the growth rate of electrostatic instability which causes the steady-state beam within the tube to break down. This growth rate was derived in monograph [16] for a non-relativistic one-dimensional beam.

3. Multisoliton solution.

We will assume that at the initial instant $\tau = 0$ there arises within the beam a spectrum of unstable harmonics

$$k_n = n k_0, \quad k_0 = 2\pi/\ell,$$

where ℓ is the length of the beam. The system of amplitude equations for $y_n = \mathcal{E}_n/\mathcal{E}_{NL}$ follows from (13.3.13) and (13.3.15):

$$\dot{y}_n = n W y_n, \quad W = \left(1 - \sum_{n=1}^{N} y_n^2 \right)^{1/2}, \tag{6.3.19}$$

[the parameters τ and $\mathcal{E}_{NL}$ were determined from formula (6.3.16)], while the relationship between the phase velocity of the wave and the field amplitude has the form

$$v_{\text{ph}} = v_L \gamma_L \left(\frac{3}{2} \frac{\Delta I}{I} \right)^{1/2} W. \tag{6.3.20}$$

In Eq. (6.3.19) let us turn to new

$$\dot{q} = q \left[1 - \sum_{n=1}^{N} y_n^2(0) q^n \right]^{1/2}, \quad q = \exp\left(2 \int_0^\tau W \, d\tau \right) \tag{6.3.21}$$

and we will specify the distribution over the harmonics for $\tau = 0$ in the form

$$y_n(0) = y_0 \exp(-an).$$

Summing the geometric progression in the right-hand side of (6.3.21), we obtain

$$\dot{Q} = 2Q \left[1 - y_0^2 \frac{Q\,(1 - Q^N)}{1 - Q} \right]^{1/2}. \tag{6.3.22}$$

As $y_0 \to 0$ the nonlinear term becomes comparable to unity only in the event that $Q \gg 1$, and it can be simplified by assuming it equal to $y_0^2 Q^N$. In this approximation, solution (6.3.22) has the form of a soliton, and it follows from (6.3.19) that

$$y_n = y_0^{1-n/N} \cosh^{-n/N} \vartheta, \tag{6.3.23}$$

where $\psi = \text{arccosh}y_N^{-1}(0) - N\tau$. The phase velocity is determined by formula (6.3.20) in which $W = \tanh\psi$.

According to (6.3.22), the largest is the N-harmonic field which has a maximum growth rate. At the instant

$$\tau_N = N^{-1}\text{arccosh}y_N^{-1}(0)$$

the field of this mode achieves a maximum value $y_N(\tau_N) = 1$, and the function W subsequently changes sign and the energy reverts to the beam. The fields of the slower harmonics which fail to increase significantly within the time τ_N, remain small with respect to the parameter $y_0^{1-n/N}$.

3. Law of momentum flux density conservation.

The study carried out above on the wave guide properties of REB without consideration of the accelerating ions which function in the role of sampling particles assumes that the energy reserve of the accelerating field is rather great. However, since it is our intention to convert the pulse of light electrons into a pulse of heavy ions, we have an interest in evaluating the effectiveness of such acceleration from the standpoint of the law of momentum flux density conservation.

The Poisson equation with consideration of the charge density of the accelerated ions

$$\rho_a = \frac{q_a N_a}{2\pi r}\,\delta(r - r_b)$$

has the form

$$\Delta\varphi = \frac{1}{2\pi r}\left(eN_e - qN_i\frac{\varphi}{\varphi} - q_a N_a\right)\delta(r - r_b). \qquad (6.3.24)$$

Completing the multiplication of (6.3.24) by $\partial\varphi/\partial z$ and integrating over the cross section of the tube, we obtain

$$\frac{d}{dz}\int_0^R\left[\left(\frac{\partial\varphi}{\partial z}\right)^2 - \left(\frac{\partial\varphi}{\partial r}\right)^2\right] rdr = 4\left(eN_e - qN_i\frac{\varphi}{\varphi} - q_a N_a\right)\left.\frac{\partial\varphi}{\partial z}\right|_{r=r_b}. \qquad (6.3.25)$$

Integrating (6.3.25) together with the equations of motion for the electron and ion beams

$$\frac{dp}{dz} = -\frac{e^2 N_e}{I}\left.\frac{\partial\varphi}{\partial z}\right|_{r=r_b}, \quad \frac{dp}{dz} = -\frac{q_a^2 N_a}{I_a}\left.\frac{\partial\varphi}{\partial z}\right|_{r=r_b}, \qquad (6.3.26)$$

we obtain the integral of motion which accounts for the contribution of the accelerating ions:

$$\frac{1}{4}\int_0^R\left[\left(\frac{\partial\varphi}{\partial z}\right)^2 - \left(\frac{\partial\varphi}{\partial r}\right)^2\right] rdr + \frac{qN_i\varphi^2}{2\varphi} + \frac{I}{e}(p - p_0) - \frac{I_a}{q_a}p_a = 0, \qquad (6.3.27)$$

where p_0 is the electron beam injection pulse, $I = -eN_e v_e$ and $I_a = q_a N_a v_a$ denote the currents of the electron and ion beams.

Without examining the means of momentum transfer from the electrons to the ions, we can assert that the acceleration will be sufficiently effective if the contribution of the last term in (6.3.27) is quite small. For the beam current that is closed to the limit, we obtain the inequality

$$q_a^{-1} I_a p_a \ll -e^{-1} I_L\,(p_0 - p_L),$$

limiting the current of accelerating ions.

§6.4. Charge density waves in a tubular REB.

It is known that the current in a tubular relativistic electron beam (REB) injected into a wave guide with conducting walls along an infinitely strong magnetic field cannot exceed a certain limit I_L [13-16]. Usually the formula for the maximum current is derived assuming that far from the injection plain the beam parameters reach their asymptotic values and the electric field is purely radial ($\partial/\partial r \gg \partial/\partial z$), i.e.,

$$\varphi = \frac{2I}{v}\Lambda, \quad I = -eNv, \quad v = c\left[1 - \left(\gamma_0 - \frac{e\varphi}{mc^2}\right)^{-2}\right]^{1/2}, \tag{6.4.1}$$

where $mc^2\gamma_0$ is the injection energy; I, N, and v are the current, linear density, and velocity of the beam; $\varphi \equiv \varphi(r_b)$; $-e$ is the electronic charge, R and r_b are wave guide and beam radii, respectively.

According to Eq. (6.4.1) the dependence of the beam current on the energy of the beam electrons has the form shown in Fig. 6.3.1, so that a solution exists for $-I \leq -I_L$, where I_L is the maximum beam current,

$$I = I_0\beta\left(\gamma_0 - \gamma\right), \quad I_L = I_0\left(\gamma_0^{2/3} - 1\right)^{3/2}, \qquad I_0 = -mc^3\left(2e\Lambda\right)^{-1}. \tag{6.4.2}$$

Here the point 1 corresponds to a higher electron kinetic energy and lower potential than at point 2.

It is found below that besides the integrals of the current and energy used in deriving Eqs. (6.4.1) and (6.4.2), the transport conditions for an REB in a cylindrical vacuum wave guide are also determined by a nonlocal integral of the momentum flux (6.3.27) which determines the "fine" energy structure of the system. In the absence of a longitudinal component of the electric field and ions the value of S is determined by the **Appendix 6.4**

$$S = \frac{mc}{e}I_0\left(\gamma_0 - \gamma\right)\left[\gamma\beta^2 + \left(\gamma_0 - \gamma\right)/2\right].$$

Here the dependence of the beam current I on the energy γ in accordance with Eq. (6.4.2) is included. This formula along Eq. (6.4.2) specifies the dependence $I(S)$ in the parametric form shown in Ref. 19]. It can be seen that the first current state corresponds to higher momentum flux than the second, i.e., $S_1 > S_2$ (for rigorous proof, see **Appendix 12.5**). At the same time, these states of the beam are not the only ones, since in the region $S_2 < S < S_1$ there is a set of periodic solutions (standing charge density waves) which continuously fill the gap between these two points. Evidently, transitions between these states are forbidden by the conservation of the momentum flux (6.3.27).

An actual physical system is not conservative and the existence of different kinds of dissipative processes is accompanied by a reduction in S, which corresponds to a $1 \rightarrow 2$ transition of the beam through all intermediate states. The collective scattering of an REB on background ions is an example of such a process (see §6.3). In paper [19] radiation damping [20] acting on beam electrons as they move in a nonuniform electric field is considered as a specific dissipative mechanism leading $1 \rightarrow 2$ transition. As a result, the beam enters state 2 which experiment show to be unstable [16]. Thus, our analysis shows that stationary currents close to the critical value may be interrupted.

1. Periodic charge density waves.

According to Ref. [15] for a semiinfinitive wave guide the Poisson equation reduces to the integral equation

$$\varphi(z) = -e \int_0^\infty G(z, z') N(z')\, dz', \tag{6.4.3}$$

where the Green function $G(z, z')$ can be represented by a series [16] or in the integral form

$$G(z, z') = \frac{4}{\pi} \sin(kz) \int_0^\infty \left[I_0 (kr_b)\, K_0 (kr_b) - \frac{K_0 (kR)}{I_0 (kR)}\, I_0^2 (kr_b) \right] \sin(kz')\, dk, \tag{6.4.4}$$

where I_0 and K_0 are modified Bessel functions of first and second kind. Here the dependence of the potential φ on the beam parameters is nonlocal and is determined by the behavior of the function $N(z)$ on the $z > 0$ semiaxis.

Since it is not possible to find an analytic solution of Eq. (6.4.3) in general form, we obtain below an asymptotic solution corresponding to $z \to \infty$ $(z \gg R)$. In fact, this procedure excludes from considerations the narrow boundary region $\delta z \simeq R$, so that the solution obtained below describe the propagation of an REB in an unbounded cylindrical wave guide, but the role of the injection plane $z = 0$ reduces to determining the current and unperturbed beam energy $mc^2\gamma_0$. The presence of rapidly oscillating functions in the integral (6.4.3) means that the main contribution to the integral is from the region, $k \simeq z^{-1}$ and the Bessel functions can be expanded in series subject to the conditions $kR \ll 1$ and $kr_b \ll 1$ [21]. In this case the Green's function can be written in the form

$$\begin{gathered} G(z, z') \simeq 2\Lambda \left(1 + L^2 \frac{d^2}{dz^2}\right) \delta(z - z'), \\ L^2 = \left(R^2 - r_b^2 - 2r_b^2\, \Lambda\right) (4\Lambda)^{-1}. \end{gathered} \tag{6.4.5}$$

After substituting Eq. (6.4.5) in Eq. (6.4.3) and using Eq. (6.4.1), we obtain a nonlinear differential equation for the beam density,

$$y'' = -y + \frac{I_0}{I} \left(\gamma_0 - \frac{y}{\sqrt{y^2 - 1}}\right) \equiv F(y), \tag{6.4.6}$$

where $y = -eNc/I$, and the primes denote differentiation with respect to $\xi = z/L$.

The above asymptotic expansion is valid if $|y''|$ is small compared to any of the terms on the right-hand side of Eq. (6.4.6). Its domain of applicability therefore coincides with the range of I and γ_0, where the function $|F(y)| \ll 1$. In addition, the omitted higher derivatives $y^{(2n)}$, $(n \geq 1)$ are small only if the function $y(\xi)$ changes sufficiently slowly: $|y''| \ll |y|/L^2$.

Given these remarks, we expand the right-hand side of Eq. (6.4.6) in a series at the point $y = y_{\min}$, $F'(y_{\min}) = 0$:

$$\begin{gathered} F(y) = F(y_{\min}) + \frac{F''(y_{\min})}{2} (y - y_{\min})^2, \\ y_{\min} = \left[1 + \left(\frac{I_0}{I}\right)^{2/3}\right]^{1/2}. \end{gathered} \tag{6.4.7}$$

Substituting Eq. (6.4.7) in Eq. (6.4.6) and using Eq. (6.3.1), we find the limiting current to be

$$y'' = \gamma_0^{1/3} \left[\left(\gamma_0^{2/3} - 1\right)^{-1/2} \left(\frac{I_L}{I_0} - 1\right) - \frac{3}{2} \left(\gamma_0^{2/3} - 1\right)^{1/2} (y - y_{\min})^2 \right]. \tag{6.4.8}$$

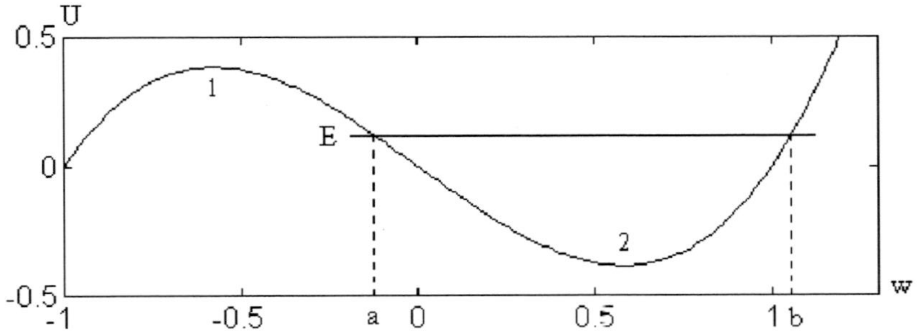

Fig. 6.4.1. The potential energy $U(w)$ of a material point as a function of the coordinate w for $I > I_L$ (σ=1). Points 1 and 2 correspond to the first and second current states; a and b are the roots of the equation $U(w)=E$.

Introducing the dimensionless quantities

$$w = \frac{\gamma_0^{1/3}}{\sqrt{2|\delta|}}\left(\frac{y}{y_{\min}} - 1\right), \quad x = \left(\frac{\gamma_0^{1/3}}{\sqrt{2|\delta|}}\right)^{1/2} \xi, \quad \delta = \frac{I_L}{I_0} - 1 \tag{6.4.9}$$

we reduce Eq. (6.4.8) to the form

$$w'' = \sigma - 3w^2, \quad \sigma = \operatorname{sign}(\delta). \tag{6.4.10}$$

Equations (6.4.9) and (6.4.10) imply that the approximations $|y''| \ll |y|/L^2$ and $|y - y_{\min}| \ll |y_{\min}|$ used here are satisfied for currents close to critical $|\delta| \ll 1$.

The first integral in Eq. (6.4.10) has the form

$$\frac{w'^2}{2} + U(w) = E, \quad U(w) = u^3 - \sigma w. \tag{6.4.11}$$

For further analysis of the solutions (6.4.11) we shall use an analogy with mechanics, assuming w to be the coordinate of a point particle with energy E and potential energy $U(w)$. The function is plotted in Fig. 6.4.1 . For a current greater than critical ($\sigma = -1$) the potential has no extrema (curve 1) and the motion of the particle is infinite. For a current below the maximum ($\sigma = 1$), the shape of the curve changes (curve 2) and the motion of the particle may be finite. The points 1 and 2 on the $U(w)$ curve in this case correspond to the two equilibrium states of the beam shown in Fig. 6.3.1. For a current equal to the maximum, points 1 and 2 merge.

It should be noted that our approximation, which is restricted to the inequality $|w| \leq 1$, is applicable only for finite motions of a particle trapped in a potential well. This corresponds to the development of nonlinear spatial oscillations in the beam density near the equilibrium value:

$$\begin{gathered} w = a + A\,\mathrm{cn}^2\left[\kappa\,(x - x_0)\,, k\right], \\ A = \kappa^2 - 3a^2/2, \quad \kappa^2 = \sqrt{1 - 3a^2/4}, \quad k = \sqrt{A/2}/\kappa, \end{gathered} \tag{6.4.12}$$

where a is the root of the equation $w - w^3 + E = 0$ $\left(-1/\sqrt{3} \leq a \leq 1/\sqrt{3}\right)$ (see Fig. 6.4.1), and x_0 is a constant of integration.

When $a \to 1/\sqrt{3}$ ($E \to 2/\sqrt{3}$), the period of the nonlinear oscillations

$$T = \sqrt{2}\int_a^b (w - w^3 + E)^{-1/2}\, dw = \frac{2}{\kappa}\, K(k), \tag{6.4.13}$$

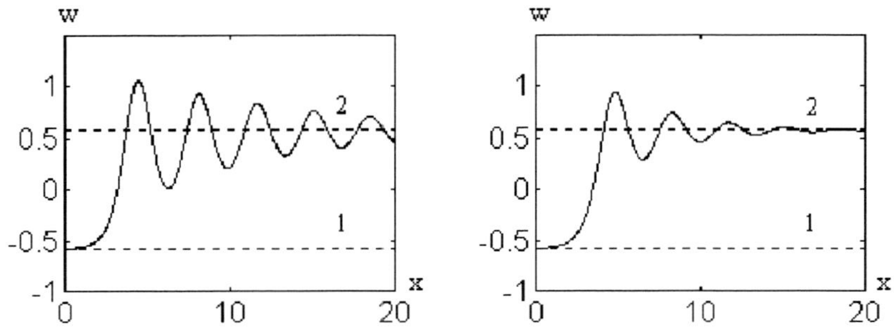

Fig. 6.4.2. Plots of the function $w(x)$ which illustrate a transition of the dissipative system from state 1 to state 2 for ν=0.2 and ν=0.5.

(b is the largest root of the function under the integral, and $K(k)$ is the complete elliptic integral of the first kind [21]), goes to infinite and there is a soliton solution

$$w = -3^{-1/2} + W,$$
$$W = \sqrt{3}\cosh^{-2}\left[\operatorname{arccosh}\left(3^{1/4}/W_0\right) - 3^{1/4}x/\sqrt{2}\right]. \tag{6.4.14}$$

Thus, without energy dissipation as $z/R \to \infty$ the current in the tubular beam either goes to one of the equilibrium states, 1 or 2, or to a periodic asymptotic relation [see Eq. (6.4.12)] which has the soliton Eq. (6.4.14) as a limiting case.

It can be shown that the integral (6.4.11) corresponds to the conservation of the momentum flux (6.3.27) along the cross section of the wave guide (see **Appendix 6.4**).

2. Radiation damping.

In the case of a monoenergetic beam in an infinite magnetic field, where the motion is uniform, the radiative correction to the Lorentz force (radiation damping force) can be derived from the general formula [20]

$$f_{\|} = -\frac{2}{3}er_0\beta\gamma\frac{\partial^2\varphi}{\partial z^2}, \tag{6.4.15}$$

where $r_0 = e^2/mc^2$ is the classical radius of the electron.

Allowance for the force (6.4.15) in the equation of motion for the beam electrons leads to the appearance in Eq. (6.4.10) of a dissipative term

$$w'' + \nu w' = 1 - 3w^2,$$
$$\nu = \frac{2r_0}{3L}\left(\frac{2}{|\delta|}\right)^{1/4}\gamma_0^{5/6}\beta_0. \tag{6.4.16}$$

For a qualitative analysis of Eq. (6.4.16), as before, we treat w as the coordinate of a point particle in a potential $U(w)$. The new term $\nu w'$, compared with that of (6.4.10), serves as a damping force and causes the oscillating particles to fall to the bottom of the potential well. As a result, the soliton solution of Eq. (6.4.10) transforms into a shock wave corresponding to a transition of the system from state 1 to state 2. With an increasing in ν, damping of the oscillations becomes greater and when $\nu \geq \nu_c$, the solution corresponding to a shock wave becomes monotonic. Plots of numerically calculated functions $w(x)$ for different values of ν are shown in Fig. 6.4.2. These plots are consistent with the above remarks.

We find ν_c by linearizing Eq. (6.4.16) near the equilibrium (state 1) value $w = -3^{-1/2} + \tilde{w}\exp(ikx)$. Solving the corresponding dispersion relation

$$k = \pm\left(2\sqrt{3} - \frac{\nu^2}{4}\right)^{1/2} + \frac{i\nu}{2} \tag{6.4.17}$$

yields $\nu_c = 2^{3/2}3^{1/4}$.

Because r_0 is small, the condition $\nu \ll 1$ is the most realistic. It is possible to find an analytic solution (6.4.16) which is asymptotic with respect to the parameter ν. Accordingly, we take the average of the derivative of E [Eq. (6.4.11)] with respect to x over the period T of the nonlinear oscillations (6.4.13), including damping [i.e., when condition (6.4.16) is satisfied]. As a result, we obtain an equation for the slow variation in the amplitude of the standing wave with position,

$$\frac{dE}{dx} = -\frac{\nu}{T}\int_0^T \left[\frac{dw_0(x,E)}{dx}\right]^2 dx, \tag{6.4.18}$$

where function $w_0(x,E)$ is specified in Eq. (6.4.12).

Introducing the quantity

$$I(E) = \int_0^T [w_0'(x,E)]^2\, dx = 2^{3/2}\int_a^b \sqrt{w - w^3 + E}\, dw \tag{6.4.19}$$

and noting that $T = dI/dE$ because of Eq. (6.4.13), we reduce Eq. (6.4.18) to the form

$$I' + \nu I = 0, \tag{6.4.20}$$

from which

$$I(E) = I(E_0)\exp(-\nu x), \tag{6.4.21}$$

where $E_0 = 2/3^{3/2}$ corresponds to the soliton solution of Eq. (6.4.10).

Substituting Eq. (6.4.12) in Eq. (6.4.19) and calculating the integral $I(E)$ by changing the variable of integration x to a in accordance with the relation

$$\begin{gathered}\operatorname{cn}\left[\kappa\left(x - x_0\right), k\right] = \cos\alpha, \\ I(E) = \kappa A^2 \int_0^\pi \sin^2 2\alpha\left(1 - k^2\sin^2\alpha\right)^{1/2} d\alpha.\end{gathered} \tag{6.4.22}$$

We then have from Eq. (6.4.21) and Eq. (6.4.22) the formula

$$2^{3/2}3^{-5/4}\kappa^5\left[2\left(k^4 - k^2 + 1\right)E(k) - (1-k^2)(2-k^2)K(k)\right] = \exp\left[-\nu\left(x - x_0\right)\right], \tag{6.4.23}$$

which establishes the functions $a(\nu x)$ and $E(\nu x) = a^3 - a$ [$K(k)$ and $E(k)$ are the complete elliptic integral of the first and second kinds [21]; A, κ, k, and a are defined by Eq. (6.4.12)].

It is shown in Ref. [19] that the analytic solution (6.4.23) approximate the exact numerical solution of Eq. (6.4.16) fairly precisely.

§6.5. Limiting current of a tubular REB.

The limiting current of a tubular REB heated to relativistic temperatures is calculated by the formula

$$\varphi(r_b, z) = -2eN_e(z)\Lambda, \quad \Lambda = \ln(R/r_b), \tag{6.5.1}$$

which follows from the Poisson equation [22]. The beam potential and running electron density obey a system of relativistic hydrodynamic equation [23]

$$\begin{gathered}\frac{\partial}{\partial t}(N\gamma)+\frac{\partial}{\partial z}(N\gamma v)=0,\\ \left(\frac{\partial}{\partial t}+v\frac{\partial}{\partial z}\right)(mG\gamma v)=e\frac{\partial\varphi}{\partial z}-\frac{1}{N\gamma}\frac{\partial}{\partial z}(NT),\\ \left(\frac{\partial}{\partial t}+v\frac{\partial}{\partial z}\right)(mc^2G-T)=\frac{T}{N}\left(\frac{\partial}{\partial t}+v\frac{\partial}{\partial z}\right)N,\end{gathered} \tag{6.5.2}$$

where $\xi = mc^2/T$, $G(\xi) = K_3(\xi)/K_2(\xi)$, $K_n(\xi)$ are modified Bessel functions [21]; and N and T are the density and temperature in the rest frame of the electrons and are related to the quantities N_e and T_e in the laboratory by the formulas

$$N_e = \gamma N, \quad T_e = T/\gamma$$

(we have retained the notations of §6.4).

Assuming in Eq. (6.5.2) that $\partial/\partial t = 0$ and integrating the equations, we find

$$\begin{gathered}N\gamma v = I/e, \quad mc^2\gamma G(\xi) - e\varphi = mc^2\gamma_0 G(\xi_0),\\ N = CTK_2(\xi)\exp\left[\xi G(\xi)\right],\end{gathered} \tag{6.5.3}$$

where C is a constant of integration.

The system of integrals (6.5.3) together with Eq. (6.5.1) permits determination of the limit beam current. Introducing the dimensionless momentum $q = \beta\gamma$, we find

$$\begin{gathered}\frac{I}{I_0} = \gamma_0 G(\xi_0)\frac{q}{\sqrt{1+q^2}} - qG(\xi),\\ q = \beta_0\gamma_0\frac{\xi}{\xi_0}\frac{K_2(\xi_0)}{K_2(\xi)}\exp\left[\xi_0 G(\xi_0) - \xi G(\xi)\right],\end{gathered} \tag{6.5.4}$$

For analysis of the stability of the REB against small perturbations $\sim \exp(ikz - i\omega t)$, we calculate the current density $\tilde{j}$ from the linearized equations (6.5.2) and substitute $\sigma_e = \tilde{j}_z/\tilde{E}_z$ ($\tilde{E}_z$ is the longitudinal electric field of the wave) into formula

$$1 = -2i\,\frac{\kappa^2\sigma_e}{\omega}\Lambda, \quad \kappa^2 = k^2 - \frac{\omega^2}{c^2}. \tag{6.5.5}$$

As a result we obtain the dispersion relation (see **Appendix 6.1**)

$$1 - \frac{I}{I_0}\frac{\kappa^2 c^3}{\gamma^3 v}\left[(\omega - kv)^2 G - \frac{mc^2 G_T'}{mc^2 G_T' - 1}\left(k - \frac{\omega v}{c^2}\right)^2\frac{T}{m}\right]^{-1} = 0. \tag{6.5.6}$$

The $I(\gamma)$ curve has a maximum, corresponding to the limiting current. The numerical calculations show that this quantity increases with increasing electron injection temperature T_e because of the increase in the "thermal" mass of the electrons, $m_T = mG(\xi)$ [see Eq. (6.5.4)]. The corresponding curves are shown in Fig. 6.5.1.

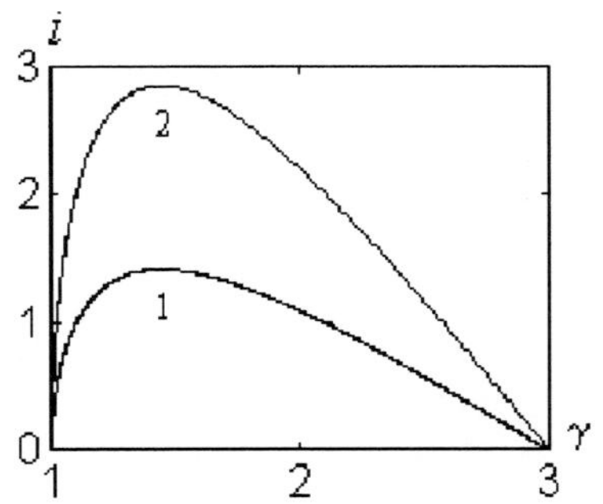

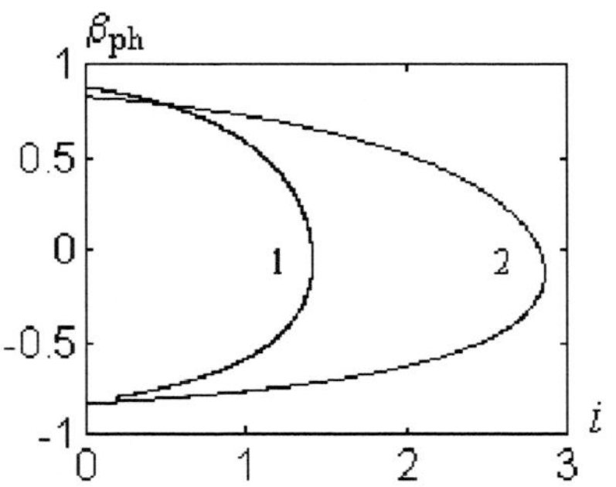

Fig. 6.5.1. Beam current $i{=}I/I_0$ versus electron energy γ (in units mc^2) for $\gamma_0{=}3$; and $\xi_0{=}$ 10 (1) and 2 (2).

Fig. 6.5.2. Phase velocity of the wave $\beta_{\rm ph}{=}v_{\rm ph}/c$ versus beam current for the same parameters of the beam.

The dependence of the phase velocity of the wave $v_{\rm ph}$ on the beam current for different temperatures found from Eq. (6.5.4) is illustrated by Fig. 6.5.2.

For nonrelativistic temperatures $\xi_0 \gg 1$ one can use the asymptotic forms of the Bessel functions

$$G = 1 + \frac{5}{2\xi}, \quad G'_T = \frac{5}{2mc^2}, \quad \xi = \xi_0 \left(\frac{q}{\beta_0\gamma_0}\right)^{2/3}$$

and symplify Eqs. (6.5.4) and (6.5.6) to [10)]

$$\frac{I_L}{I_0} = q_L^3 + \frac{5}{2}\frac{\gamma_0^{2/3} q_L}{\xi_0}\left[1 - \left(\frac{\beta_c}{q_L}\right)^{2/3}\right],$$
$$\beta_{\rm ph} = q_L\left[\frac{3}{2}\left(1 - \frac{I}{I_L}\right)\right]^{1/2}, \quad q_L = \left(\gamma_0^{2/3} - 1\right)^{1/2}, \qquad (6.5.7)$$

The latter expression (6.5.7) obtained by expanding (6.5.8) in the small parameter $1 - I/I_L \ll 1$ corresponds to a phase velocity of the slow wave, $\beta_{\rm ph} = v_{\rm ph}/c$, in a monochromatic relativistic electron beam nearly critical current I_L [17].

In the ultrarelativistic limit $\xi_0 \ll 1$ (at small values of the arguments of the Bessel functions) formulas (6.5.4) and (6.5.6) become

$$\frac{I_L}{I_0} = \frac{4}{\xi_0}\left[\frac{\gamma_0 q}{\sqrt{1+q^2}} - \left(\beta_0\gamma_0 q^2\right)^{1/3}\right], \quad \xi = \xi_0\left(\frac{q}{\beta_0\gamma_0}\right)^{1/3},$$
$$1 - \frac{I}{I_0}\frac{\kappa^2 c^3 \xi}{4\gamma^3 v}\left[(\omega - kv)^2 - \frac{c^3}{3}\left(k - \frac{\omega v}{c^2}\right)^2\right]^{-1} = 0. \qquad (6.5.8)$$

From first equation (6.5.8) under the condition $I'(q_L) = 0$ we obtain the relation

$$\frac{1}{(1+q_L^2)^{3/2}} - \frac{2}{3}\left(\frac{\beta_0}{q_L\gamma_0^2}\right)^{1/3} = 0, \qquad (6.5.9)$$

10) In this approximation relations (6.5.3) agree with the formulas of nonrelativistic hydrodynamics [24].

which determinies the limiting value of the momentum q_L of the beam and and which has a solution for $\gamma_0 > 3/2$. The asymptotic solution of (6.5.8) and (6.5.9) for $\gamma_0 \gg 1$ takes the form

$$q_L = \gamma_L = \left(\frac{3}{2}\right)^{3/8} \gamma_0^{1/4}, \quad \frac{I_L}{I_0} = \frac{4\gamma_0}{\xi_0},$$
$$\beta_{\text{ph}} = \left(\frac{2}{3}\right)^{1/8} \gamma_0^{1/4} \left[2\left(1 - \frac{I}{I_L}\right)\right]^{1/2}. \tag{6.5.10}$$

The second formula (6.5.10) is obtained by expanding (6.5.8) in the small parameter $1 - I/I_L \ll 1$ and is the analog of (6.5.7) at ultrarelativistic temperatures. The $\beta_{\text{ph}}(I)$ curve is more gradual than in the case of cold beam because of the increase of the limiting current by a factor $4/\xi_0$ and the stretching of the longitudinal scale [see Fig. (6.5.1)].

§6.6. Acceleration of an ion bunch by a slow modulated-phase cyclotron wave in a REB.

In the method of autoresonance acceleration of the ions the phase velocity of a slow cyclotron wave

$$V = \frac{\omega}{\omega + \omega_B} v, \tag{6.6.1}$$

[ω is the wave frequency, $\omega_B = eB_0/mc\gamma$ is the cyclotron frequency of the beam electrons, $e > 0$, B is the external magnetic field, $v = \beta c$ is the beam velocity, and $\gamma = (1 - \beta^2)^{-1/2}$ is the relativistic factor] is presumed to be increased by a reduction in the external longitudinal magnetic field along the acceleration length [3].

In this section we demonstrate the possibility of accelerating an individual ion bunch along the relativistic electron beam (REB) propagating along a spatially uniform wave system. The necessary increase in the phase velocity of the wave is achieved by means of time-dependence modulation of the REB at the beginning of the acceleration. In contrast to Ref. [3], where the frequency was fixed, in our case the acceleration results from an increase in the frequency $\omega|_{z=0} = \omega(t)$ at the entrance to the system in a constant magnetic field.

We consider a REB, of finite radius (see §3.2), propagating along an external magnetic field in a vacuum. The nonlinear equation for the beam radius R in a strong magnetic field in a strong magnetic field $\omega_B^2 \gg \omega_b^2$ ($\omega_b^2 = 4\pi n_b e^2/m\gamma^3$ is the Langmuir frequency and n_b is the beam electron density) has the form

$$\left(\frac{\partial}{\partial t} + v\frac{\partial}{\partial z}\right)^2 R + \frac{\omega_B^2}{4}\left(1 - \frac{R_0^4}{R^4}\right) R = 0, \tag{6.6.2}$$

where R_0 is the equilibrium beam radius.

Moving to the new variable $t, z \to t' = t - z/v$, $z' = z$,

$$(\partial/\partial t + v\partial/\partial z)^2 R = d^2R/dz'^2,$$

after integrating (6.6.2) over z, we obtain

$$R^2(\psi) = R_0^2\left[\sqrt{1 + a^2(t')} + a(t')\cos\psi\right],$$
$$\psi = \frac{\omega_B}{v} z - \alpha(t'), \tag{6.6.3}$$

where the amplitude $a(t')$ and phase $\alpha(t')$ are arbitrary functions of the parameter t'.

The accelerating field along the beam axis can be reconstructed from the results of §3.2,

$$E_z = \frac{I}{R^2 v}\left(\frac{\partial}{\partial z} + \frac{v}{c^2}\frac{\partial}{\partial t}\right) R^2,$$

where $I = -\pi R_0^2 e n_b v$ is the beam current. In the field of the wave with constant amplitude, $a = const$, the equation of motion of the ion has the form

$$\frac{d}{dt}\left(\Gamma \dot{Z}\right) = -\frac{qaI}{Mv^2}\left(\omega_B + \frac{1}{\gamma^2}\frac{\partial \alpha}{\partial t}\right)\frac{R_0^2 \sin\psi}{R^2(\psi)}, \tag{6.6.4}$$

where $q > 0$ and M are the ion charge and mass, and $\Gamma = \left(1 - \dot{Z}^2/c^2\right)$.

For a synchronous particle $Z = Z_s(t)$ and $\dot{Z} = V$, moving in a fixed phase $\psi = \psi_s$, the equation of motion has the form

$$\dot{V} = -\frac{qaI\omega_B}{Mv^2\Gamma^3}\frac{\left(1 - vV/c^2\right)}{(1 - V/v)}\frac{R_0^2}{R^2}\sin\psi_s. \tag{6.6.5}$$

Integrating (6.6.5), we find a implied relation for the functions $Z_s(t)$ and $V(t)$

$$\begin{aligned} A\omega_B t &= \Gamma_0 - \Gamma + \gamma\frac{v}{c}\left[W(V_0) - W(V)\right], \\ A\omega_B Z_s &= \Gamma_0 V_0 - \Gamma V + c\gamma\left[W(V_0) - W(V)\right], \\ W = \arcsin&\left(\frac{v - V}{c - Vv/c}\right), \quad A = -\frac{qI}{Mvc^2}\frac{R_0^2 a \sin\psi_s}{R^2(\psi_s)}, \end{aligned} \tag{6.6.6}$$

where V_0 is the initial velocity of the ion. After some time the particle velocity $V(t)$ reaches the beam velocity v (Fig. 6.6.1). Assuming $V_0 = 0$ and $V = v$ in (6.6.6), we find the acceleration time T and length L:

$$A\omega_B T = 1 - \frac{c}{\gamma v}\arcsin\frac{v}{c}, \quad A\omega_B L = c\gamma\left(\arcsin\frac{v}{c} - \frac{v}{c}\right). \tag{6.6.7}$$

The phase velocity V can be found as a function of t and z from Eqs. (6.6.6), by going over from the Lagrangian variables t and $Z_s(t)$ to the Eulerian variable t and z. Introducing the function $t' = t - Z_s(t)/v$ into our discussion and then omitting the subscript, we obtain

$$A\omega_B t' = \Gamma_0\left(1 - \frac{V_0}{v}\right) - \Gamma\left(1 - \frac{V}{v}\right) + \frac{c}{\gamma v}\left[W(V) - W(V_0)\right]. \tag{6.6.8}$$

The wave phase velocity is shown in Fig. 6.6.2 as a function of the coordinate at the subsequent moments. After some time the curve $V(z)$ shifts to the right with the beam velocity $v > V$, so that the accelerated particles is displaced into the region of the high phase velocities and its velocity increases to the beam velocity.

The present acceleration mechanism occurs as a result of phase modulation $\alpha(t, z)$, i.e., excitation of a wave with frequency and wave number in the beam. Indeed, differentiating the second of Eqs. (6.6.3) with respect to time at $\psi = \psi_s$ and $z = Z_s(t)$, we find

$$\frac{\partial \alpha}{\partial t} = \omega_B \frac{V}{v - V}. \tag{6.6.9}$$

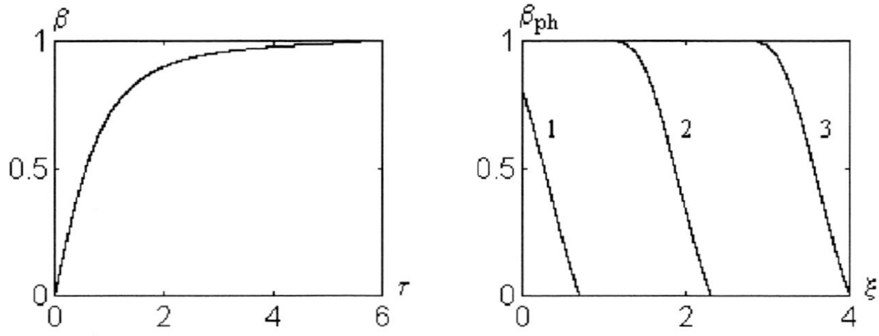

Fig. 6.6.1. Velocity $\beta = V/c$ of a synchronous particle as a function of time $\tau = A\omega_B t$ for $\gamma=10$ and $V_0=0$.

Fig. 6.6.2. Phase velocity $\beta_{\rm ph}$ as a function of coordinate $\xi = A\omega_B z/v$ at the subsequent moments: $\tau=$ 0.7 (1); 2.4 (2); 4 (3) for $\gamma=10$.

When the wave has a fixed phase velocity Eq. (6.6.9) coincides with (6.6.1) if $\omega = \partial\alpha/\partial t =$const; the modulation mode (6.6.8), therefore, corresponds to a frequency of beam modulation at the entrance to the system $z = 0$; $\omega(t) = \partial\alpha/\partial t$.

When ions are accelerated in unneutralized electron beams the accelerated particles have simultaneous phase and radial stability [26]. For small deviations from the trajectory of a synchronous particle $Z_1 = Z(t) - Z_s$, upon linearizing Eq. (6.6.4) we obtain

$$\ddot{Z}_1 + \delta \dot{Z}_1 + \Omega_{\rm ph}^2 Z_1 = 0, \tag{6.6.10}$$

where

$$\Omega_{\rm ph}^2 = \omega_B^2 \frac{qaI}{Mv^3\Gamma^3} \frac{R_0^4}{R^4} \frac{1 - vV/c^2}{(1 - V/v)^2} \left[\frac{qaI \sin^2 \psi_s}{M\Gamma^3\gamma^2 v^3 (1 - V/v)^2} + a + \sqrt{1 + a^2} \cos \psi_s \right],$$

$$\delta = 3\Gamma^2 \frac{V\dot{V}}{c^2},$$

and $V(t)$ is defined by (6.6.5).

The region of stable phases $\Omega_{\rm ph}^2 > 0$ is given by the inequalities

$$-\sqrt{1 + a^2(1 - f)^2} - \sqrt{1 + a^2} < fa \cos \psi_s < \sqrt{1 + a^2(1 - f)^2} - \sqrt{1 + a^2},$$
$$f = -\frac{2qI}{M\Gamma^3\gamma^2 v^3 (1 - V/v)^2}. \tag{6.6.11}$$

At the same time ($\dot{V} > 0$) autophasing occurs [26], since during acceleration the phase spread of the particles

$$\psi - \psi_s = \frac{\omega_B}{v - V} Z_1$$

decreases. The radial stability condition

$$\Omega_r^2 = -\frac{2qI}{Mv\Gamma R^2(\psi_s)} > 0 \tag{6.6.12}$$

is satisfied independently of the particle phase.

In conclusion, we note that the group velocity of the wave changes when the dispersion-induced spreading of the accelerated packets of waves is taken into account. From the formulas of §3.1 it follows that

$$\Delta v_g = v - v_g = \frac{\omega_b^2}{\omega_B k_\perp^2}\left(\frac{k_z}{\gamma^2} + \frac{v\omega_B}{c^2}\right). \tag{6.6.13}$$

(the following order with respect to the small parameter, $k_z R \ll 1$, is taking into account). Therefore, the condition of our approximation (6.6.2),

$$T\Delta v_g \ll k_z^{-1},$$

is satisfied only in the case of an ultra-relativistic electron beam:

$$\frac{M}{m\gamma^2}\frac{e}{qa} \ll 1. \tag{6.6.14}$$

This acceleration mechanism presumes that only one bunch is present in the acceleration length; accordingly, a train of accelerating pulses must be produced at the system entrance to ensure continuous operation. The modulation frequency of the beam can be changed rapidly in this case by using a nonlinear two-level system [31].

§6.7. Propagation of a bounded high-current REB in a low-density plasma.

A small addition of ions to a relativistic electron beam

$$n_i = en_b/q\gamma^2, \tag{6.7.1}$$

(where n_b and $v = \beta c$ are the density and velocity of the beam, -*e* and q are the charges of the electron and ion, and γ is the relativistic factor) is known to lead to the establishment of radial equilibrium, under conditions where the Coulomb repulsive force is balanced by the self-magnetic field of the beam [27].

A low-density plasma $n_p \leq n_b$ in the path of beam propagation can be used to produce radial equilibrium of a REB through partial neutralization of the beam charge by the ions. As we will show below, the field of the excess electron charge ejects "light" plasma electrons (with mass m) from the bulk of the beam, while "heavy" beam electrons (with an effective mass $m_b = m\gamma^3$) are focused by the charge of the plasma ions [28]. However, the dynamics of the process is such that when condition (6.7.1) is fulfilled, the radial oscillations of the REB with ions occur. The nature of these oscillations depends on the relativistic factor γ which determines the ratio of the mass m_b of the electrons to the mass M of the ions.

1. Charge neutralization of a beam.

To describe the reaction of a REB with a plasma, we will use the system of hydrodynamic equations for the beam and for the plasma electrons and ions. We will assume that the characteristic length of the longitudinal perturbations L significantly exceeds the beam radius R:

$$L \gg R, \quad \frac{\partial}{\partial r} \gg \frac{\partial}{\partial z} \tag{6.7.2}$$

and the transverse motion of particles with constant longitudinal velocity dominates.

A self-similar solution exists in this approximation

$$v_{r\alpha}(t,r) = \frac{r}{R_\alpha}\frac{dR_\alpha}{dt}, \qquad n_\alpha(t,r) = n_{o\alpha}\frac{R_{o\alpha}^2}{R_\alpha^2}\,\vartheta(r-R_\alpha), \tag{6.7.3}$$

$$\frac{d}{dt} = \frac{\partial}{\partial t} + v_{o\alpha}\frac{\partial}{\partial z}, \qquad \vartheta(r-R_\alpha) = \begin{cases} 1, & r \le R_\alpha, \\ 0, & r > R_\alpha, \end{cases}$$

$n_\alpha(t,\mathbf{r})$ and $v_\alpha(t,\mathbf{r})$ are the density and velocity of the beam particle α, $R_\alpha(t,z)$ is the radius of the beam or of the plasma electrons and ions:

$$\frac{d^2R_\alpha}{dt^2} = \frac{e_\alpha}{m_\alpha\gamma_\alpha}\frac{1}{r}\,(E_r - \beta_\alpha B_\varphi)\,R_\alpha \tag{6.7.4}$$

(here e_α, m_α, and γ_α are the charge, mass, and relativistic factor of particles of species α).

The radial electric field E_r and the azimuthal magnetic field B_φ satisfy the equations

$$\begin{aligned} \frac{1}{r}\frac{\partial}{\partial r}(rE_r) &= 4\pi\sum_\alpha e_\alpha n_{o\alpha}\frac{R_{o\alpha}^2}{R_\alpha^2}\vartheta(r-R_\alpha), \\ \frac{1}{r}\frac{\partial}{\partial r}(rB_\varphi) &= 4\pi\sum_\alpha e_\alpha\beta_\alpha n_{o\alpha}\frac{R_{o\alpha}^2}{R_\alpha^2}\vartheta(r-R_\alpha). \end{aligned} \tag{6.7.5}$$

The radial coordinate r eliminated from equations (6.7.4) and (6.7.5) if E_r, $B_\varphi \sim r$.

Obviously, such a solution exists only in the region $r \le (R_\alpha)_{\min}$, where $(R_\alpha)_{\min}$ is the shortest of the radii. Therefore, the model in question can be used to describe the beam if the radii of plasma electrons R_e and of plasma ions R_i exceed the beam radius R when oscillations appear (see Chap. 3). Assuming these conditions to be fulfilled, we obtain

$$\begin{aligned} \frac{1}{R}\frac{d^2R}{dt^2} &= \frac{\omega_b^2}{2}\left[\frac{R_0^2}{R^2} + \gamma^2\frac{n_p}{n_{ob}}R_p^2\left(\frac{1}{R_e^2} - \frac{q}{eR_i^2}\right)\right], \\ \frac{1}{R_e}\frac{d^2R_e}{dt^2} &= \frac{\omega_b^2}{2}\gamma^3\left[\frac{R_0^2}{R^2} + \frac{n_p}{n_{ob}}R_p^2\left(\frac{1}{R_e^2} - \frac{q}{eR_i^2}\right)\right], \\ \frac{1}{R_i}\frac{d^2R_i}{dt^2} &= -\mu\frac{\omega_b^2}{2}\left[\frac{R_0^2}{R^2} + \frac{n_p}{n_{ob}}R_p^2\left(\frac{1}{R_e^2} - \frac{q}{eR_i^2}\right)\right], \end{aligned} \tag{6.7.6}$$

where $\omega_b^2 = 4\pi e^2 n_{ob}/m\gamma^3$, $\mu = qm\gamma^3/eM$, R_p and n_p are the initial radius and density of the plasma, and R_0 and n_{ob} are the initial radius and density of the beam.

When conditions (6.7.1) for a spatially homogeneous beam $\partial/\partial z = 0$ is met in the dimensionless variables

$$X = \frac{R}{R_0}, \quad Y = \frac{R_e}{R_p}, \quad Z = \left(\frac{e}{q}\right)^{1/2}\frac{R_i}{R_p}, \quad T = \frac{\omega_b t}{\sqrt{2}}$$

system (6.7.6) assumes the form:

$$\begin{aligned} \ddot{X} &= \left(\frac{1}{X^2} + \frac{1}{Y^2} - \frac{1}{Z^2}\right)X, \quad \ddot{Y} = \gamma\left(\frac{\gamma^2}{X^2} + \frac{1}{Y^2} - \frac{1}{Z^2}\right)Y, \\ \ddot{Z} &= -\mu\left(\frac{1}{X^2} + \frac{1}{\gamma^2 Y^2} - \frac{1}{\gamma^2 Z^2}\right)Z. \end{aligned} \tag{6.7.7}$$

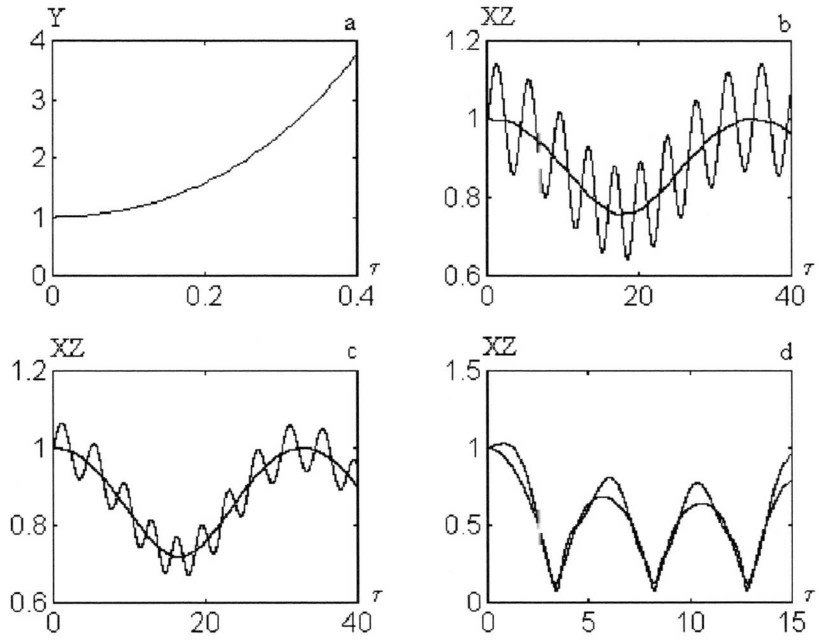

Fig. 6.7.1. Plasma electrons radius Y (a), plasma ions radius Z and beam radius X (b) as a function of τ, γ=3. Synchronous electron-ion oscillations when γ=5 (c) and γ=7 (d).

The results of numerical integration of Eq. (6.7.7) with the initial conditions

$$X(0) = Y(0) = Z(0) = 1, \quad \dot{X}(0) = \dot{Y}(0) = \dot{Z}(0) = 0 \tag{6.7.8}$$

are represented in Fig. 6.7.1. After rapid escape of plasma electrons, radial collapse of the REB takes place together with the ions that partially neutralize its charge.

The singularity is eliminated if the ion temperature is finite and the kinetic pressure gradient impedes the compression of the ion channel to zero radius. According to §3.2, allowance for the temperature gradient

$$T_i = T_0 \frac{R_p^2}{R_i^2}\left(1 - \frac{r^2}{R_i^2}\right)$$

(here T_0 is the initial temperature on the channel axis $r = 0$) leads to the following modification of the third equation in the system (6.7.7):

$$\ddot{Z} = -\mu\left(\frac{1}{X^2} + \frac{1}{\gamma^2 Y^2} - \frac{1}{\gamma^2 Z^2}\right) Z + \frac{\delta}{Z^3}, \tag{6.7.9}$$

where $\delta = 4T_0/M\omega_p^2 R_p^2$.

The results of numerical integration of Eq. (6.7.9), together with the first equations of system (6.7.7), are represented in Fig. 6.7.1. The appearance of two time scales that correspond to electron oscillations with the Langmuir frequency ω_b and to ion oscillations with frequency $\omega_i = \sqrt{\mu}\,\omega_b$ is a characteristic feature of the process of beam neutralization by ions. As the relativistic mass of the electrons (the parameter γ) increases, the oscillations of the REB and the ion core become synchronized (Fig. 6.7.1c,d).

2. Transverse electron-ion instability.

Over time $t \geq \omega_b^{-1}$, plasma electrons are expelled from the beam (see Fig. 1.1.1a), and the beam with background ions trapped by the electron Coulomb field is described by the first equation of system (6.7.6) and by equation (6.7.9). If the beam radius R is small relative to wavelength

$$kR \ll 1, \quad k = \omega_b/v, \quad \omega_b = \sqrt{4\pi e^2 n_b/\gamma^3}, \tag{6.7.10}$$

the field of the oscillations is directed across the beam radius and under the condition of electron-ion instability defocuses the electrons and ions.

In order for electron-ion system to be stable, condition (6.7.1) must be supplemented with the analogous condition of ion radial equilibrium. Setting the right side of equation (6.7.9) equal to zero, we find the ion temperature

$$T_{eq} = \pi e q n_{ob} \beta^2 R_p^2, \tag{6.7.11}$$

at which the kinetic pressure gradient balances the focusing electric field.

We consideration for equations (6.7.1) and (6.7.11), the system of equations that describes the motion of a REB in the ion channel in the plasma has the form

$$\begin{gathered}\left(\frac{\partial}{\partial t} + v\frac{\partial}{\partial z}\right)^2 R = \frac{\omega_b^2}{2}\left(\frac{R_0^2}{R^2} - \frac{R_p^2}{R_i^2}\right) R, \\ \frac{\partial^2 R_i}{\partial t^2} = \frac{\omega_i^2}{2}\left(-\frac{R_0^2}{R^2} + \frac{R_p^2}{\gamma^2 R_i^2} + \beta^2 \frac{R_p^4}{R_i^4}\right) R_i,\end{gathered} \tag{6.7.12}$$

where $\omega_i^2 = 4\pi e q n_{ob}/M$.

Let us investigate the stability of system (6.7.12) by considering small deviations from the equilibrium position

$$R = R_0(1+y), \quad R_i = R_0(1+\rho), \quad |y| \ll 1, \quad |\rho| \ll 1.$$

For perturbations of the type $\sim \exp(i\omega t - ikz)$, the dispersion equation follows from system (6.7.12):

$$1 - \frac{\omega_b^2}{(\omega - kv)^2} - \frac{\omega_i^2}{\omega^2 - \omega_i^2} = 0. \tag{6.7.13}$$

For a not-too-relativistic beam with

$$\mu = \omega_i^2/\omega_b^2 = qm\gamma^3/eM \ll 1,$$

when the inequalities $kv \gg \omega \gg \omega_i$ are fulfilled, Eq. (6.7.13) yields

$$2\frac{\omega^3}{\omega_b^3} - \left(1 - \frac{\omega_b^2}{k^2v^2}\right)\frac{\omega^2}{\omega_b^2} + \mu = 0. \tag{6.7.14}$$

Complex roots of (6.7.14), which correspond to unstable solutions, exist if

$$1 - \frac{\omega_b^2}{k^2v^2} < 3\mu^{1/3},$$

while the maximum growth rate corresponds to

$$\omega = -\frac{1+i\sqrt{3}}{2^{4/3}}\sqrt{\mu}\,\omega_b, \quad k = \frac{\omega_b}{v}. \tag{6.7.15}$$

The estimates of linear theory yield $\rho \approx \mu^{1/3} y$, and hence when $\mu^{1/3} \ll 1$ the ions remain linear $|\rho| \ll 1$ in the nonlinear stage of the instability $y \simeq 1$. In this approximation system (6.7.12) simplifies:

$$\begin{gathered} \left(\frac{\partial}{\partial t} + v\frac{\partial}{\partial z}\right)^2 X = \frac{\omega_b^2}{2}\left(\frac{1}{X^2} - 1 + 2\rho\right) X, \\ \frac{\partial^2 \rho}{\partial t^2} = -\frac{\omega_i^2}{2}\left(\frac{1}{X^2} - 1\right), \quad X = \frac{R}{R_0}. \end{gathered} \tag{6.7.16}$$

Since the inequalities

$$v\frac{\partial}{\partial z} \simeq \omega_b \gg \frac{\partial}{\partial t} \simeq \mu^{1/3}\omega_b$$

hold, we have as a first approximation

$$\frac{d^2 X}{d\xi^2} = \frac{1}{2}\left(\frac{1}{X} - X\right), \quad \xi = \frac{\omega_b z}{v}. \tag{6.7.17}$$

By analogy with mechanics, the first integral of this equation

$$\frac{1}{2}\left(\frac{dX}{d\xi}\right)^2 + U(X) = E, \quad U(X) = -\ln\sqrt{X} + \left(\frac{X}{2}\right)^2 \tag{6.7.18}$$

describes the periodic motion of a particle in the potential well between the turning points $U(X_1) = U(X_2) = 0$. The oscillation period

$$L = \sqrt{2}\int_{X_1}^{X_2} [E - U(X)]^{-1/2}\, dX.$$

is a function of the energy E.

Near the minimum potential energy $a^2 = 2E - 1/2 \ll 1$, we can fulfill the expansion of the right-hand side of equation (6.7.17) in series in $y = X - 1$, $|y| \ll 1$. As a result we obtain the following asymptotic equation

$$\begin{gathered} y'' + y - \frac{y^2}{2} + \frac{y^3}{2} = 0, \\ y(\xi) = a\cos\psi + \frac{a^2}{4} - \frac{a^2}{12}\cos 2\psi, \quad \psi(\xi) = \xi\left(1 + \frac{a^2}{6}\right)^{1/2} + \varphi, \end{gathered} \tag{6.7.19}$$

where the amplitude a and phase φ are constant.

To the next order in the small parameter $\mu^{1/3} \ll 1$, the solution remains in the form of Eq. (6.7.19), but the amplitude and phase φ are time dependent:

$$y(t,z) = \frac{1}{4}|A|^2 + \operatorname{Re}\left(Ae^{i\xi} - \frac{A^2}{12}e^{2i\xi}\right), \tag{6.7.20}$$

where $A(t) = a(t)\exp i\varphi(t)$.

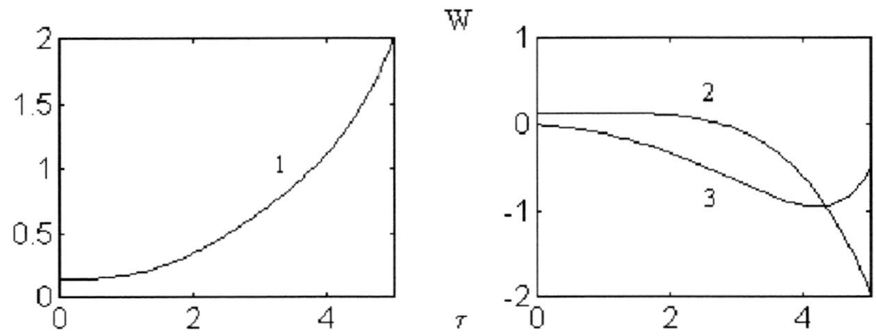

Fig. 6.7.2. Evolution of the complex field amplitude versus time: $|W|$ (1), $\mathrm{Re}\,W|$ (2) and $\mathrm{Im}\,W|$ (3), γ=3.

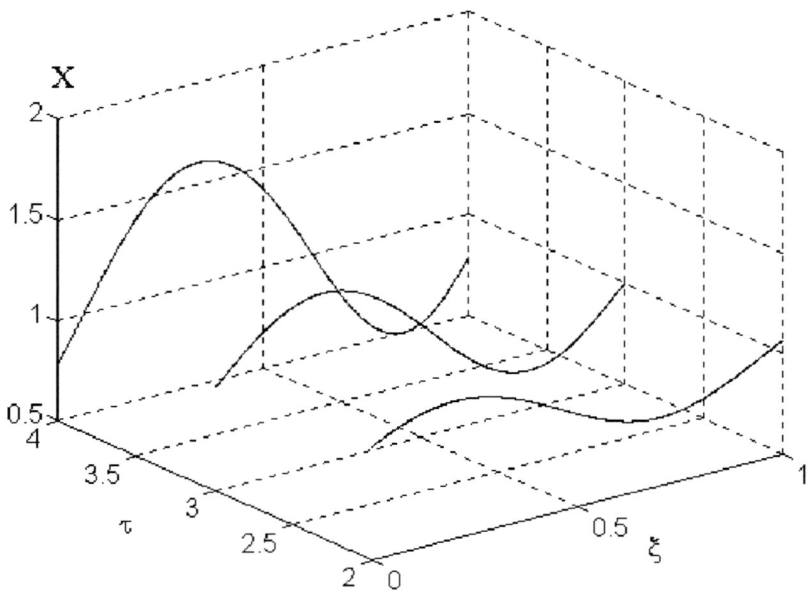

Fig. 6.7.3. Evolution of the beam radius under the action of the transverse field under conditions of the electron-ion instability.

A nonlinear equation for the complex function $A(t)$ equivalent to (6.7.16) in the limit of small nonlinearity $|A|^2 \ll 1$ can be obtained from (6.7.14) by using the following substitution:

$$\omega = i\frac{d}{dt}, \quad \frac{\omega_b^2}{v^2} \to k_{NL}^2 = \frac{\omega_b^2}{v^2}\left(1 + \frac{|A|^2}{6}\right).$$

In the dimensionless variables

$$W^2 = A^2/6\mu^{1/3}, \quad \tau = \mu^{1/3}\omega_b t, \quad \Delta = \mu^{-1/3}(1 - \omega_b^2/k^2v^2)$$

we obtain

$$2iW''' - [(\Delta - |W|^2)W]'' - W = 0. \tag{6.7.21}$$

The numerical solution of equation (6.7.21) for the maximally unstable mode (6.7.15) is shown in Fig. 6.7.2. We see that the small nonlinearity does not stabilize the increase in the amplitude of the oscillations, and the instability enters the strongly nonlinear regime $a \simeq 1$. Fig. 6.7.3 gives a qualitative idea of the spatial distribution of the oscillations. According to this figure, an increase in amplitude is accompanied by strong modulation of ions and by the separation of the ion channel into individual bunches with a spatial period $L \simeq 2\pi v/\omega_b$. The instability is aperiodic, and the phase velocity of the perturbation wave is small, $v_{\rm ph} \simeq \mu^{1/3} v$, so that the picture of the field is nearly static in the laboratory coordinate system.

Let us note that the condition of applicability of the envelope model $kR \ll 1$ limits the beam current through the following inequality:

$$I = \pi R^2 e n_b v \ll mc^3/4e(\gamma\beta)^3.$$

3. Transport of an electron beam in a plasma channel in a longitudinal magnetic field.

In this section we show that an external longitudinal magnetic field B, which inhibits transverse motion of the beam and plasma particles, causes damping of the REB oscillations in the ion channel. Since the gyrofrequency $\omega_{B\alpha} = e_\alpha B/m_\alpha c$ for all species is inversely proportional to the particle mass $m_\alpha = m\gamma_\alpha$, in the process of neutralizing the REB and expelling electrons from the beam interior the effect of the magnetic field is felt primarily by the non-relativistic plasma electrons.

The growth rate of an electron-ion instability decreases and the process of exciting low-frequency oscillations slows as the magnetic field strength rises. However, the instability is not stabilized by the nonlinearity when the field amplitude is small, and as in an unmagnetized plasma, the REB is strongly modulated with a spatial period on the order of the instability wavelength.

A particle with charge e_α in a longitudinal magnetic field experiences a Lorentz force $e_\alpha \mathbf{v}_\alpha \times \mathbf{B}/c$, and in the equations (6.7.6) for the radii of the unmagnetized beam and plasma particles there are additional terms $-\left(\omega_{B\alpha}^2 R_\alpha/4\right)\left(1 - R_{o\alpha}^4/R_\alpha^4\right)$, taking into account this effect (see §3.2). Consequently, the system of equations (6.7.7) is altered and becomes [29]

$$\begin{aligned}
\ddot{X} &= \left[\frac{1}{X^2} + \frac{1}{Y^2} - \frac{1}{Z^2} - \frac{\kappa}{2}\left(1 - \frac{1}{X^4}\right)\right] X, \\
\ddot{Y} &= \left[\gamma\left(\frac{\gamma^2}{X^2} + \frac{1}{Y^2} - \frac{1}{Z^2}\right) - \frac{\kappa\gamma^2}{2}\left(1 - \frac{1}{Y^4}\right)\right] Y, \qquad (6.7.22)\\
\ddot{Z} &= -\mu\left[\frac{1}{X^2} + \frac{1}{\gamma^2 Y^2} - \frac{1}{\gamma^2 Z^2} + \frac{\kappa\mu^2}{2\gamma^4}\left(1 - \frac{1}{Z^4}\right)\right] Z + \frac{\delta}{Z^3},
\end{aligned}$$

where $\kappa = \omega_B^2/\omega_b^2$ and we have retained the notations of Eq. (6.7.7)

By integrating (6.7.22) numerically we obtain results illustrating the effect of the longitudinal magnetic field on the process by which the REB is neutralized in a hydrogen plasma with $mq/Me = 1/1840$ for different values of the parameter κ is shown in Fig. 6.7.4. In contrast with the above similar calculations for an unmagnetized plasma, where the plasma electrons were essentially completely expelled from the interior of the beam, the fairly strong magnetic field is partly balanced by the repulsive radial Coulomb field and plasma electrons are restrained from escaping.

Below we assume that the plasma electrons leave the beam interior $Y^2 \gg X^2$, $Y^2 \gg Z^2$, and the ion channel is in equilibrium with the BEB. Under this conditions the length over which the beam is transported is determined by the inverse growth rate of the transverse

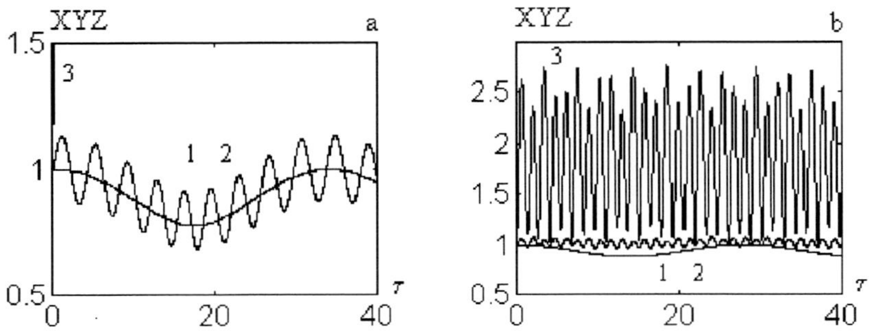

Fig. 6.7.4. Transverse oscillations of the beam with $\gamma = 3$ and plasma in a longitudinal magnetic field: κ=0.1 (a) and κ=6 (b). Oscillating curves 1 correspond to the beam radius, and curves 2 and 3 to the ion and electron radii.

electron-ion instability, modified with respect to that of section 2 by the presence of the longitudinal magnetic field, which retains particles from leaving their equilibrium position and slow down the process by which the oscillations are exited.

We drop the terms $\sim Y^{-2}$ in Eq. (6.7.22) for the beam and ion radii and, following section 2, linearize the ion terms: $Z = 1 + \rho$, $|\rho| \ll 1$. Assuming also that the ions remains unmagnetized because of their large mass $\omega_{Bi}^2 \ll \mu\omega_b^2$, we find the following system of nonlinear partial differential equations:

$$\left(\frac{\partial}{\partial\tau} + \frac{\partial}{\partial\xi}\right)^2 X = \frac{1}{2}\left(\frac{1}{X} - X\right) - \frac{\kappa}{4}\left(X - \frac{1}{X^3}\right) + \rho X,$$
$$\frac{\partial^2\rho}{\partial\tau^2} = -\frac{\mu}{2}\left(\frac{1}{X^2} - 1\right), \quad \xi = \frac{\omega_b z}{v}, \tag{6.7.23}$$

which describe the growth of transverse electron-ion oscillations of the ion channel with the trapped REB in a longitudinal magnetic field.

In the absence of the ions we have $\rho = \partial/\partial\tau = 0$ and the asymptotic solution of (6.7.23) can be found by setting $X = 1 + y$ and expanding the nonlinear terms in powers of $|y| \ll 1$. As a result we obtain the equation

$$y'' + (1+\kappa)y - (1+3\kappa)\frac{y^2}{2} + (1+5\kappa)\frac{y^3}{2} = 0, \tag{6.7.24}$$

where the prime denotes differentiation with respect to ξ. The solution of (6.7.24) takes the form

$$y(\xi) = a\cos\psi + \frac{a^2}{4}\frac{1+3\kappa}{1+\kappa}\left(1 - \frac{1}{3}\cos 2\psi\right),$$
$$\Psi = k^2(a)\xi + \varphi, \quad k^2 = 1 + \kappa + \frac{a^2}{6}\frac{1+6\kappa}{1+\kappa}, \tag{6.7.25}$$

where a and φ are the amplitude and phase of a linear oscillator.

When ions are included in the next approximation in the problem

$$\eta = \left(\frac{\omega_b\omega_i}{\Omega_b^2}\right)^{2/3} = \left[\frac{\mu}{(1+\kappa)^2}\right]^{1/3} \ll 1, \tag{6.7.26}$$

the solution again has the form (6.7.25), but the amplitude and phase depend on time.

For small perturbations we find from (6.7.23) the dispersion relation

$$2i\frac{\omega^3}{\Omega_b^3}-\left(1-\frac{\Omega_b^2}{k^2v^2}\right)\frac{\omega^2}{\Omega_b^2}+\frac{\omega_i^2\omega_b^2}{\Omega_b^4}=0,$$
$$\Omega_b=\omega_B^2+\omega_b^2,\quad \omega_i^2=4\pi eqn_b/M\gamma^2,\quad \left|1-\Omega_b^2/k^2v^2\right|\ll 1. \tag{6.7.27}$$

where ω_i takes into account Eq. (6.7.1).

For mode with the largest growth rate it follows from (6.7.27) that

$$\omega=-\frac{1+i\sqrt{3}}{2^{4/3}}\eta\Omega_b,\quad k=\frac{\Omega_b}{v}. \tag{6.7.28}$$

Comparing (6.7.28) and (6.7.15) we see that in a longitudinal magnetic field the growth rate $\operatorname{Im}\omega$ decreases in proportion to $(\omega_b/\Omega_b)^{1/3}$.

The nonlinear equation for the amplitude and phase of the oscillations in the weakly nonlinear approximation can be recovered from Eqs. (6.7.25) and (6.7.27). Going over to the complex amplitude $A(t)=a(t)\exp[i\varphi(t)]$ we can represent (6.7.25) in the form

$$y(t,z)=\operatorname{Re}\left[Ae^{ikz}+\frac{|A|^2}{4}\frac{1+3\kappa}{1+\kappa}\left(1-\frac{1}{3}e^{2ikz}\right)\right]. \tag{6.7.29}$$

Then, setting

$$\omega=i\frac{d}{dt},\quad \frac{\Omega_b^2}{v^2}\to\frac{\Omega_b^2}{v^2}\left[1+\frac{|A|^2}{6}\frac{1+6\kappa}{(1+\kappa)^2}\right]$$

in (6.7.27) and introducing the dimensionless variables

$$\tau_\kappa=\eta\Omega_b t,\quad W_\kappa^2=\frac{|A|^2}{6\eta}\frac{1+6\kappa}{(1+\kappa)^2},\quad \Delta_\kappa=\frac{1}{\eta}\left(1-\frac{\Omega_b}{k^2v^2}\right)$$

we find the equation which coincides with (6.7.21) and generalizes the above results to the case $\kappa>0$. On other hand, including the external longitudinal magnetic field does not change the qualitative picture of how the beam radius changes in time under the action of the electron-ion instability. This follows from the formal similarity of above equation and the equation for unmagnetized plasma.

For the magnetized beam with $\kappa\gg1$ the steady solution of Eq. (6.7.23) takes the form

$$X(\tau)=\left[\sqrt{1+a^2}+a\cos\left(\sqrt{\kappa}\xi+\varphi\right)\right]^{1/2}, \tag{6.7.30}$$

where a and φ are constants of integration. Including the ions causes these quantities to become time-dependent in agreement with (6.7.21). From (6.7.30) it follows that for $a^2\gg1$ the amplitude of the oscillations of the function $X(\tau,\xi)$ varies in the range

$$(2a)^{-1/2}\le X(\tau,\xi)\le(2a)^{1/2}.$$

A drift in the phase $\varphi(\tau)$ is accompanied by a displacement of the maxima and minima of $X(\tau,\xi)$ parallel to $\dot z=\eta v\simeq(\mu\kappa^{-2})^{1/3}v$.

Appendix 6.1

Substituting (6.2.9) in Eq. (6.2.7), we rewrite this equation in the form

$$w'' = \frac{\beta_0(1+w)}{\sqrt{\beta_0^2+2w+w^2}} + \frac{1}{\beta_0^2}\left[\left(1+\frac{\nu}{2}\right)^2 - \beta_0^2\right] w - 1. \tag{A6.1.1}$$

In the limit of a small ion density, $\nu \ll 1$, we can fulfill the expansion of the right-hand side of equation (A6.1.1) in series in w

$$\frac{\beta_0(1+w)}{\sqrt{\beta_0^2+2w+w^2}} \simeq 1 + w\left(1-\frac{1}{\beta_0^2}\right)\left(1-\frac{3w}{2\beta_0^2}\right) \tag{A6.1.2}$$

to simplify the above equation.

Substituting (A6.1.2) in (A6.1.1) and omitting the small terms of the subsequent order of smallness in parameter ν, we find the equation

$$\beta_0^2\, w'' - \nu w - \frac{3}{2\beta_0^2\gamma_0^2}\, w^2 = 0, \tag{A6.1.3}$$

where $\gamma_0^{-2} = 1-\beta_0^2$.

In this approximation, an asymptotic solution of Eq. (A6.1.1) has the form of soliton, and it follows from (A6.1.3) that

$$w = -\nu\beta_0^2\gamma_0^2 \cosh^{-2}\left(\frac{\sqrt{\nu}}{2\beta_0}\xi\right), \quad \nu\gamma_0^2 \ll 1. \tag{A6.1.4}$$

Appendix 6.2

The small electromagnetic perturbations $\sim \exp(ikz - i\omega t)$ satisfy the system of Maxwell equations:

$$\begin{aligned} ikE_r - \frac{dE_z}{dr} &= i\frac{\omega}{c}B_\varphi, \quad kB_\varphi = \frac{\omega}{c}E_r, \\ \frac{1}{r}\frac{d}{dr}(rB_\varphi) &= -i\frac{\omega}{c}E_z + \frac{4\pi}{c}\tilde{j}, \end{aligned} \tag{A6.2.1}$$

where $\tilde{j} = \sigma(r)E_z$ is a component of the beam current density. Excluding the electric E_r and magnetic B_φ fields, we obtain the following equation

$$\frac{1}{r}\frac{d}{dr}\left(r\frac{dE_z}{dr}\right) = \kappa^2\left[1 + i\frac{4\pi}{\omega}\sigma(r)\right]E_z, \tag{A6.2.2}$$

where $\kappa^2 = k^2 - \omega^2/c^2$.

We assume that the charge density of the the tubular REB is equal to

$$\sigma(r) = \sigma(\omega,k)(2\pi r)^{-1}\delta(r-r_b)$$

where r_b is the beam radius. Correspondingly, the solution of Eq. (A6.2.2) takes the form:

$$E_z^i = AI_0(\kappa r), \quad E_z^e = BI_0(\kappa r) + CK_0(\kappa r), \tag{A6.2.3}$$

where E_z^i and E_z^e correspond to the regions $r < r_b$ and $r > r_b$, respectively; I_0 and K_0 are modified Bessel functions [21], and A, B C are the constants of integration.

Eliminating C from (A6.2.3) by means of the boundary condition $E_z^e(R) = 0$, we find

$$E_z^e = B\left[I_0(\kappa r) - \frac{I_0(\kappa R)}{K_0(\kappa R)}K_0(\kappa r)\right]. \tag{A6.2.4}$$

Substituting the first solution (A6.2.3) and (A6.2.4) in the following boundary conditions

$$\begin{gathered} E_z^i(r_b) = E_z^e(r_b), \\ \left(\frac{dE_z^i}{dr} - \frac{dE_z^e}{dr}\right)_{r=r_b} = -2i\,\frac{\kappa^2}{\omega r_b}\sigma(\omega, k)E_z^i(r_b), \end{gathered} \tag{A6.2.5}$$

we get a system of equations for the constants A and B, and their condition of solvability arrive at the dispersion relation [14]:

$$\frac{I_0(\kappa R)}{I_0(\kappa r_b)} = -2i\,\frac{\kappa^2}{\omega}\sigma(\omega, k)\left[I_0(\kappa R)K_0(\kappa r_b) - I_0(\kappa r_b)K_0(\kappa R)\right]. \tag{A6.2.6}$$

At the long-wave limit

$$I_0(\kappa R) \simeq 1, \quad K_0(\kappa R) \simeq -\ln(\kappa R), \quad \kappa R \ll 1$$

(A6.2.6) changes to the following equation

$$1 = -2i\,\frac{\kappa^2}{\omega}\,\sigma(\omega, k)\,\Lambda, \quad \Lambda = \ln\frac{R}{r_b}. \tag{A6.2.7}$$

We use a system of equations (6.5.2) to calculate an electron conductivity in a case of the thermal beam. Substituting

$$\begin{gathered} v(t, z) = v + \tilde{v}, \quad N(t, z) = N + \tilde{N}, \quad T(t, z) = T + \tilde{T}, \\ \tilde{j} = -e\left(N\gamma^3\tilde{v} + \gamma v\tilde{N}\right), \quad G = G(T) + G_T'\tilde{T}, \end{gathered}$$

and linearizing the above equations, we obtain a set of relations for the small perturbations

$$\begin{gathered} \tilde{T} = \frac{T/N}{mc^2G_T' - 1}\tilde{N}, \quad \tilde{N} = \frac{k - \omega v/c^2}{\omega - kv}N\gamma^2\tilde{v}, \\ \tilde{j} = -\frac{\omega}{\omega - kv}eN\tilde{v}. \end{gathered} \tag{A6.2.8}$$

The dependence on the electric field is determined by the equation of motion:

$$-im(\omega - kv)\left(G\gamma^3\tilde{v} + \gamma vG_T'\tilde{T}\right) = -eE_z - i\,\frac{k}{\gamma}mc^2G_T'\tilde{T}. \tag{A6.2.9}$$

Comparing (A6.2.8) and (A6.2.9), we find

$$\sigma_e = \frac{ie^2N_e\omega}{m\gamma^3}\left[(\omega - kv)^2G - \frac{mc^2G_T'}{mc^2G_T' - 1}\left(k - \frac{\omega v}{c^2}\right)^2\frac{T}{m}\right]^{-1}. \tag{A6.2.10}$$

where $\sigma_e = \tilde{j}/E_z$ and $N_e = N\gamma$.

In the limit case $T \to 0$ $(G = 1)$, the dispersion relation reduces to the form

$$1 = \frac{2e^2\kappa^2 N_e \Lambda}{m\gamma^3 (\omega - kv)^2} = 0, \tag{A6.2.11}$$

[see formulas (A6.2.7) and (A6.2.10)]. Substituting $I = -eN_e v$ and $I = I_0\beta(\gamma_0 - \gamma)$ in (A6.2.11) and solving the equation, we find a phase velocity of the slow wave, $\beta_{\rm ph} = \omega/ck$, as a function of the beam energy $\gamma = (1 - \beta^2)^{-1/2}$:

$$\beta_{\rm ph} = \frac{\beta - \sqrt{\beta^2 - (1+h)(\beta^2 - h)}}{1+h}, \quad h = \frac{\gamma_0 - \gamma}{\gamma^3}, \tag{A6.2.12}$$

which changes between the limits $\beta_{\rm ph}(0) = (1 - 1/\gamma_0)^{-1/2}$ and $\beta_{\rm ph}(\gamma_0) = \beta_0$. A function $\beta_{\rm ph}(I)$ is represented in Fig. 6.3.1.

Appendix 6.3

To calculate the ionic conductivity for the thermal beam, $\sigma_i = \tilde{j}_i/E_z$, we represent a distribution function in the form:

$$f = f_0 + \tilde{f}, \quad f_0 \gg |\tilde{f}|,$$
$$f_0 = A\sqrt{u_m^2 - u^2}, \quad \frac{\partial \tilde{f}}{\partial t} + u\frac{\partial \tilde{f}}{\partial z} + \frac{q}{M}E_z\frac{\partial f_0}{\partial u} = 0. \tag{A6.3.1}$$

where u_m is the velocity of the trapped ions at the bottom of the well, φ, and $A = -N_i M/\pi q\varphi$.

A ion current density, $\tilde{j}_i$, is determined by the component of the distribution function, $\tilde{f}$. For monochromatic perturbations $\sim \exp(ikz - i\omega t)$ we find

$$\sigma_i = -i\frac{q^2}{M}\int_{-u_m}^{u_m} u\frac{\partial f_0}{\partial u}\frac{du}{\omega - ku}. \tag{A6.3.2}$$

Let us introduce the dimensionless variable $y = u/u_m$ and $\alpha = \omega/ku_m$ and change (A6.3.2) by means of integration by parts

$$\sigma_i = -iA\frac{q^2\omega}{Mk^2}\left[\pi - \alpha\int_{-1}^{1}\frac{y\,dy}{(\alpha - y)\sqrt{1 - y^2}}\right].$$

Calculating an integral in the right-hand side, one can rewrite the above formula in the form:

$$\sigma_i = -i\pi A\frac{q^2\omega}{Mk^2}\left[1 - \alpha g(\alpha)\right], \tag{A6.3.3}$$

where function

$$g(\alpha) = \begin{cases} (\alpha^2 - 1)^{-1/2}, & \alpha > 1, \\ i(1 - \alpha^2)^{-1/2}, & \alpha < 1 \end{cases}$$

takes into account a complex phase of the integral.

From formulas (A6.2.7), (A6.2.10), and (A6.3.3) we obtain the dispersion equation for a tubular REB considering the contribution of the captured ions

$$1 = 2\Lambda\kappa^2 \left[\frac{e^2 N_e}{m\gamma^3 (\omega - kv)^2} + \frac{qN_i}{k^2\varphi} [1 - \alpha g(\alpha)] \right]. \tag{A6.3.4}$$

Eliminating a beam velocity from Eq. (A6.3.4) by means of relations $N_e = -I/ev$ and $I = I_0\beta(\gamma_0 - \gamma)$, we can fulfill the expansion of the right-hand side of this equation in series in

$$1 - I/I \ll 1, \quad qN_i v_L/|I_L|, \quad I = I_L - qN_i v_L \tag{A6.3.5}$$

(we have retained the notation of Eqs. (6.3.1), (6.3.10), (6.3.13) and (6.4.2)). As a result the dispersion equation reduces to the form:

$$2\,\frac{v_{\text{ph}}}{v_L} = -\alpha g(\alpha) a + b,$$
$$v_{\text{ph}} = \frac{\omega}{k}, \quad \alpha = \frac{v_{\text{ph}}}{v_L}\frac{1}{\sqrt{2\mu}}, \quad \mu = \frac{q}{e}\frac{m\gamma_0}{M}. \tag{A6.3.6}$$
$$a = \frac{qN_i v_L}{|I_L|}, \quad b = \gamma_0^{1/3}\left[6\left(1 - \frac{I}{I}\right)\right]^{1/2}.$$

In the absence of ions, $a = 0$, Eq. (A6.3.6) goes over to Ref. [17]. For $a > 0$ dispersion relation is an algebraic equation of fourth order with respect to v_{ph}. Since its solution in general form is quite awkward, it is more convenient to examine its asymptotic form.

In the limit case $\alpha \gg 1$, the degree of the equation (A6.3.6) is lowered:

$$z^3 + \Delta z^2 + \mu = 0,$$
$$z = \left(\frac{2}{a}\right)^{1/2}\frac{v_{\text{ph}}}{v_L}, \quad \Delta = \frac{a - b}{(4a)^{1/3}}. \tag{A6.3.7}$$

Complex roots of Eq. (A6.3.7), which correspond to instability of REB in a cold ion plasma [12], appear for $\Delta > \Delta_c = -3\sqrt{\mu/4}$. Asymptotic solutions of this equation are

$$z = \begin{cases} \left(-1 \pm i\sqrt{3}\right)\mu^{1/3}/2, & |\Delta| \ll \mu^{1/3}, \\ \pm i\sqrt{\mu/\Delta}, & \Delta \gg \mu^{1/3}. \end{cases} \tag{A6.3.8}$$

In the region of the small phase velocities $\alpha \ll 1$, when the thermal motion of the ions becomes important, a different asymptotic form of equation (A6.3.6) is valid:

$$\frac{v_{\text{ph}}}{v_L} = \left(1 \pm \frac{ia}{\sqrt{8\mu}}\right)^{-1}\frac{b}{2}. \tag{A6.3.9}$$

In the transition region $\alpha \simeq 1$ and $a \gg \sqrt{8\mu}$ the solution (A6.3.6) has a simple form:

$$\frac{v_{\text{ph}}}{v_L} = \pm\left(\frac{2\mu}{a^2/b^2 - 1}\right)^{1/2}, \quad a > b. \tag{A6.3.10}$$

In the limit cases $a - b \ll a$ and $a \gg b$, formula (A6.3.10) goes over to (A6.3.8) and (A6.3.9).

Appendix 6.4

We shall show that the integral (6.4.11) corresponds to the conservation of the momentum flux through the beam cross section (6.3.27),

$$mc^2N\gamma\beta^2 + \frac{1}{4}\int_0^R \left[\left(\frac{\partial\varphi}{\partial r}\right)^2 - \left(\frac{\partial\varphi}{\partial z}\right)^2\right] rdr = S \qquad (A6.4.1)$$

in the absence of trapped ions. Integrating the term $\sim (\partial\varphi/\partial r)^2$ in Eq. (A6.4.1) by parts and using Poisson's equation

$$\Delta\varphi = 2eNr^{-1}\delta(r - r_b)$$

and the conservation of energy $\gamma - \gamma_0 = -e\varphi/mc^2$, we transform this equation to the form

$$mc^2N\left[\gamma\beta^2 + \frac{1}{2}(\gamma - \gamma_0)\right] + \frac{1}{4}\int_0^R \left[\varphi\frac{\partial^2\varphi}{\partial z^2} - \left(\frac{\partial\varphi}{\partial z}\right)^2\right] rdr = S. \qquad (A6.4.2)$$

The integral (A6.4.2) is calculated in the long-wavelength approximation

$$\varphi(r, z) = -2eN(z)\begin{cases}\ln(R/r_b), & r \le r_b \\ \ln(R/r), & r > r_b.\end{cases}$$

After reducing it to dimensionless form, $y = -eNc/I$, we obtain

$$\begin{aligned}&\frac{1}{2}\left(y'^2 + y^2\right) - \frac{I}{I_0}\left(\gamma_0 y - \sqrt{y^2-1}\right)\\ &-\frac{y}{2}\left[y'' + y - \frac{I_0}{I}\left(\gamma_0 - \frac{y}{\sqrt{y^2-1}}\right)\right] = -\frac{c^2S}{2I^2\Lambda}.\end{aligned} \qquad (A6.4.3)$$

The expression in the braces on the left of Eq. (A6.4.3) is equal to zero because of Eq. (6.4.6), and remaining expression is the same as first integral Eq. (6.4.6) [i.e., the same as Eq. (6.4.11) as $y \to y_{\min}$].

Appendix 6.5

Using Eqs. (6.4.1), Eqs. (6.4.1) and (A6.4.2), we calculate the derivative dI/dS to find

$$\frac{dI}{dS} = \frac{dI/d\beta}{dS/d\beta} = \frac{eI_0}{mc}\frac{1}{\gamma\beta} > 0. \qquad (A6.4.4)$$

Since at the point $I = I_L$ the momentum fluxes in the first and second states are the same, $S_1 = S_2 = S_L$, and because of Eq. (A6.4.2)

$$\frac{dI_1}{dS_1} > \frac{dI_2}{dS_2}, \qquad (A6.4.5)$$

we find (since $S_{1,2} < S_L$) that $I_2(S_2)$ decreases more rapidly than $I_1(S_1)$ as S decreases, and that $S_1 > S_2$ for given value of I (see Fig.1b in Ref. [19]).

CHAPTER 7

ELECTROSTATIC PERTURBATION COLLISIONAL DAMPING IN A WEAKLY IONIZED PLASMA

The heating of plasma by the field of rectangular electric pulse with the time less than relaxation time of electron distribution function is investigated on the basis of the kinetic equation with Davydov collisional integral [1-3]. Analytic solutions are obtained for transient electron distribution during and after pulse. The distribution function is used for analyzing evolution of electronic ionization coefficient.

Electric-field attenuation is considered for a homogeneous weakly ionized plasma consequent on elastic collisions between electrons and neutral particles. The frequency and decrement for small Langmuir oscillations are determined by the ratio between the frequency of those collisions and the plasma frequency. The background electron distribution is altered in a strong electric field and the treatment becomes nonlinear. The solution in that case shows that the field energy is completely converted to electron thermal energy.

§7.1. Heating of a weakly ionized plasma by a short electric-field pulse.

A dynamic equilibrium is set up in a weakly ionized plasma in a steady electric field $\mathbf{E}$, which has a non-Maxwellian electron velocity distribution, where the electrons acquire energy from the field and lose it by elastic collisions with neutral atoms (Druvestein distribution [4]).

Since a characteristic time,

$$t_T \simeq \left(\frac{ml}{eE}\right)^{1/2} \left(\frac{M}{m}\right)^{3/4}, \tag{7.1.1}$$

[l is the mean free path of an electron in the plasma, $-e$ and m are the charge and mass of the electron, and M is the mass of the neutral particle] is quite substantial, the distribution does not have time attain a stationary state in a short electric pulse $t < t_T$, and the plasma heating is essentially nonstationary.

If the frequency of arising in a plasma short electric pulses is large enough, the processes of ionization and recombination do not have time to change a plasma density when the oscillations of electron distribution function occurs. Plasma-field system (using in a plasma-gas laser [5]) enters the quasi-stationary regime at which the heating and cooling of plasma electrons are not accompanied by the considerable alteration of the plasma density. Analytic solutions are obtained below for transient electron distribution during and after short electric pulse [6].

1. Acceleration of electrons.

We consider a kinetic equation for the electron distribution function $f(t,\mathbf{v})$ for a homogeneous weakly ionized plasma

$$\frac{\partial f}{\partial t} + \frac{e}{m}E\frac{\partial f}{\partial \mathbf{v}} = I(f), \tag{7.1.2}$$

where $I(f)$ is the integral of elastic collisions between electrons and neutrals, and the electric pulse is rectangular

$$\mathbf{E}(t) = \begin{cases} \mathbf{E}, & t \le t_T \\ 0, & t > t_T \end{cases} .$$

We assume that a characteristic time of plasma heating is much less than an inverse time of the elastic collisions. Putting $I(f) = 0$ in the right-hand side of Eq. (7.1.2) and carrying out the integration, we obtain

$$f(t, bfv) = \frac{n_e}{(\pi v_T^2)^{3/2}} \exp\left[-\frac{v_\perp^2 + (v_z - V)^2}{v_T^2}\right]. \tag{7.1.3}$$

At the initial time $t = 0$ this distribution is a Maxwellian distribution with a thermal velocity $v_T = \sqrt{T_n/m}$ (T_n is the temperature of neutral particles). For $t > 0$ an electron gas (which is accelerated as a whole) is moving with the velocity $V = eEt/m$ along an external electric field $\mathbf{E}$.

To find a distribution function $f(t, v)$, we put $v_\perp = v\cos\vartheta$ and $v_z = v\sin\vartheta$. After averaging Eq. (7.1.3) over the angles

$$\Big\langle f(t,\mathbf{v})\Big\rangle = \frac{1}{4\pi}\int f(t,\mathbf{v})\, d\Omega, \quad d\Omega = 2\pi \sin\vartheta\, d\vartheta,$$

we obtain

$$\Big\langle f(t,\mathbf{v})\Big\rangle = \frac{n_e}{2\pi^{3/2} v_T v V} \sinh\left(\frac{2vV}{v_T^2}\right) \exp\left(-\frac{v^2 + V^2}{v_T^2}\right). \tag{7.1.4}$$

2. Heating of electrons.

In this section we show that the elastic collisions between electrons and neutrals in the presence of an external electric field causes the diffusive evolution of the electron distribution function in a velocity space. This stage of plasma heating is analyzed with the help of the Davydov kinetic equation [1-3]:

$$f(t,\mathbf{v}) = F_0(t,v) + \frac{\mathbf{v}}{v}\mathbf{F}_1(t,v),$$

$$\frac{\partial F_0}{\partial t} = \frac{1}{v^2}\frac{\partial}{\partial v}\left(\frac{ev^2}{3m}\mathbf{E}\mathbf{F}_1 + \frac{mv^4}{Ml}F_0 + \frac{v^3 T_n}{Ml}\frac{\partial F_0}{\partial v}\right), \tag{7.1.5}$$

$$\frac{\partial \mathbf{F}_1}{\partial t} + \frac{v}{l}\mathbf{F}_1 = \frac{e}{m}\mathbf{E}\frac{\partial F_0}{\partial v},$$

in which a series expansion of initial Eq. (7.1.2) in the ratio between the mass of electron and ion is used.

We assume that the mean free path of an electron l does not depend on the velocity. Introducing the dimensionless variables

$$\begin{gathered}\tau = t/t_E, \quad t_E = \sqrt{ml/eE}, \quad u = v/v_E, \quad v_E = \sqrt{eEl/m}, \\ \mu = m/M, \quad F = F_0/n_e, \quad F_E = \mathbf{E}\mathbf{F}_1/En_e,\end{gathered} \tag{7.1.6}$$

we rewrite the equations (7.1.5) in the form

$$\begin{gathered}\frac{\partial F}{\partial \tau} = \frac{1}{u^2}\frac{\partial}{\partial u}\left(\frac{u^2}{3}F_E + \mu u^4 F + \frac{\mu}{2}\frac{v_T^2}{v_E^2}u^3\frac{\partial F}{\partial u}\right), \\ \frac{\partial \mathbf{F}_E}{\partial \tau} + u\mathbf{F}_E = \frac{\partial F}{\partial u},\end{gathered} \tag{7.1.7}$$

where t_E and v_E are the characteristic time and velocity of the elastic-collision evolution, respectively.

The system of equations (7.1.7) is obtained under conditions of $\tau > 1$ when the relaxation time exceeds the collisional one. For $\tau \gg 1$ and $u \simeq 1$ the second equation has an asymptotic solution

$$F_E \simeq \frac{1}{u}\frac{\partial F}{\partial u}. \tag{7.1.8}$$

Substituting (7.1.8) in the first equation (7.1.7), we find

$$\frac{\partial F}{\partial \tau} = \frac{1}{u^2}\frac{\partial}{\partial u}\left(\frac{u}{3}\frac{\partial F}{\partial u} + \mu u^4 F + \frac{\mu}{2}\frac{v_T^2}{v_E^2}u^3\frac{\partial F}{\partial u}\right). \tag{7.1.9}$$

To proceed, we assume that the conditions $\mu v_T^2 \ll v_E^2$ and $\mu u^4 \ll 1$ are fulfilled. Thereafter omitting these terms in Eq. (7.1.9), we make use of the substitution $\tilde{u} = 2u^{3/2}/3$ to rewrite this equation as a diffusion equation in the cylindrical coordinate system. We use an integral Hankel transformation [7] with respect to u to solve the equation:

$$F(\tau, u) = \frac{1}{\tau - \tau_0}\int_0^\infty F(\tau_0, u)\exp\left[-\frac{u^3 + x^3}{3(\tau - \tau_0)}\right] I_0\left[\frac{2(ux)^{3/2}}{3(\tau - \tau_0)}\right] x^2\, dx, \tag{7.1.10}$$

In the limit $\tau \to \tau_0$, one can use the asymptotic form of a modified Bessel function $I_0(x)$,

$$\begin{gathered}I_0(x) \simeq \frac{e^x}{\sqrt{2\pi x}}, \quad x \gg 1, \\ \lim_{\epsilon \to 0}\frac{1}{\sqrt{\pi\epsilon}}\exp\left[-\frac{(x-a)^2}{\epsilon}\right] = \delta(x - a),\end{gathered} \tag{7.1.11}$$

to verify an initial condition:

$$\lim_{\tau \to \tau_0} F(\tau, u) = F(\tau_0, u).$$

At a qualitative level, we can use Eq. (7.1.4) (which found for $\tau \ll 1$) as initial condition for Eq. (7.1.10) in the limit $v_T \to 0$:

$$F(\tau_0, u) = \frac{1}{4\pi v_E^3 u^2}\delta(u - \tau_0). \tag{7.1.12}$$

Substituting (7.1.12) in (7.1.10), we find

$$F(\tau,u)=\frac{1}{4\pi v_E^3(\tau-\tau_0)}\exp\left[-\frac{u^3+\tau_0^3}{3(\tau-\tau_0)}\right]I_0\left[\frac{2(u\tau_0)^{3/2}}{3(\tau-\tau_0)}\right]. \tag{7.1.13}$$

The electrons are heated rapidly with time, $u\simeq\tau^{1/3}$, and the distribution function loses the memory of the original velocity distribution. In the limit $\tau\gg\tau_0$, at small values of the argument of the Bessel function, $\sqrt{\tau_0^3/\tau}\ll 1$, formula (7.1.13) becomes

$$F(\tau,u)\simeq\frac{1}{4\pi v_E^3\tau}\exp\left(-\frac{u^3}{3\tau}\right). \tag{7.1.14}$$

The small term μu^4F in Eq. (7.1.9) must be taken into account for $\tau_1\simeq\mu^{-3/4}\gg 1$ when the distribution function attains its asymptotic. Putting $\partial F/\partial t=0$, we find a stationary Druvestein distribution [4]:

$$F_\infty=\frac{n_e}{\pi v_E^3}\left(\frac{3\mu}{4}\right)^{3/4}\left[\Gamma\left(\frac{3}{4}\right)\right]^{-1}\exp\left(-\frac{3}{4}\mu u^4\right) \tag{7.1.15}$$

in which the average value of energy for plasma electrons is $T_\infty=eEl/\sqrt{3\mu}$.

3. Cooling of electrons.

Thereafter the electric field is switch off at the time t, the elastic collisions between electrons and neutrals lead to a cooling of electrons to neutral temperature. The first equation (7.1.7) in the absence of the field, $F_E=0$, is used below to analyze this process:

$$\frac{\partial F}{\partial\tau}=\frac{\mu}{u^2}\frac{\partial}{\partial u}\left(u^4F+\frac{v_T^2}{2v_E^2}u^3\frac{\partial F}{\partial u}\right). \tag{7.1.16}$$

Omitting small term $v_T^2/v_E^2\ll u^2$ in the right-hand side of Eq. (7.1.9), one can represent a solution in the form

$$F(\tau,v)=\frac{1}{u^4}\Phi\left(\frac{1}{u}-\mu\tau\right), \tag{7.1.17}$$

where Φ is an arbitrary function.

We use a dynamic function (7.1.14) at the moment t to concrete Φ. Comparing (7.1.14) and (7.1.17), we find the expression

$$F(\tau,u)=\frac{1}{4\pi v_E^3\tau q^4}\exp\left(-\frac{u^3}{3\tau q^3}\right),$$
$$q=1-\mu u(\tau-\tau),\quad \tau=t/t_E,\quad \tau>\tau, \tag{7.1.18}$$

which satisfies the normalization condition

$$4\pi v_E^3\int_0^\infty F(\tau,u)u^2\,du=1.$$

According to (7.1.18), the density of plasma electrons decreases exponentially with time uniformly over an entire length of the velocity space $u>0$. As a result, electrons displace

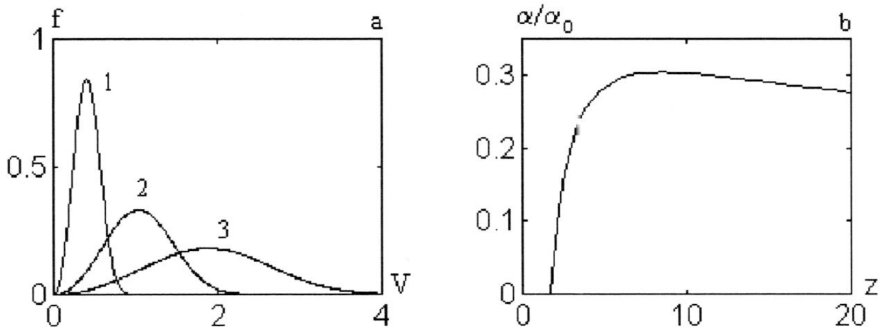

Fig. 7.1.1a. Electron distribution function $f(V)$ at successive times z=0.1 (1); 1.67 (2) и 10 (3).

Fig. 7.1.1b. Ionization coefficient α/α_0 as a function of the parameter $z=3\tau(v_e/v_i)^3$.

in the range of small energies $u^2 \simeq v_T^2/v_E^2$ where last term in Eq. (7.1.16) must be taken into account. After a time,

$$t - t \simeq l/\mu v_T, \tag{7.1.19}$$

the electron temperature decreases monotonically to the asymptotic value T_n (Maxwellian temperature of neutrals).

4. Evolution of ionization coefficient.

The transient electron distribution function (7.1.14) can be used for analyzing evolution of electronic ionization coefficient [8]

$$\alpha(t) = N \int v\sigma(v) F_0(t,v)\, dv. \tag{7.1.20}$$

We assume that the velocity dependence of the elastic-scattering cross section is defined by the formula

$$\sigma(v) = \pi \left(\frac{e^2}{mv^2}\right)^2 \left(\frac{v^2}{v_i^2} - 1\right)^2, \tag{7.1.21}$$

where N is the density of neutral particles, and the velocity v_i corresponds to the ionization threshold.

Substituting (7.1.14) and (7.1.21) in (7.1.20), we obtain

$$\alpha(z) = \frac{\alpha_0}{z} \int_{1/z}^{\infty} \left[\left(\frac{z^2}{x}\right)^{1/3} - \frac{1}{x}\right] \exp(-x)\, dx,$$

$$\alpha_0 = \frac{\pi n_e e^4}{m^2 v_i^3}, \quad z = 3\tau \left(\frac{v_E}{v_i}\right)^3. \tag{7.1.22}$$

The integral in Eq. (7.1.22) can be expressed in terms of incomplete gamma-function $\Gamma\left(2/3,\, z^{-1}\right)$ and exponential integral $\mathrm{Ei}\left(-z^{-1}\right)$ [9].

Time-evolution of the distribution function (7.1.14) in the dimensionless variables

$$f(z,V) = \frac{4\pi}{3} v_i^3 V^2 F = \frac{V^2}{z} \exp\left(-\frac{V^3}{z}\right), \quad V = \frac{v}{v_i} \tag{7.1.23}$$

is represented in Fig. 7.1.1a. Curve 2 corresponds to $\alpha \geq 0$ when the ionization process starts.

The function $\alpha(z)/\alpha_0$ in Fig. 7.1.1b illustrates the dependence of $\alpha(\tau)$ on the electric field and ionization potential. A maximum of the curve $\alpha_{\mathrm{m}} = 0.3\,\alpha_0$ does not depend from the field value E.

Note that the condition of applicability of an elastic-collision model ($n_p = \text{const}$) limits the time of electric pulse through the following inequality: $t \ll \alpha_{\mathrm{m}}^{-1}$.

§7.2. Small damped oscillations.

The electric field remained constant during the velocity-distribution relaxation; that approximation applies only if there is an external energy source that compensates for the energy loss from the field in accelerating the electrons or if the plasma density is low enough, when one can neglect the polarization field.

Here we consider the heating of a weakly ionized plasma by an electric field pulse in the self-consistent field approximation. The relaxation characteristics are derived from the following simple model. The plasma lies between two nonconducting planes, whose separation is much greater than the electron mean free path and the Debye length, so the plasma can be taken as homogeneous. The system is in an external electric field E_0, which does not penetrate into the plasma because the charges are separated and a polarization field $E_p = -E_0$ arises. At the start, the electric field is switched off, and the Langmuir oscillations arise in the plasma as a result of E_p. The elastic electron-neutral collisions damp the ordered oscillations, and the energy is converted to electron thermal energy. If the field amplitude is small, the oscillations are linear, and their frequency and decrement are dependent on the ratio of the elastic-collision frequency to the plasma frequency. In a strong field, the electron distribution is substantially distorted and the treatment becomes nonlinear. In the latter case, we have solved the nonlinear partial differential equation system, which enables one to examine how the field energy is converted to thermal form.

We consider a self-consistent system consisting of Davydov kinetic equation for the electrons (7.1.5) and an equation for the electric field, which is an analog of Poisson's equation for a homogeneous plasma:

$$\frac{\partial \mathbf{E}}{\partial t} = 4\pi e \int \frac{\mathbf{v}}{v} (\mathbf{v}\mathbf{F}_1)\, dv, \qquad \operatorname{curl} \mathbf{E} = 0. \tag{7.2.1}$$

The mean free path of an electron, $l(v) = 1/N\sigma(v)$, in Eqs. (7.1.5) depends on the elastic-scattering cross section between electrons and neutrals. As we envisage a weakly ionized plasma, we take the cross section as [2]

$$\sigma = \sigma_0 \begin{cases} 1, & v \le v_{\mathrm{m}}, \\ v_{\mathrm{m}}/v, & v > v_{\mathrm{m}}. \end{cases} \tag{7.2.2}$$

When there is no field ($\mathbf{E} = \mathbf{F}_1 = 0$), the electrons are in Maxwellian equilibrium with the neutrals

$$F_0(v) = \frac{n_e}{(\pi v_T^2)^{3/2}} \exp\left(-\frac{v^2}{v_T^2}\right), \tag{7.2.3}$$

in which $v_T^2 = \sqrt{2T_n/m}$ and n_e are the electron's thermal velocity and density.

Convenient variables are

$$\mathcal{E} = \frac{E}{4\pi e n_e l_0}, \quad F = \frac{(\omega_p l_0)^3}{n_e} F_0, \quad F_E = \frac{(\omega_p l_0)^3}{n_e} \frac{\mathbf{E}\mathbf{F}_1}{E},$$

$$u = \frac{v}{\omega_p l_0}, \quad \omega_p^2 = \frac{4\pi e^2 n_e}{m}, \quad l_0 = \frac{1}{N\sigma_0}, \quad \tau = \omega_p t,$$

$$g(u) = \frac{l_0}{l(u)}, \quad \mu = \frac{m}{M}$$

where (7.1.5) and (7.2.1) become

$$\frac{\partial F}{\partial \tau} = \frac{1}{u^2}\frac{\partial}{\partial u}\left[\frac{u^2 g(u)}{3}\mathcal{E}F_E + \mu u^4 g(u)F + \frac{\mu}{2}\frac{v_T^2}{(\omega_p l_0)^2}u^3\frac{\partial F}{\partial u}\right], \tag{7.2.4}$$

$$\frac{\partial F_E}{\partial \tau} + ug(u)F_E = \mathcal{E}\frac{\partial F}{\partial u}, \quad \frac{\partial \mathcal{E}}{\partial \tau} = \frac{4\pi}{3}\int_0^\infty F_E\, u^3\, du. \tag{7.2.5}$$

We assume that the electric field is sufficiently weak and that

$$\mathcal{E}F_E \ll \mu u^2 F, \tag{7.2.6}$$

so the energy acquired by an electron in the field is small by comparison with that lost by elastic collision with a neutral atoms, and the background distribution is close to the Maxwellian.

We use a Laplace transformation with respect to time to solve (7.2.5); we assume that the initial conditions $\mathcal{E}(0) = \mathcal{E}_0$ and $F_E = 0$ are obeyed to get

$$\mathcal{E}(\tau) = \frac{\mathcal{E}_0}{2\pi i}\int_{\sigma - i\infty}^{\sigma + i\infty}\frac{\exp(p\tau)\, dp}{p - \dfrac{4\pi}{3}\displaystyle\int_0^\infty \frac{u^3\, du}{p + g(u)u}\frac{\partial F}{\partial u}}, \tag{7.2.7}$$

in which the integration path lies to the right of any singularities in the integrand. The residue theorem indicates that the (7.2.7) integral has a value determined by the points at which the denominator becomes zero:

$$p = -\frac{\beta}{\zeta^5}\left[\int_0^{u_\mathrm{m}}\frac{u^4\exp(-u^2/\zeta^2)}{p+u}\, du + \int_{u_\mathrm{m}}^\infty \frac{u^4\exp(-u^2/\zeta^2)}{p+u_\mathrm{m}}\, du\right], \tag{7.2.8}$$

in which $\zeta = v_T/\omega_p l_0$, $u_\mathrm{m} = v_\mathrm{m}/\omega_p l_0$, and $\beta = 8/3\sqrt{\pi}$. This relation defines the Langmuir oscillation frequency and damping in such a plasma and is an analog of the dispersion equation for a collision-free plasma. We examine (7.2.8) for two limiting cases.

For $v_T \ll v_\mathrm{m}$, the cross section is constant, $\sigma(v) \simeq \sigma_0$, for most of the electrons (7.2.3) and one can pass to the asymptote $u_m \to \infty$ in (7.2.8):

$$p = -\frac{\beta}{\zeta^5}\int_0^\infty \frac{u^4\exp(-u^2/\zeta^2)}{p+u}\, du. \tag{7.2.9}$$

The roots of (7.2.9) are dependent on ζ, which is the ratio of the mean elastic-collision frequency to the plasma frequency.

The asymptotic solutions to (7.2.9) are

$$p = -\frac{\beta}{2}\begin{cases}\zeta - (\beta^2 - 2)\zeta^3 \mp \dfrac{2i}{\beta}\left[1 + \left(\dfrac{3\beta^2}{2} - 5\right)\dfrac{\zeta^2}{4}\right], & \zeta \ll 1,\\[2ex] 1 + \dfrac{2}{3\zeta^2}, & \zeta \gg 1.\end{cases} \tag{7.2.10}$$

Numerical integration in the region $\zeta \simeq 1$ is based on representing (7.2.9) as an equivalent differential equation for the complex function $p(\zeta)$:

$$\frac{dp}{d\zeta} = \frac{p}{\zeta} - \frac{p^2\,\zeta}{p^3 + (1 - 3\zeta^2/2)p - \beta\zeta}. \tag{7.2.11}$$

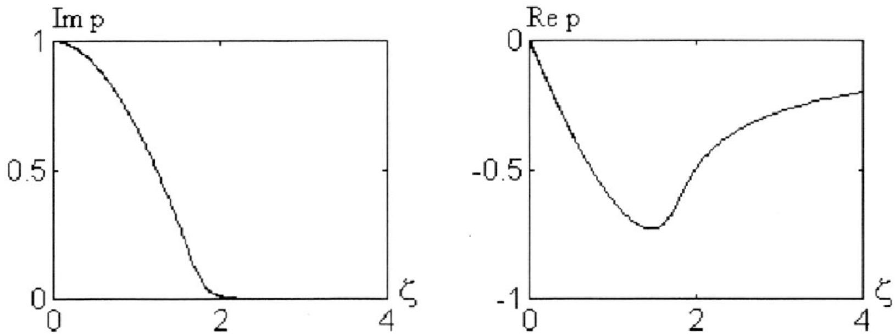

Fig. 7.2.1. Oscillation frequency $\mathrm{Im}\,p$ and damping decrement $\mathrm{Re}\,p$ as functions of ζ.

We put $p = a + ib$ and separate the real and imaginary parts in (7.2.11) to get the following nonlinear differential-equation system:

$$\frac{da}{d\zeta} = \frac{a}{\zeta} - \frac{Q_1}{Q_3}\zeta, \quad \frac{db}{d\zeta} = \frac{b}{\zeta} - \frac{Q_2}{Q_3}\zeta, \tag{7.2.12}$$

where

$$Q_1 = a\left(a^2+b^2\right)^2 + a\left(1-3\zeta^2/2\right)\left(a^2+b^2\right) - \beta\zeta\left(a^2-b^2\right),$$
$$Q_2 = -b\left(a^2+b^2\right)^2 + b\left(1-3\zeta^2/2\right)\left(a^2+b^2\right) - 2\beta\zeta ab,$$
$$Q_3 = \left(a^2+b^2\right)^3 + \left(1-3\zeta^2/2\right)\left(a^2+b^2\right)\left[1-3\zeta^2/2+2\left(a^2-b^2\right)\right]$$
$$-2\beta\zeta a\left(1-3\zeta^2/2+a^2-3b^2\right)+\beta^2\zeta^2.$$

The initial conditions here are defined by the first asymptote in (7.2.10). Fig. 7.2.1 shows numerical integration results for (7.2.12). For $\zeta \ll 1$, the collisions have little effect, and weakly damped Langmuir oscillations are present $a \ll b$. At ζ increases, so does the decrement, since the energy lost by the electrons is proportional to the number of collisions. However, $\zeta \gg 1$, the time between collisions $\Delta t \simeq l_0/v_T$ is reduced so much that an electron does not have time to acquire considerable energy in the electric field over the length l_0, and the energy transfer from the field to the electrons via the neutral particles becomes less effective. At the same time, as ζ increases, there is a rapid fall in the oscillation frequency because the collisions disrupt the ordered electron motion.

Condition (7.2.6) with (7.2.5) and (7.2.9) can be put as $|\mathcal{E}|^2\mathrm{Re}\,p \ll \mu\zeta^3$, so from (7.2.10) we have in dimensional variables that

$$\sqrt{\mu}\,T_n \gg \begin{cases} e|E|d, & d \ll l_0 \\ e|E|l_0, & d \gg l_0 \end{cases}, \tag{7.2.13}$$

in which $d = v_T/\omega_p$ is the Debye electron radius. The left side corresponds to the energy lost by an electron on collision with a neutral and the right to the energy acquired by it between collisions in a weakly collisional plasma ($d \ll l_0$) or a highly collisional one ($d \gg l_0$).

If $v_T \gg v_{\mathrm{m}}$, the cross section is $\sigma(v) \sim 1/v$ and elastic collision frequency remains constant: $\nu = N\sigma(v)v = v_{\mathrm{m}}/l_0$. We pass to the asymptote $u_{\mathrm{m}} \to 0$ in (7.2.8), which corresponds to discarding terms of order $(u_{\mathrm{m}}/\zeta)^5$), to get

$$p_\pm = -u_{\mathrm{m}}/2 \pm \sqrt{u_{\mathrm{m}}^2/4 - 1}. \tag{7.2.14}$$

Eq. (7.2.14) implies that the u_{m} dependence for the decrement and frequency is qualitatively the same as found above for u_T for a constant electron mean free path. As u_{m}

increases from zero to two, the decrement increases, while the frequency decreases and the process becomes aperiodic. Any further increase in u_{m} leads to the damping decreasing in proportion to $1/u_{\mathrm{m}}$, since for $u_{\mathrm{m}} > 2$ the damping is determined by the real root of (7.2.14), p_+, less in absolute value.

§7.3. Nonlinear electron-distribution relaxation in a strong field.

If (7.2.6) is violated, one has to correct for the change in background distribution, and the treatment becomes nonlinear. The nonlinear equation derived below involves the assumption that the electron mean free path is constant. Also, (7.1.5) has been derived on the assumption that

$$F_E \ll F, \tag{7.3.1}$$

which restricts the field.

For $\zeta \ll 1$, where collisions have little effect and weakly damped Langmuir oscillations arise, we omit the uF_E term in (7.2.5) and put $\tau \simeq 1$ to get from (7.3.1) that $\mathcal{E} \ll \zeta$ ($E^2/4\pi \ll n_e T$), which means that the permissible field energy density is small by comparison with the electron thermal-energy density and the change in the background distribution is slight. Therefore, large amplitude Langmuir oscillations cannot be described in the (7.1.5) approximation.

If $\zeta \gg 1$, the applicability range for the Davydov equations expands, since the collisions rapidly rearrange the ordered electron motion to the thermal form and prevent ordered collective collisions energy from accumulating. This limiting case is examined below in the nonlinear approximation.

When the electric field is switch on, a current arises in a plasma, and F_E alters; we therefore integrate (7.2.5) with constant $\mathcal{E}$ and F to get

$$F_E = \frac{\mathcal{E}}{u}\frac{\partial F}{\partial u}\left[1 - \exp(-u\tau)\right]. \tag{7.3.2}$$

Within the time for a few collisions $\tau > \zeta^{-1}$, F attains the asymptotic value

$$F_E \simeq \frac{\mathcal{E}}{u}\frac{\partial F}{\partial u} \tag{7.3.3}$$

and the subsequent behavior is determined by the variations in $\mathcal{E}$ and F.

Eq. (7.3.1) and (7.3.3) imply a constraint on the field $\mathcal{E} \ll \zeta^2$, while the inequality $\mathcal{E} \gg \sqrt{\mu}\zeta^2$ enables one to omit the $(\mu/u^2)\partial(u^4 F)/\partial u$ term in (7.2.4), so the condition for our approximation is

$$\sqrt{\mu}\,\zeta^2 \ll \mathcal{E} \ll \zeta^2, \tag{7.3.4}$$

and (7.2.4) and (7.2.5) simplify to

$$\frac{\partial F}{\partial \tau} = \frac{\mathcal{E}^2}{u^2}\frac{\partial F}{\partial u}\left(\frac{u}{3}\frac{\partial F}{\partial u}\right), \quad \frac{\partial \mathcal{E}}{\partial \tau} = \frac{4\pi}{3}\mathcal{E}\int_0^\infty u^2\frac{\partial F}{\partial u}\,du. \tag{7.3.5}$$

The solution to first equation (7.3.5) [satisfying (7.1.3) at $t = 0$] has been derived §7.1. Putting in Eq. (7.1.10)

$$\tau - \tau_0 \to \theta = \int_{\tau_0}^{\tau} \mathcal{E}^2\,d\tau',$$

we get

$$F(\theta, u) = \frac{1}{\sqrt{\pi}\theta\zeta^3}\int_0^\infty \exp\left[-\frac{x^2}{\zeta} - \frac{u^3 + x^3}{3\theta}\right] I_0\left[\frac{2(ux)^{3/2}}{3\theta}\right] x^2\,dx, \tag{7.3.6}$$

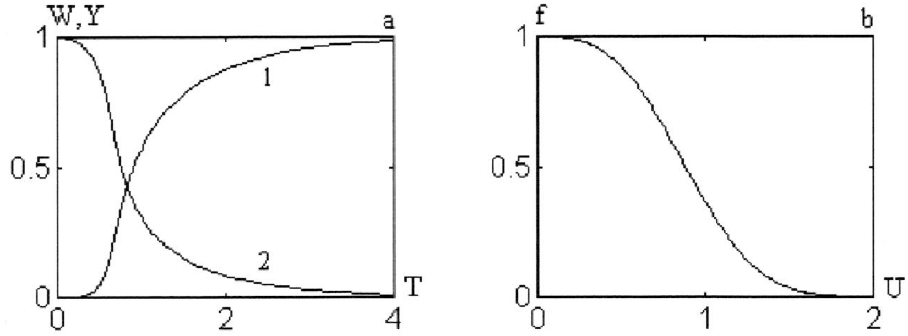

Fig. 7.1.1a. Temperature Y (1) and field W (2) as functions of time T.
Fig. 7.1.1b. Asymptote of electron distribution $f(U)$ for maximum temperature $Y=1$.

where $\tau_0 \geq \zeta^{-1}$ is the time corresponding to the start of the nonlinear heating.

In sufficiently strong field, the electrons are heated rapidly, and starting at time $\theta^{1/3} \gg \zeta$, the distribution loses the memory of the original velocity distribution, so one can use the asymptote for the integral in (7.3.6):

$$F(\theta, u) = \frac{1}{4\pi\theta} \exp\left(-\frac{u^3}{3\theta}\right). \tag{7.3.7}$$

The mean electron energy is

$$\overline{u^2} = 4\pi \int_0^\infty F u^4 \, du = \alpha\theta^{2/3}, \tag{7.3.8}$$

where $\alpha = \left(2/3^{1/3}\right) \Gamma\left(2/3\right)$.

We substitute (7.3.7) into (7.3.5) to get a nonlinear differential equation containing total derivatives with respect to time:

$$\ddot{\theta} + \frac{2\alpha}{3} \theta^{-1/3} \dot{\theta} = 0 \tag{7.3.9}$$

We integrate (7.3.9) to get

$$\dot{\theta} + \alpha\,\theta^{2/3} = \mathcal{E}_0^2, \tag{7.3.10}$$

which reflects the conservation law for the electron and field energies.

Eq. (7.3.10) implies that within a time $\tau_{NL} \simeq \mathcal{E}_0$, $\theta(\tau)$ increases monotonically to the asymptotic value $\theta_{\rm m} = \alpha^{-3/2}\mathcal{E}_0^3$, while field energy is completely converted to electron thermal energy $\mathcal{E}_{\rm min} = 0$ and $\overline{u^2} = \mathcal{E}_0^2$. We integrate (7.3.10) and use the fact that $\dot{\theta} = \mathcal{E}^2$ to get the time dependence of the field in inexplicit form:

$$\tau = \frac{3\mathcal{E}_0}{2\alpha^{3/2}} \left[\ln\left(\frac{1+\sqrt{1-W}}{1-\sqrt{1-W}}\right) - 2(1-W)\right], \tag{7.3.11}$$

where $W = \mathcal{E}^2/\mathcal{E}_0^2$.

Convenient variables are

$$Y = (1-W)^{3/2}, \quad Y = \theta/\theta_{\rm m}, \quad T = (2\alpha^{3/2}/3\mathcal{E}_0)\tau,$$
$$f(U) = 4\pi\theta_{\rm m} F = \frac{1}{Y} \exp\left(-\frac{U^3}{Y}\right), \quad U^3 = \frac{u^3}{3\theta_{\rm m}}. \tag{7.3.12}$$

to represent above solution in Fig. 7.3.1.

This nonlinear solution applies also for stronger fields $\mathcal{E}_0 \simeq \zeta^2$, but there (7.3.1) is violated, so one cannot consider the initial electron heating stage (7.3.2), while Eqs. (7.3.5) are applicable above times $\tau \simeq \mathcal{E}_0^{-1/2}$, when the electrons have already been greatly heated by the field $u \gg \zeta$. At the same time, one has to incorporate the constraint $\mathcal{E}_0^2 \ll \mu^{-1}$ for which the (7.2.4) $(\mu/u^2)\partial(u^4 F)/\partial u$ term remains small.

The application range for this nonlinear theory $\zeta \leq \mathcal{E} \ll \zeta^2$ in the dimension variables is equal to $\sqrt{4\pi n_e T} \leq E \ll NT\sigma e^{-1}$.

REFERENCES

To Introduction

1. Alfven H., Wernholm P. *A new type of accelerator.* Arkiv Fysic, **5**, 175, (1952).

2. Veksler V.I. *Coherent principle of acceleration of charged particles.* Proc. Symp. CERN, **1**, 80, (1956).

3. Budker G.I. *Relativistic stabilized electron beam.* Proc. Symp. CERN, **1**, 68, (1956); Atomnaya Energiya (Atomic Energy), **3**, 9, (1956).

4. Fainberg Ya.B. *The use of plasma wave guides as accelerating structure.* Proc. Symp. CERN, **1**, 84, 1956; Fainberg Ya.B. Atomnaya Energiya (Atomic Energy), **6**, 431, (1959).

5. Bennet W.H. *Magnetically self-focusing streams.* Phys. Rev., **45**, 890, (1934); *Self-focussing streams.* Phys. Rev., **98**, 1584, (1955).

6. Ivanov A.A., Rudakov L.I. *Powerful electron beam in a plasma.* Zh. Eksp. Teor. Fiz., **58**, 1332, (1970).

7. Ivanov A.A. *Physics of Highly Nonequilibrium Plasmas*, Atomizdat, Moscow, 1977.

8. Davidson R.C. *Theory of Nonneutral Plasmas*, W.A.Benjamin, Inc., Advanced Book Program Reading, Massachusets, 1974.

9. Rukhadze A.A., Bogdankevich L.S., Rosinskii S.E., Rukhlin V.G. *Physics of High-Current Relativistic Electron Beams*, Atomizdat, Moscow, 1980.

10. Fainberg Ya.B. Proc. VII Internat. Conference on Accelerators, Erevan, 1969.

11. Krasovitskii V.B. *Radial self-focusing of an electron beam due to the two-stream instability in a plasma.* Zh. Eksp. Teor. Fiz., **56**, 1253, (1969) [Sov. Phys. JETP, **29**, 674, (1969)]; *On the nonlinear theory of the beam-plasma instability.* Zh. Eksp. Teor. Fiz., **64**, 1597, (1973) [Sov. Phys. JETP, **37**, 809, (1973)];

12. Gabovich M.D., Levitskii S.M., Soloshenko I.A. *Excitation of surface plasma waves in an ionic-beam plasma and a radial compression of neutralized ion beam.* Pis'ma Zh. Tech. Fiz., **1**, 416, (1975) [JTP Lett. **1**, No. 9 , (1975)].

13. Winterberg F. *Electrostatic self-focusing of intense relativistic electron beams in dense plasma.* Bull. Amer. Phys. Soc., No.11, 1453, (1970); Atomkernenergie, **22**, 142, (1973).

14. Krasovitskii V.B. *Radiation pressure at the surface of a beam of coherently radiating electrons.* Zh. Tekh. Fiz., **47**, 10, (1977) [Sov. Phys. Tekh. Phys., **22**, 4, (1977)].

15. Bliokh Yu.P. *Electromagnetic instability of a relativistic electron beam.* Fiz. Plasmy, **2**, 285, (1976); Bliokh Yu.P., Onischenko I.N., Fainberg Ya.B. *Nonlinear theory of self-compression of a relativistic electron beam exposed to coherent electromagnetic radiation.* Fiz. Plasmy, **9**, 389, (1983).

16. Krasovitskii V.B. *Nonlinear radial self-focusing of an electron beam in a plasma.* Pis'ma Zh. Eksp. Teor. Fiz., **9**, 679, (1969) [JETP Lett. **9**, 422, (1969)].

17. Simonenko L.S., Soloshenko I.A. *Self-focusing of a negative ion beam due to the excitation of ionic plasma oscillations.* Fiz. Plasmy, **1**, 635, (1975) [Sov. J. Plasma Phys., , No. 4, (1977)].

18. Gladkii A.M., Kovalenko V.P., Yushmanov P.N. *The properties of plasma waves exited by the injected electron bunches.* Pis'ma Zh. Eksp. Teor. Fiz., **24**, 533, (1976).

To Chapter 1

1. Krasovitskii V.B. *Nonlinear radial self-focusing of an electron beam in a plasma.* Pis'ma Zh. Eksp. Teor. Fiz., **9**, 679, (1969) [JETP Lett. **9**, 422, (1969)].

2. Landau L.D, Lifshits. *Electrodynamics of Continuous Media*, Pergamon, Oxford, 1984.

3. Kovalenko V.P., Yushmanov P.N. *Stationary nonlinear waves in a plasma with a modulated electron beam.* Fiz. Plasmy, **3**, 1284, (1977) [Sov. J. Plasma Phys., **3**, 714, (1977)].

4. Dorofeenko V.G., Krasovitskii V.B. *Self-focusing of a modulated electron beam in a plasma.* Ukr. Fiz. Zh., **29**, 395, (1984).

5. Gradshteyn I.S., Ryshik I.M. *Tables of Integrals, Series, and Products*, Academic Press, New York, 1965.

6. Bohm D., Gross E.P. *Theory of plasma oscillations. A. Origin of medium-like behavior.* Phys. Rev. **75**, 1851, 1864, 1949.

7. Dorofeenko V.G., Krasovitskii V.B. *Electrostatic equilibrium of a modulated electron beam in a plasma.* Izv. Vyssh. Uchebn. Zaved., Radiofiz., **27**,1043 (1984) [Radiophys. Qu. Electr., **27**, 735, (1984)].

8. Krasovitskii V.B., Naguchev O.Yu. *Equilibrium electron-ion bunches in a plasma.* Fiz. Plasmy, **20**, 405, (1994) [Sov. J. Plasma Phys., **20**, , (1994)].

9. Krasovitskii V.B., Kurilko V.I., Strzhemechnii M.A. *Nonlinear theory of interaction between modulated beam and plasma.* Atomnaya Energiya (Atomic Energy), **24**, 545, (1968).

To Chapter 2

1. Akhiezer A.I., Akhiezer I.A., Polovin R.V., Sitenko A.G., Stepanov K.N. *Plasma Electrodynamics* [in Russian], Nauka, Moscow, 1974.

2. Aleksandrov A.F., Bogdankevich L.S., Rukhadze A.A. *Principles of Plasma Electrodynamics*, Springer, New York, 1984.

3. Dorofeenko V.G., Krasovitskii V.B. *Electrostatic equilibrium of a modulated electron beam in a plasma.* Izv. Vyssh. Uchebn. Zaved., Radiofiz., **27**,1043 (1984) [Radiophys. Qu. Electr., **27**, 735, (1984)].

4. Krasovitskii V.B., Osmolovskii S.I. *Self-focusing of relativistic electron clumps in a heated plasma.* Fiz. Plasmy, **18**, 1623, (1992) [Sov. J. Plasma Phys., **18**, 843, (1992)].

5. Miller M.A. *Motion of charge particles in a high-frequency electromagnetic field.* Izv. Vyssh. Uchebn. Zaved., Radiofiz., **1**, 110, (1958); Gaponov A.V., Miller M.A. Zh. Eksp. Teor. Fiz., **34**, 242, (1958) [Sov. Phys. JETP, **7**, 168, (1958)].

6. Talanov V.I. *Self-focusing of electromagnetic waves in a nonlinear media.* Izv. Vyssh. Uchebn. Zaved., Radiofiz., **7**, 564, (1964).

7. Krasovitskii V.B., Fomin G.V. *Transport of electron bunches in a magnetized plasma.* Fiz. Plasmy, **17**, 1400, (1991) [Sov. J. Plasma Phys., **17**, 811, (1991)].

8. Dorofeenko V.G. *Electrostatic equilibrium of a modulated electron beam in a magnetized plasma.* Izv. Vyssh. Uchebn. Zaved., Fizika, No. 3, p. 3, (1985).

9. Kuklin V.M. *One dimensional charged particles bunch moving through plasma.* Ukr. Fiz. Zh., **31**, 853, (1986); Kuklin V.M., Moiseev S.S., Panchenko I.P., et al. *Dynamics and emission of charged three-dimensional clumps in a plasma.* Reprint of Inst. of Space Research, 1619, Moscow, 1989; Kuklin V.M., Panchenko I.P. *Nonlinear structure formation in dissipative media.* Plasma Phys. Rep., **20**, 733 (1994).

To Chapter 3

1. Dorofeenko V.G., Krasovitskii V.B. *Self-focusing of relativistic electron bunches in a plasma.* Zh. Eksp. Teor. Fiz., **99**, 498, (1991) [Sov. Phys. JETP, **72**, 278, (1991)].

2. Dorofeenko V.G., Krasovitskii V.B., Moiseev S.S., Naguchev O.Yu. *Transport of relativistic electron bunches in a dense plasma.* Zh. Tekh. Fiz., **61**, 21, (1991); [Sov. Phys. Tekh. Phys., **36**, 732, (1991)].

3. Aleksandrov A.F., Bogdankevich L.S., Rukhadze A.A. *Principles of Plasma Electrodynamics*, Springer, New York, 1984.

4. Sudan R.N. In: *Basic Plasma Physics.* Suppl. to V. 2 (eds. Galeev A.A. and Sudan R.N.), North-Holland, 1984.

5. Kruer W.L., Dawson J.M., Sudan R.N. *Trapped particle instability.* Phys.Rev.Lett., **23** ,838 (1969).

6. Akhiezer A.I., Demutskii V.P., Polovin R.V. *On the theory of plasma waves with trapped electrons.* Fiz. Plasmy, **3**, 1079 (1977) [Sov. J. Plasma Phys., **3**, 597, (1976)].

7. Kadomtsev B.B. *Collective Processes in Plasmas*, Pergamon, Oxford, 1976.

8. Abramowitz, M. and Stegun, I. A. (Editors) *Handbook of Mathematical Functions with Formulas, Graphs, and Mathematical Tables*, Dover, New York, 1972.

9. Krasovitskii V.B., Naguchev O.Yu., Fomin G.V. *Spatial distribution of the induced charge and current in a plasma containing a deeply modulated electron-density distribution.* Izv. Vyssh. Uchebn. Zaved., Radiofiz., **37**,1273 (1994) [Radiophys. Qu. Electr., **37**, 822, (1994)].

10. Kovalenko V.P. *Stationary BGK-waves in a dissipative plasma.* Fiz. Plasmy, **6**, 864 (1980) [Sov. J. Plasma Phys., **6**, (1980)].

11. Krasovitskii V.B., Fomin G.V. *Electromagnetic shock wave with trapped relativistic electrons in a collisional plasma.* Zh. Eksp. Teor. Fiz., **101**, 858, (1992) [Sov. Phys. JETP, **74**, 458, (1992)];

12. Landau L.D, Lifshits E.M. *Mechanics*, Pergamon, New York.

13. Batishchev O.V., Krasovitskii V.B., Naguchev O.Yu., Sigov Yu.S. *Self-focusing of a modulated ribbon REB in dense plasma.* Fiz. Plasmy, **19**, 738, (1993) [Plasma Phys. Rep., **19**, 379, (1993)].

14. Hockney R.G., Eastwood J.W. *Computer Simulation Using Particles*, McGraw-Hill, New York, 1981.

15. Sigov Yu.S. *Numerical Methods of the Kinetic Theory of Plasmas* [in Russian], MFTI, Moscow, 1984.

16. Birdsall C.K., Langdon A.B. *Plasma Physics Via Computer Simulation*, McGraw-Hill, New York, 1985.

17. Batishchev O.V., Belova N.G., Sigov Yu.S. *Preprint No. 21, Keldish Institute of Applied Mathematics*, Moscow, 1987.

18. Kuzelev M.V., Rukhadze A.A. *Electrodynamics of High-Density Electron Beams in a Plasma* [in Russian], Nauka, Moscow, 1990.

19. Krasovitskii V.B. *On the nonlinear theory of the interaction of a modulated relativistic electron beam with a plasma.* Izv. Vyssh. Uchebn. Zaved., Radiofiz., **13**, 1902 (1970) [Radiophys. Qu. Electr., **13**, , (1970)].

20. Krasovitskii V.B. *Excitation of a regular plasma wave by a modulated beam with high energy density.* Pis'ma Zh. Eksp. Teor. Fiz., **15**, 346, (1972) [JETP Lett. **15**, 244, (1972)].

21. Fainberg Ya.B., Shapiro V.D., Shevchenko V.I. *On the nonlinear theory of the interaction of a monochromatic relativistic electron beam with a plasma.* Zh. Eksp. Teor. Fiz., **57**, 966, (1969) [Sov. Phys. JETP, **30**, 528, (1970)].

To Chapter 4

1. Gorev V.V., Kingsep A.S., Rudakov L.I. *High Langmuir turbulence in a plasma.* Izv. Vyssh. Uchebn. Zaved., Radiofiz., **19**, 691, (1976).

2. Bachin I.V., Krasovitskii V.B., Krymskii A.M. *Nonlinear stationary space-charge waves in beam-plasma systems.* Fiz. Plasmy, **8**, 593, (1982) [Sov. Plasma Phys., **8**, 336, (1982)].

3. Mokhov Yu.V., Chukbar K.V. *Nonlinear stationary waves in beam-plasma systems.* Fiz. Plasmy, **10**, 1097, (1984).

4. Dorofeenko V.G., Krasovitskii V.B. *Solitary Langmuir pulse in a plasma with an electron beam.* Fiz. Plasmy, **13**, 1090, (1987) [Sov. Plasma Phys., **13**, 628, (1987)].

5. Bachin I.V., Krasovitskii V.B. *Stationary Langmuir waves with trapped particles.* Izv. Vyssh. Uchebn. Zaved., Radiofiz., **27**, 550, (1984) [Radiophys. Qu. Electr., **27**, 369, (1984)].

6. Bachin I.V., Krasovitskii D.V. *Langmuir solitons in a plasma with an admixture of resonant atoms.* Fiz. Plasmy, **13**, 490, (1987) [Sov. Plasma Phys., **13**, 279, (1987)].

7. Landau L.D. *On the oscillations of electron plasma.* Zh. Eksp. Teor. Fiz., **16**, 574, (1948)

8. Krasovitskii D.V. *Nonlinear spatial damping of Langmuir pulses in a plasma.* Izv. Vyssh. Uchebn. Zaved., Radiofiz., **29**, 1310, (1986) [Radiophys. Qu. Electr., **29**, 983, (1986)].

9. Talanov V.I. *Self-focusing of electromagnetic waves in a nonlinear media.* Izv. Vyssh. Uchebn. Zaved., Radiofiz., **7**, 564, (1964).

10. Vedenov A.A., Rudakov L.I. *On the interaction of waves in a continuous media.* Dokl. Akad. Nauk SSSR, **159**, 767, (1964)

11. Kadomtsev B.B. *Collective Processes in Plasmas*, Pergamon, Oxford, 1976.

12. Krasovitskii V.B., Kurilko V.I. *Theory of amplification of longitudinal waves by a charge particle beam in a nonlinear plasma.* Zh. Eksp. Teor. Fiz., **51**, 445, (1966) [Sov. Phys. JETP, **24**, 300, (1967)].

13. Gradshteyn I.S., Ryshik I.M. *Tables of Integrals, Series, and Products*, Academic Press, New York, 1965.

14. Fraiman G.M. *Collisionless damping of nonlinear plasma oscillations.* Zh. Eksp. Teor. Fiz., **64**, 1589, (1973) [Sov. Phys. JETP, **64**, , (1973)].

15. Krasovitskii V.B., Kurilko V.I. *Propagation and excitation of electromagnetic waves in a nonlinear medium.* Zh. Eksp. Teor. Fiz., **48**, 353, (1965) [Sov. Phys. JETP, **21**, 232, (1965)].

16. Krasovitskii D.V. *Electromagnetic solitons in two-level system.* Zh. Tekh. Fiz., **54**, 1798, (1984) [Sov. Phys. Tekh. Phys., **29**, 1048, (1984)].

17. Korn G.A., Korn T.M. (editors) *Mathematical Handbook*, McGraw-Hill, New York, 1968.

18. Abramowitz, M. and Stegun, I. A. (Editors) *Handbook of Mathematical Functions with Formulas, Graphs, and Mathematical Tables*, Dover, New York, 1972.

19. Dawson J.M. *On Landau Damping.* Phys. Fluids, **4**, 869, (1961).

20. McCall S.L., Hahn S.L. *Self-induced transparency by pulsed coherent light.* Phys. Rev. Lett., **22**, 908, (1967).

21. O'Neil T. *Collisionless damping of nonlinear plasma oscillation.* Phys. Fluids, **8**, 2255, (1965).

22. Shapiro V.D., Shevchenko V.I. *Wave-particle interaction in nonequilibrium media.* Izv. Vyssh. Uchebn. Zaved., Radiofiz., **19**, 767, (1976).

To Chapter 5

1. Kadomtsev B.B. *Collective Processes in Plasmas*, Pergamon, Oxford, 1976.

2. Lighthill M.J., Inst. Math. Appl., **1**, 269, (1965); Proc. Roy. Soc., **A229**, 28, (1967).

3. McCall S.L., Hahn E.L. *Self-induced transparency by pulsed coherent light.* Phys. Rev. Lett., **22**, 908, (1967); *Self-induced transparency.* Phys. Rev., **183**, 457, (1969).

4. Krasovitskii D.V. *Electromagnetic solitons in two-level system.* Zh. Tekh. Fiz., **54**, 1798, (1984) [Sov. Phys. Tekh. Phys., **29**, 1048, (1984)].

5. Bachin I.V., Krasovitskii V.B. *Propagation of electromagnetic solitons in nonequilibrium dispersive media.* Zh. Eksp. Teor. Fiz., **79**, 472, (1980) [Sov. Phys. JETP, **52**, 237, (1980)].

6. Bachin I.V. *Solitary helicon waves in a nonequilibrium nonlinear plasma.* Ukr. Fiz. Zh., **26**, 11, (1981).

7. Bachin I.V., Krasovitskii V.B. *Electromagnetic solitons with trapped particles in an ion-beam plasma.* Fiz. Plasmy, **9**, 277, (1983) [Sov. J. Plasma Phys., **9**, 162, (1983)].

8. Krasovitskii V.B., Prudskikh V.V. *Solitary electromagnetic pulses in a retarding system with an electron-ion beam.* Izv. Vyssh. Uchebn. Zaved., Radiofiz., **39**, 850, (1996) [Radiophys. Qu. Electr., **39**, 562, (1996)].

9. Krasovitskii D.V. *Resonant electromagnetic solitons in a plasma.* Fiz. Plasmy, **12**, 1394, (1986) [Sov. J. Plasma Phys., **12**, 809, (1986)].

10. Krasovitskii V.B., Prudskikh V.V. *Autoresonance soliton in a plasma.* Fiz. Plasmy, **20**, 564, (1994). [Sov. J. Plasma Phys., **20**, 505, (1994)].

11. Istomin Ya.N., Karpman V.V. *Nonlinear evolution of a quasi-monochromatic packet of spiral waves in a plasma.* Zh. Eksp. Teor. Fiz., **63**, 1698, (1972) [Sov. Phys. JETP, **63**, , (1972)].

12. Bohm D., Gross E.P. *Theory of plasma oscillations. A. Origin of medium-like behavior.* Phys. Rev. **75**, 1851, 1864, (1949).

13. Korn G.A., Korn T.M. (editors) *Mathematical Handbook for Scientists and Engineers*, McGraw-Hill, New York, 1968.

14. Kamke E. *DifferentialGleichungen*, Leipzig, 1959.

15. Krasovitskii V.B., Kurilko V.I. *Propagation and excitation of electromagnetic waves in a nonlinear medium.* Zh. Eksp. Teor. Fiz., **48**, 353, (1965) [Sov. Phys. JETP, **21**, 232, (1965)].

To Chapter 6

1. Veksler V.I. *Coherent principle of acceleration of charged particles.* Proc. Symp. CERN, **1**, 80, (1956); Usp. Fiz. Nauk., **66**, 99, (1958) [Sov. Phys.-Usp., **1**, 54, (1958)].

2. Krasovitskii V.B. *Nonlinear theory of collective acceleration by ions by a relativistic electron beam.* Zh. Eksp. Teor. Fiz., **59**, 176, (1970) [Sov. Phys. JETP, **32**, 98, (1971)].

3. Sloan M.L., Drummond W.E. *Autoresonant accelerator concept.* Phys. Rev. Lett., **31**, 1234, (1973).

4. Indikul V.P., Panchenko I.P., Shapiro V.D., Shevchenko V.I. *On the theory of autoresonant acceleration of ions by a high-density relativistic electron beam.* Fiz. Plasmy, **2**, 775, (1976). **22**. Shapiro V.D., Shevchenko V.I. *Wave-particle interaction in nonequilibrium media.* Izv. Vyssh. Uchebn. Zaved., Radiofiz., **19**, 767, (1976).

5. Fainberg Ya.B., Butz V.A., Egorov A.M., Ivanov B.I., Onishchenko I.N. *Investigation of acceleration by charge density waves in plasma-filled wave guides and charge particle beams.* In: *Plasma Electronics* [in Russian], Naukova Dumka, Kiev, 1989.

6. Sprangle P., Drobot A.D., Mancheimer W.M. *Collective ion acceleration in a converging wave guide.* Phys. Rev. Lett., **36**, 1180, (1976).

7. Limar A.G., Hijnyak N.A., Belikov V.V. *Acceleration by ions by a modulated electron beam.* Pis'ma Zh. Eksp. Teor. Fiz., **1**, 615, (1975).

8. Fainberg Ya.B., Hijnyak N.A., Belikov V.V. *Methods of acceleration of heavy particles with the help of charge density waves in a plasma and vacuum.* Usp. Fiz. Nauk., **127**, 331, (1979).

9. Krasovitskii V.B. *Solitary pulse solution of longitudinal waves in a plasma.* Zh. Tekh. Fiz., **37**, 493, (1967) [Sov. Phys. Tekh. Phys., **12**, 354, (1967)].

10. Bohm D., Gross E.P. *Theory of plasma oscillations. A. Origin of medium-like behavior.* Phys. Rev. **75**, 1851, 1864, (1949).

11. Stix T.H. *The theory of plasma waves*, McGraw-Hill, New York, 1962.

12. Buneman O. *Dissipation of currents in ionized media.* Phys. Rev. **115**, 503, (1959).

13. Bogdankevich L.S. Rukhadze A.A. *Stability of relativistic electron beams in a plasma and problem of limiting currents.* Usp. Fiz. Nauk., **103**, 609, (1971) [Sov. Phys. Usp., **14**, 163, (1971)].

14. Brejzman B.N., Rytov D.D. *Powerful relativistic electron beams in a plasma and in a vacuum (theory).* Nucl. Fus. **14**, 873, (1974).

15. Rytov D.D. *On the limiting current of a relativistic electron beam.* Zh. Tekh. Fiz., **47**, 709, (1977) [Sov. Phys. Tekh. Phys., **22**, , (1977)].

16. Miller R. *Introduction to the Physics of High Current Charged Particle Beams* [Russian translation], Mir, Moscow, 1984.

17. Briggs R.J. *Space-charge waves on a relativistic, unneutralized electron beam and collective ion acceleration.* Phys. Fluids, **19**, 1257, (1976).

18. Krasovitskii V.B. *Nonlinear evolution of a slow electron-ion wave in a tubular relativistic electron beam.* Izv. Vyssh. Uchebn. Zaved., Radiofiz., **31**, 1400, (1988) [Radiophys. Qu. Electr., **29**, 1012, (1989)].

19. Dorofeenko V.G., Krasovitskii V.B. *Nonlinear stationary charge density waves in an annular relativistic electron beam.* Zh. Tekh. Fiz., **63**, 130, (1993) [Sov. Phys. Tekh. Phys., **38**, 700, (1993)].

20. Landau L.D, Lifshits E.M. *The Classical Theory of Fields*, Pergamon, New York, 1976.

21. Abramowitz, M. and Stegun, I. A. (Editors) *Handbook of Mathematical Functions with Formulas, Graphs, and Mathematical Tables*, Dover, New York, 1972.

22. Krasovitskii V.B. *Limiting current of a hollow relativistic electron beam.* Zh. Tekh. Fiz., **58**, 1090, (1988) [Sov. Phys. Tech. Phys., **33**, 639 (1988)].

23. Dzhavakhishvili D.I., Tsintsadze N.L. *Transport phenomena in an ionized ultrarelativistic plasma.* Zh. Eksp. Teor. Fiz., **64**, 1314, (1973) [Sov. Phys. JETP, **37**, 666, (1974)].

24. Braginskii S.I., *Transport phenomena in a plasma.* In: *Reviews of Plasma Physics.* Vol. 1 (ed. by M.A.Leontovich), Consultants Bureu, New York, 1963.

25. Bachin I.V., Dorofeenko V.G., Krasovitskii V.B. *Acceleration of an ion bunch by a slow modulated-phase cyclotron wave in a relativistic electron beam.* Fiz. Plasmy, **16**, 91, (1990) [Sov. J. Plasma Phys., **16**, 53, (1990)].

26. Fainberg Ya.B. *On the problems of charge particle acceleration in a plasma.* In: *Linear Accelerator Theory and Design* [in Russian], Gosatomizdat, Moscow, 1962.

27. Ivanov A.A. *Physics of Highly Non-equilibrium Plasmas*, Atomizdat, Moscow, 1977.

28. Krasovitskii V.B., Naguchev O.Yu., Osmolovskii S.I., Fomin G.V. *Propagation of a high-current relativistic electron beam in a low-density plasma.* Fiz. Plasmy, **17**, 445, (1991) [Sov. J. Plasma Phys., **17**, 261, (1991)].

29. Krasovitskii V.B., Naguchev O.Yu., Fomin G.V. *Transport of high-power REB in a plasma channel in a longitudinal magnetic field.* Fiz. Plasmy, **18**, 1233, (1992) [Sov. J. Plasma Phys., **18**, 637, (1992)].

30. Dolgopolov V.V., Kirichenko Yu.V., Romanov S.S., Tkach Yu.V. *Numerical simulation by particles of a collective acceleration of ions by a relativistic electron beam.* Reprint of Kharkov Phys. and Tech. Inst., KPTI 95-4, Kharkov, 1995.

31. Krasovitskii V.B. *Acceleration of a modulated beam of charge particle in a medium with inverted population level.* Zh. Tekh. Fiz., **61**, 1093, (1971) [Sov. Phys. Tech. Phys., **16**, 864, (1971)].

To Chapter 7

1. Davidov B.I. *On the theory of electron motion in gases and semiconductors.* Zh. Eksp. Teor. Fiz., **7**, 1069, (1937).

2. Golant V.E., Zhilinskii A.P., Sakharov S.A. *The Principles of Plasma Physics* [in Russian], Atomizdat, Moscow, 1977.

3. Akhiezer A.I., Akhiezer I.A., Polovin R.V., Sitenko A.G., Stepanov K.N. *Plasma Electrodynamics* [in Russian], Nauka, Moscow, 1974.

4. Druyvesteyn M.J. Physica, **10**, 61, (1930); **14**, 1003, (1934).

5. Petrash G.G. *Impulse Gas Lasers.* Usp. Fiz. Nauk., **105**, 645, (1971); Reshetnyak S.A., Shelepin L.A. *Kinetic of a plasma electron distribution in an electric field.* Fiz. Plasmy, **3**, 859, (1977) [Sov. J. Plasma Phys., **3**, 637, (1977)].

6. Krasovitskii D.V., Papakin V.F. *Heating of a weakly ionized plasma by a short electric-field pulse.* Ukr. Fiz. Zh., **30**, 63, (1985).

7. Bateman H., Erdelyi A. *Higher Transcendental Functions*, McGraw-Hill, New York, 1953.

8. Smirnov B.M. *Physics of Weakly Ionized Gas*, Nauka, Moscow, 1972.

9. Janke E., Emde F., Losch F. *Tables of Higher Functions*, McGraw-Hill, New York, 1960.

10. Krasovitskii D.V., Papakin V.F. *Electrostatic-perturbation collisional damping in a weakly ionized plasma.* Izv. Vyssh. Uchebn. Zaved., Radiofiz., **31**, 1186, (1988) [Radiophys. Qu. Electr., **31**, 841, (1988)].

INDEX

A

B

C

D

E

F

G

H

I

K

Q

R

S

T

U

V

W